Newnes
Electrical Pocket Book

Newnes
Electrical Pocket Book

Edited by E. A. Reeves
DFH (Hons), CEng, MIEE

Twenty-second edition

Newnes
An imprint of Butterworth-Heinemann
Linacre House, Jordan Hill, Oxford OX2 8DP
A division of Reed Educational and Professional Publishing Ltd

 A member of the Reed Elsevier plc group

OXFORD   JOHANNESBURG   BOSTON
MELBOURNE   NEW DELHI   SINGAPORE

First published by George Newnes Ltd 1937
Twenty-first edition 1992
Twenty-second edition 1995
Reprinted 1996

British Library Cataloguing in Publication Data
Newnes Electrical Pocket Book. – 22Rev.ed
  I. Reeves, E. A.
  621.3

ISBN 0 7506 2566 X

Library of Congress Cataloguing in Publication Data
Newnes electrical pocket book / edited by E.A. Reeves. – 22nd ed.
  p.   cm.
  Includes index.
  ISBN 0 7506 2566 X (pbk.)
  1. Electric engineering–Handbooks, manuals, etc.   I. Reeves, E.
  A.
  TK151.N52   1955   95–20183
  621.3–dc20   CIP

Printed in Great Britain by The Bath Press, Bath

Contents

Preface

Although the twenty-first edition of the Pocket Book was only published in 1992 a number of important changes have occurred which must be reflected in this book. First of all the IEE Wiring Regulations have now become BS 7671, with the title 'Regulations for Electrical Installations'. Furthermore six of the seven Guidance Notes, referred to in the last edition, have now been published (for list of these see page 278), requiring extensive modifications to the existing text of Chapter 12. The issuing, in December 1994 of Amendment No. 1 to BS 7671 has further modified the original text of the 16th edition of the IEE Wiring Regulations. Since 1992 the UK has come into line with Europe, changing its public supply voltage from 240/415 V to 230/400 V, effective from 1 January 1995. The limits allowed on these new voltages are sufficient to include our present voltages, so that they can remain as they are; at the present time there is therefore no likelihood of them actually being altered. At the same time the growing importance of the so-called electronic office has given rise to the development of versatile wiring systems to supply computers, fax machines, data equipment and other electronic aids, as well as conventional business machines. The importance of these wiring systems is reflected by the addition of another chapter (Chapter 22) under the title of 'Cable Management Systems'.

Another addition has been Chapter 23 dealing with hazardous areas and equipment suitable for installing in such areas. With the increasing emphasis on safety and the growing importance of the subject it was felt only right to recognize this by covering it in this new edition. In the chemical, mining, petrochemical industries and many others combustible materials can form explosive atmospheres which will, under certain conditions, cause an explosion. Two of these conditions are the generation of a spark or high temperatures from electrical equipment (i.e. the carcase of an electric motor). Therefore when electrical equipment is used in a hazardous atmosphere it has to conform to strict regulations and the new chapter indicates what the requirements are. The remaining text in the Pocket Book is unaltered and therefore that part of the Preface relating to it is reproduced from the twenty-first edition. As stated in that edition of the Pocket Book, The IEE Wiring Regulations (now BS 7671) is quite different from its predecessor, the fifteenth edition. One of the major changes is that many of the Appendices of the fifteenth edition do not appear in the new edition but are published separately as Guidance Notes. At the time of writing this preface, six of the seven Guidance Notes are available, and much information from them is contained in Chapter 12, so that the heading of it is not strictly true.

An important difference in BS 7671 is the inclusion of Part 6, 'Special installations or locations', the original Part 6 of the fifteenth edition 'Inspection and testing' now being Part 7. For this reason we have included lengthy extracts from Part 6 covering locations containing a bath tub or shower basin, swimming pools, hot air saunas, construction site installations, agricultural and horticultural premises, restrictive conductive locations, earthing requirements for the installation of equipment having high earth leakage currents, electrical installations in caravans, motor caravans and caravan sites, and highway power supplies and street furniture. Space is reserved for future use and also for marinas.

The system of numbering the Regulations is rather more complicated than for the fifteenth edition; the first digit signifies a part, the second a chapter, the third a section and the subsequent digits the Regulation number. For example Section 413 is made up as follows: Part 4 – Protection for safety, Chapter 41– Protection against electric shock, Section 413 – Protection against indirect contact.

The second important change since 1989 is that the full details of the privatization of the UK supply industry are now known. At the time of going to press with the previous edition we only knew the two bodies in England as Big G and Little G with Big G including all the nuclear plants. This is now different, with nuclear power remaining as two separate entities known as Nuclear Electric and Scottish Nuclear, both owned by the Government, Big G becoming National Power, and Little G, PowerGen. The transmission system is also a separate company known as the national grid. Area electricity boards have been formed into separate distribution companies. Details of these organizations are given in the updated Chapter 7.

Other changes that have occurred during the last few years include the growing importance of building automation systems and work on superconductivity. At the same time many minor changes have occurred and so these have also been incorporated as necessary. For example, publication of a new British Standard in 1986 (too late for inclusion in that edition) had radically altered the classification of insulating materials in the upper temperature regions. The concept of Groups Y, A, and E etc. no longer exists, but a table in another British Standard, BS 5691 Part 2, lists materials and their tests which to some extent tie up with the old group concept.

Many of the tabulated figures in the lighting chapter have been modified to take into account latest practice, and advantage has been taken to include details of the recently introduced high frequency electronic ballast which provides silent, instant, flicker-free starting for single and two fluorescent lamps up to 1,800 mm long and both 26 mm and 38 mm diameter.

Updating of the chapter on switchgear and protection has involved replacement of some circuits and mention of the latest microprocessor-controlled overcurrent and earth-fault relays. The restriction placed on current balance protection schemes when applied to transformers is now included, in which a harmonic biased relay has to be employed rather than an unbiased design which is suitable for use on generator protective schemes.

Since 1986 the 4,000 MW Drax power station has been commissioned, Berkeley and Hunterston A nuclear station have ceased generating while Bradwell nuclear station will be subject to an annual review. Britain's first pressurized water reactor (PWR) station is at Sizewell, with an installed capacity of 1.200 MW plant being estimated to cost £2.03 billion (1987 cost). It became fully operational by January 1995. Wind generation is becoming more fashionable and three demonstration wind parks are to be built, each with twenty-five wind turbines having a combined capacity of 24 MW.

Another area of fast growth is in Building Automation Systems (BASs), which rely heavily on computers, microprocessors and other electronic artifices. The majority of BASs are based on the relatively inexpensive personal computer, with only the largest systems employing a mini-computer. Intelligence and stand alone capability are today included in individual controllers rather than being concentrated in a central processing unit or in field processing units.

E. A. Reeves

Introduction

The chief function of any engineer's pocket book is the presentation in convenient form of facts, tables and formulae relating to the particular branch of engineering dealt with.

In the case of electrical engineering, it is essential that the engineer should have a clear understanding of the methods by which the various formulae are derived in order that he can be quite certain that any particular formula is applicable to the conditions which he is considering. This applies with particular force in the case of alternating current work.

The first section of the Pocket Book is, therefore, devoted to the theoretical groundwork upon which all the practical applications are based. This covers symbols, fundamentals, electrostatics and magnetism.

When an engineer is called upon to deal with any particular type of electrical apparatus, for example, a protective relay system, a thermostatically controlled heating system, or industrial switchgear and control gear, the first requirement is that he shall understand the principles upon which these systems operate. In order to provide this information, much space has been devoted in the various sections to clear descriptions of the circuits and principles which are used in the different types of electrical apparatus.

The inclusion of technical descriptions, together with the essential data embodied in the tables, will be found to provide the ideal combination for those engineers engaged on the utilisation side of the industry, where many different types of equipment and electrical appliances, ranging from semiconductor rectifiers to electrode steam boilers, may have to be specified, installed and maintained in efficient operation.

An extensive summary of the sixteenth edition of the 'IEE Regulations for Electrical Installations' (now BS 7671) is contained in Chapter 12. The layout and content are markedly different to the previous editions and for those personnel

1

working in electrical contracting it is important that they obtain their own copy of the Regulations. One of the most important changes is the exclusion of many of the Appendices which have been published as separate Guidance Notes (see page 278). Another change is the inclusion of a new Part 6, 'Special installations or locations'. More is said about these in the Preface and Chapter 12.

1

Fundamentals and theory

FUNDAMENTALS

Current. The term 'current' is used to denote the rate at which electricity flows. In the case of a steady flow the current is given by the quantity of electricity which passes a given point in one second. The magnitude of the current depends not only upon the electromotive force but also upon the nature and dimensions of the path through which it circulates.

Ohm's Law. Ohm's law states that the current in a d.c. circuit varies in direct proportion to the voltage and is inversely proportional to the resistance of the circuit. By choosing suitable units this law may be written

$$\text{Current} = \frac{\text{Electromotive force}}{\text{resistance}}$$

The commercial units for these quantities are

Current – the ampere	(A)
Electromotive force – the volt	(V)
Resistance – the ohm	(Ω)

Using the symbols I, V and R to represent the above quantities in the order given, Ohm's law can be written

$$I = \frac{V}{R}$$

or

$$V = I \times R$$

The law not only holds for a complete circuit, but can be applied for any part of a circuit provided care is taken to use the correct values for that part of the circuit.

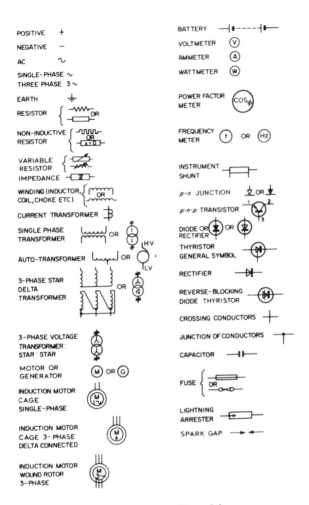

Figure 1.1
Graphical Symbols – BS3939

4

Resistivity. The resistivity of any material is the resistance of a piece of material having unit length and unit sectional area. The symbol is ρ and the unit is the ohm metre. The resistivity of a material is not usually constant but depends on the temperature. Table 1.1 shows the resistivity (with its reciprocal, conductivity) of the more usual metals and alloys.

Table 1.1   Resistivities at 20°C

Material	Resistivity Ohm metres		Conductivity Siemens per metre	
Silver	1.64	$\times 10^{-8}$	6.10	$\times 10^7$
Copper (annealed)	1.72	$\times 10^{-8}$	5.8	$\times 10^7$
Gold	2.4	$\times 10^{-8}$	4.17	$\times 10^7$
Aluminium (hard)	2.82	$\times 10^{-8}$	3.55	$\times 10^7$
Tungsten	5.0	$\times 10^{-8}$	2.00	$\times 10^7$
Zinc	5.95	$\times 10^{-8}$	1.68	$\times 10^7$
Brass	6.6	$\times 10^{-8}$	1.52	$\times 10^7$
Nickel	6.9	$\times 10^{-8}$	1.45	$\times 10^7$
Platinum	11.0	$\times 10^{-8}$	9.09	$\times 10^6$
Tin	11.5	$\times 10^{-8}$	8.70	$\times 10^6$
Iron	10.15	$\times 10^{-8}$	9.85	$\times 10^6$
Steel	19.9	$\times 10^{-8}$	5.03	$\times 10^6$
German Silver	16–40	$\times 10^{-8}$	6.3 $-$ 2.5	$\times 10^6$
Platinoid	34.4	$\times 10^{-8}$	2.91	$\times 10^6$
Manganin	44.0	$\times 10^{-8}$	2.27	$\times 10^6$
Gas carbon	0.005		200	
Silicon	0.06		16.7	
Gutta-percha	2	$\times 10^7$	5	$\times 10^{-8}$
Glass (soda-lime)	5	$\times 10^9$	2	$\times 10^{-10}$
Ebonite	2	$\times 10^{13}$	5	$\times 10^{-14}$
Porcelain	2	$\times 10^{13}$	5	$\times 10^{-14}$
Sulphur	4	$\times 10^{13}$	2.5	$\times 10^{-14}$
Mica	9	$\times 10^{13}$	1.1	$\times 10^{-14}$
Paraffin-wax	3	$\times 10^{16}$	3.3	$\times 10^{-17}$

Resistance of a conductor. The resistance of a uniform conductor with sectional area A and length l is given by

$$R = \rho \frac{l}{A}$$

The units used must be millimetres and square millimetres if ρ is in ohm millimetre units.

Temperature coefficient. The resistance of a conductor at any temperature can be found as follows:

$$R_t = R_o (1 + \alpha t)$$

R_t = resistance at temperature $t°C$

R_o = resistance at temperature 0°C

The coefficient α is called the temperature coefficient and it can be described as the ratio of the increase in resistance per degree C rise in temperature compared with the actual resistance at 0°C.

The coefficient for copper may be taken as 0.004. The increase in resistance for rise of temperature is important, and for many calculations this point *must* be taken into account.

Power. Power is defined as the rate of doing work. The electrical unit of power (P) is the *watt* (abbreviation W), and taking a steady current as with d.c.,

$$1 \text{ W} = 1 \text{ V} \times 1 \text{ A}$$

or

$$\text{W} = \text{V} \times \text{A}$$

or in symbols

$$P = V \times I$$

(For alternating current, see page 14)
Note: 1 kW = 1000W

Energy. Energy can be defined as power × time, and electrical energy is obtained from

$$\text{Energy} = VIt$$

where t is the time in seconds.

The unit obtained will be in joules, which is equivalent to 1 ampere at 1 volt for 1 second. The practical unit for energy is the kilowatt hour and is given by

$$\frac{\text{watts} \times \text{hours}}{1000} = \text{kWh}$$

Energy wasted in resistance. If we pass a current I through resistance R, the volt drop in the resistance will be given by

$$V = IR$$

The watts used will be VI, therefore the power in the circuit will be $P = VI = (IR) \times I = I^2R$.

This expression (I^2R) is usually known as the copper loss or the I^2R loss.

SI units. The SI system uses the metre as the unit of length, the kilogram as the unit of mass and the second as the unit of time. These units are defined in BS 3763 'The International System (SI) Units'.

SI units are used throughout the next of this book and include most of the usual electrical units. With these units however the permittivity and permeability are constants. They are:

Permittivity ϵ_0 = 8.85×10^{-12} farad per metre
Permeability μ_0 = $4\pi \times 10^{-7}$ henry per metre

These are sometimes called the electric and magnetic space constants respectively. Materials have relative permittivity ϵ_r and relative permeability μ_r, hence ϵ_r and μ_r for vacuum are unity.

ELECTROSTATICS

All bodies are able to take a *charge* of electricity, and this is termed static electricity. The charge on a body is measured by means of the force between two charges, this force following the inverse square law (i.e. the force is proportional to the product of the charges and inversely proportional to the square of the distance between them). This may be written

$$F = \frac{q_1 q_2}{4\pi\epsilon_0 d^2} \quad \text{N}$$

where q_1 and q_2 are the charges in coulombs and d the distance in metres – the space in between the charges being either air or a vacuum with a permittivity ϵ_0. N is Newtons.

If the two charged bodies are separated by some other medium the force acting may be different, depending on the relative permittivity of the *dielectric* between the two charged bodies. The relative permittivity is also termed the dielectric constant.

In this case the force is given by

$$F = \frac{q_1 q_2}{4\pi\epsilon_r\epsilon_0 d^2} \quad \text{N}$$

where ϵ_r is the contant for the particular dielectric. For air or a vacuum the value of ϵ_r is unity.

Intensity of field. There is an electrostatic field due to any charged body and the *intensity* of this field is taken as the force on unit charge.

The intensity of field at any given point due to an electrostatic charge is given by

$$E = \frac{q}{4\pi\epsilon_0 d^2} \quad \text{V/m}$$

7

Note: The ampere is the defined unit. Hence a coulomb is that quantity of charge which flows past a given point of a circuit when a current of one ampere is maintained for one second.

The value of the ampere, adopted internationally in 1948, is defined as that current which, when flowing in each of two infinitely long parallel conductors in a vacuum, separated by one metre between centres, causes each conductor to have a force acting upon it of 2×10^{-7} N/m length of conductor. The definition on the basis of the silver voltameter is now obsolete.

Dielectric flux. The field due to a charge as referred to above is assumed to be due to imaginary *tubes of force* similar to magnetic lines of force, and these tubes are the paths which would be taken by a free unit charge if acted on by the charge of the body concerned.

By means of these tubes of force we get a *dielectric flux-density* of so many tubes of force per square metre of area. For our unit we take a sphere 1 m radius and give it unit charge of electricity. We then get a dielectric flux density on the surface of the sphere of unit = one tube of force per square centimetre. The total number of tubes of force will be equal to the surface area of the sphere = 4π. For any charge q at a distance r the dielectric flux density will be

$$D = \frac{q}{4\pi r^2} \text{ C/m}^2$$

We have seen that the intensity of field or electric force at any point is

$$E = \frac{q}{4\pi\epsilon_0\epsilon_r r^2}$$

so that this can also be stated as $E = D/\epsilon_r\epsilon_0$.

Electrostatic potential. The potential to which a body is raised by an electric charge is proportional to the charge and the *capacity* of the body – so that $C = Q/V$, where V is the potential and C the capacity. The definition of the capacity of a body is taken as the charge or quantity of electricity necessary to raise the potential by one volt. This unit of potential is the work done in joules, in bringing unit charge (1 coulomb) from infinity to a point at unit potential.

Capacitance. The actual measurement of capacity is termed *capacitance*, and for practical purposes the unit is arranged for use

8

with volts and coulombs. In this case the unit of capacitance is the farad, and we get $C = Q/V$, where C is in farads, Q is in coulombs and V in volts.

The farad is a rather large unit, so that we employ the microfarad $= 10^{-6}$ of a farad or 1 picofarad $= 10^{-12}$ of a farad.

CAPACITORS

The capacity of a body is increased by its proximity to earth or to another body and the combination of the two is termed a capacitor. So long as there is a potential difference between the two there is a capacitor action which is affected by the dielectric constant of the material in between the two bodies.

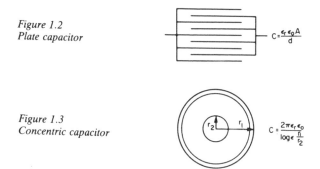

Figure 1.2
Plate capacitor

$C = \dfrac{\epsilon_r \epsilon_0 A}{d}$

Figure 1.3
Concentric capacitor

$C = \dfrac{2\pi \epsilon_r \epsilon_0}{\log_\epsilon \dfrac{r_1}{r_2}}$

Flat plate capacitor. Flat plate capacitors are usually made up of metal plates with paper or other materials as a dielectric. The rating of a plate capacitor is found from

$$C = \epsilon_r \epsilon_0 A/d \text{ farads}$$

where A is the area of each plate and d the thickness of the dielectric. For the multi-plate type we must multiply by the number of actual capacitors there are in parallel.

Concentric capacitor. With electric cables we get what is equivalent to a concentric capacitor with the outer conductor or casing of radius r_1 m. and the inner conductor of radius r_2 m. If now the dielectric has a constant of ϵ_r, the capacity will be (for 1m length)

$$C = \frac{2\pi\epsilon_r\epsilon_0}{\log\epsilon \ (r_1/r_2)} \text{ farad per metre}$$

Values of ϵ_r for different materials

Air	1
Paper, Pressboard	2
Cotton tape (rubbered)	2
Empire cloth	2
Paper (oiled)	2
Shellac	3
Bakelite	6
Paraffin-wax	3
Mica	7
Porcelain	7
Glass	7
Marble	8
Rubber	2.5
Ebonite	2.5
Gutta-percha	4
Polyethylene	2.3

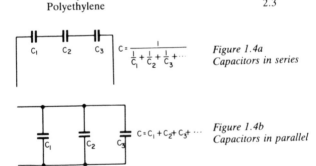

$$C = \frac{1}{\dfrac{1}{C_1} + \dfrac{1}{C_2} + \dfrac{1}{C_3} +\cdots}$$

Figure 1.4a
Capacitors in series

$$C = C_1 + C_2 + C_3 +\cdots$$

Figure 1.4b
Capacitors in parallel

THE MAGNETIC CIRCUIT

Electromagnets. Magnetism is assumed to take the form of *lines of force* which flow round the magnetic circuit. This circuit may be a complete path of iron or may consist of an iron path with one or more air-gaps. The transformer is an example of the former and a generator the latter.

The lines of force are proportional to the *magneto-motive-force* of the electric circuit and this is given by

$$\text{mmf} = \frac{IN}{10} \text{ ampere turns}$$

where I is the current in amperes and N the number of turns in the coil or coils. This mmf is similar in many respects to the emf of an electric circuit and in the place of the resistance we have the *reluctance* which may be termed the resistance of the magnetic circuit to the passage of the lines of force. The reluctance is found from

$$\text{Reluctance} = S = \frac{l}{A\mu_r\mu_0} \text{ At/Wb}$$

where l is the length of the magnetic circuit in millimetres, A is the cross-section in square millimetres and $\mu_r\mu_0$ is the permeability of the material. The permeability is a property of the actual magnetic circuit and not only varies with the material in the circuit but with the number of lines of force actually induced in the material if that material is iron.

The actual flux induced in any circuit is proportional to the ratio $\dfrac{\text{mmf}}{\text{reluctance}}$ and so we get

$$\text{Total flux} = \phi = \frac{\text{mmf}}{S} \text{ Wb}$$

The relative permeability μ_r is always given as the ratio of the number of lines of force induced in a circuit of any material compared with the number of lines induced in air for the same conditions. The permeability of air is taken as unity and so permeability can be taken as the magnetic conductivity compared with air.

Taking the formula for total flux given above, we can combine this by substituting values for mmf and S, giving

$$\text{Total flux} = \phi = \frac{\mu_r\mu_0 INA}{10l} \text{ Wb}$$

11

Having obtained the total flux, we can obtain the flux density or number of lines per square metre of cross-section as follows:

$$\text{Flux density} = B = \frac{\phi}{A} \text{ tesla (T)}$$

The tesla is one weber per square metre.

Where there is an air-gap it will be found that there is a certain amount of magnetic leakage and the actual flux in the air-gap will be smaller than that in the iron. The ratio between these two is given by the leakage coefficient which

$$= \frac{\text{flux in air-gap}}{\text{flux in iron}}$$

Ampere-turns per metre (At/m). In order to deal with complex magnetic circuits such as generators, motors, etc., it is more convenient to take the various sections of the magnetic circuit separately, and for this purpose it is useful to have the ampere-turns required per metre to give a fixed flux density. Taking our complete formula above for total flux, we get

$$B = \frac{\phi}{A} = \mu_r\mu_0 \frac{IN}{10l} = \mu_r\mu_0 H$$

so that the permeability and flux density are linked by the expression $\frac{IN}{10l} = H$ which is called the *magnetising force* and it will be seen that this is equal to the ampere-turns per unit length (i.e. metre).

The relation between B and H is usually given by means of a $B - H$ curve (Figure 1.6), but by using a different scale the actual value of ampere-turns per metre required can be read off. This scale is also shown in Figure 1.6.

Hysteresis. If a piece of iron is gradually magnetised and then slowly demagnetised it will be found that when the current is reduced to zero there is still some residual magnetism or remanence and the current has to be reversed to annul the flux. This is shown in Figure 1.7 where the complete curve of magnetisation is shown by the circuit ABCDEF. This lagging of the flux behind the magnetising force is termed *hysteresis* and during a complete *cycle* as shown by the figure ABCDEF there is definite loss called the *hysteresis loss*. Frequency is expressed in hertz (Hz) so that 1 Hz = 1 cycle/second.

12

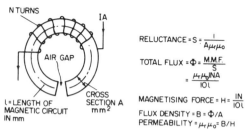

$$\text{RELUCTANCE} = S = \frac{l}{A\mu_r\mu_0}$$

$$\text{TOTAL FLUX} = \Phi = \frac{\text{M.M.F.}}{S}$$

$$= \frac{\mu_r\mu_0 INA}{10l}$$

$$\text{MAGNETISING FORCE} = H = \frac{IN}{10l}$$

$$\text{FLUX DENSITY} = B = \Phi/A$$
$$\text{PERMEABILITY} = \mu_r\mu_0 = B/H$$

Figure 1.5. The magnetic circuit

Figure 1.6
The B–H curve

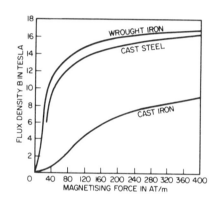

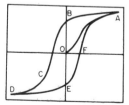

ABCDEF = HYSTERESIS LOOP
OB = REMANENCE
HYSTERESIS LOSS IN JOULES/
cu cm / CYCLE AND IN WATTS/CYCLE
$W = nfB^{1.6} \times 10^{-1}$ PER CU METRE

Figure 1.7
Hysteresis loss

In an alternating current machine this loss is continuous and its value depends on the materials used.

$$\text{Watts lost per cubic metre} = nf\text{B}_{\text{max.}}^{1.6} \times 10^{-1}$$

where n is a coefficient depending on the actual material, the temperature, the flux density and the thickness of the laminations, f = frequency and $\text{B}_{\text{max.}}$ is the maximum flux-density.

Typical values for n are:

Cast steel	0.003 to 0.012
Cast iron	0.011 to 0.016
Soft iron	0.002
Silicon steel	
0.2%	0.0021
3%	0.0016
4.8%	0.00076

Magnetic paths in series.   Where the magnetic path is made up of several different parts, the total reluctance of the circuit is obtained by adding the reluctance of the various sections. Taking the ring in Figure 1.5. the total reluctance of this is found by calculating the reluctance of the iron part and adding the reluctance of the air-gap. The reluctance of the air-gap will be given by

$$\frac{l}{\mu_0 A}$$

The value of $\mu_0 = 4\pi \times 10^{-7}$ H/m

A.C. THEORY

Alternating currents.   Modern alternators produce an emf which is for all practical purposes sinusoidal (i.e. a sine curve), the equation between the emf and time being

$$e = E_{\text{max}} \sin \omega t$$

where e = instantaneous voltage
E_{max} = maximum voltage
ωt = angle through which the armature has turned from neutral.

Taking the frequency as f hertz, the value of ω will be $2\pi f$, so that the equation reads

$$e = E_{\text{max}} \sin (2\pi f)t$$

The graph of the voltage will be as shown in Figure 1.8.

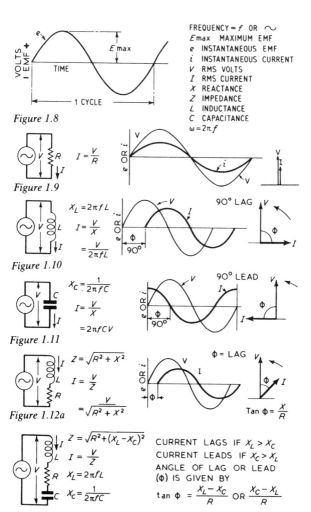

Figure 1.8

FREQUENCY = f OR $\sim$
Emax   MAXIMUM EMF
e INSTANTANEOUS EMF
i INSTANTANEOUS CURRENT
V RMS VOLTS
I RMS CURRENT
X REACTANCE
Z IMPEDANCE
L INDUCTANCE
C CAPACITANCE
$\omega = 2\pi f$

$I = \dfrac{V}{R}$

Figure 1.9

$X_L = 2\pi f L$

$I = \dfrac{V}{X}$

$\quad = \dfrac{V}{2\pi f L}$

90° LAG

Figure 1.10

$X_C = \dfrac{1}{2\pi f C}$

$I = \dfrac{V}{X}$

$\quad = 2\pi f C V$

90° LEAD

Figure 1.11

$Z = \sqrt{R^2 + X^2}$

$I = \dfrac{V}{Z}$

$\quad = \dfrac{V}{\sqrt{R^2 + X^2}}$

$\phi = $ LAG

$\text{Tan } \phi = \dfrac{X}{R}$

Figure 1.12a

$Z = \sqrt{R^2 + (X_L - X_C)^2}$

$I = \dfrac{V}{Z}$

$X_L = 2\pi f L$

$X_C = \dfrac{1}{2\pi f C}$

CURRENT LAGS IF $X_L > X_C$
CURRENT LEADS IF $X_C > X_L$
ANGLE OF LAG OR LEAD
(ϕ) IS GIVEN BY

$\tan \phi = \dfrac{X_L - X_C}{R} \text{ OR } \dfrac{X_C - X_L}{R}$

Figure 1.12b

15

Average or mean value. The average value of the voltage will be found to be 0.636 of the maximum value for a perfect sine wave, giving the equation

$$E_{Av} = 0.636E_{max}$$

The mean value is only of use in connection with processes where the results depend on the current only, irrespective of the voltage, such as electroplating or battery-charging.

RMS (root-mean-square) value. The value which is to be taken for power purposes of any description is the rms value. This value is obtained by finding the square-root of the mean value of the squared ordinates for a cycle or half-cycle. (See Figure 1.8).

$$E_{RMS} = E_{max} \times \frac{1}{\sqrt{2}} = 0.707E_{max}$$

This is the value which is used for all power, lighting and heating purposes, as in these cases the power is proportional to the square of the voltage.

A.C. circuits

Resistance. Where a sinusoidal emf is placed across a pure resistance the current will be in phase with the emf, and if shown graphically will be in phase with the emf curve.

The current will follow Ohm's law for d.c. i.e. $I = V/R$,

Where V is the rms value of the applied emf or voltage, and R is the resistance in ohms – the value of I will be the rms value. (See Figure 1.9.)

Inductance. If a sinusoidal emf is placed across a pure inductance the current will be found to be $I = V/[(2\pi f)L]$ where V is the voltage (rms value), f is the frequency and L the inductance in henries, the value of I being the rms value. The current will lag behind the voltage and the graphs will be as shown in Figure 1.10, the phase difference being 90°. The expression $(2\pi f)L$ is termed the inductive reactance (X_L).

Capacitance. If a sinusoidal emf is placed across a capacitor the current will be $I = (2\pi f).CV$, where C is the capacitance in farads, the other values being as above. In this case the current

16

leads the voltage by 90°, as shown in Figure 1.11. The expression $1/[(2\pi f)C]$ is termed the capacitive reactance (X_C) and the current is given by

$$I = \frac{V}{X_C}$$

Resistance and inductance in series.   In this circuit, shown in Figure 1.12, the current will be given by

$$I = \frac{V}{\sqrt{(R^2 + X_L^2)}}$$

where X_L is the reactance of the inductance ($X_L = (2\pi f)L$). The expression $\sqrt{(R^2 + X_L^2)}$ is called the impedance (Z), so that $I = V/Z$. The current will lag behind the voltage, but the angle of lag will depend on the relative values of R and X_L – thè angle being such that $\tan \phi = X_L/R$ (ϕ being the angle as shown in Figure 1.12a.

Resistance and capacitance in series.   For this circuit the current will be given by

$$I = \frac{V}{\sqrt{(R^2 + X_C^2)}}$$

where X_C is the reactance of the capacitance ($1/(2\pi fC)$). The current will lead the voltage and the angle of lead will be given by $\tan \phi = X_C/R$.

Resistance, inductance and capacitance in series. The impedance (Z) of this circuit will be $Z = \sqrt{[R^2 + (X_L - X_C)^2]}$ with $I = V/Z$, and the phase difference will be either

$$\tan \phi = \frac{X_L - X_c}{R} \text{ or } \frac{X_c - X_L}{R}$$

whichever is the higher value. (Here X_L = inductive reactance and X_c = capacitance reactance.)

The latest IEC recommendation is that the terms inductance and capacitance can be dropped when referring to reactance, provided that reactance due to inductance is reckoned positive and to capacitance, negative.

Currents in parallel circuits.   The current in each branch is calculated separately by considering each branch as a simple

17

circuit. the branch currents are then added *vectorially* to obtain the supply current by the following method:

Resolve each branch-current vector into components along axes at right angles (see Figure 1.13), one axis containing the vector of the supply emf. This axis is called the *in-phase* axis; the other axis at 90° is called the *quadrature* axis. Then the supply current is equal to

$$\sqrt{[(\text{sum of in-phase components})^2 + }$$
$$(\text{sum of quadrature components})^2]$$

and
$$\cos \phi = \frac{\text{sum of in-phase components}}{\text{supply current}}$$

Thus if $I_1, I_2, \ldots,$ denote the branch-circuit currents, and $\phi_1, \phi_2, \ldots,$ their phase differences, the in-phase components are $I_1 \cos \phi_1, I_2 \cos \phi_2,$ etc. and the quadrature components are $I_1 \sin \phi_1, I_2 \sin \phi_2,$ etc. Hence the line or supply current is

$$I = \sqrt{(I_1 \cos \phi_1 + I_2 \cos \phi_2 + \ldots)^2 + }$$
$$(I_1 \sin \phi_1 + I_2 \sin \phi_2 + \ldots)^2$$

and
$$\cos \phi = \frac{I_1 \cos \phi_1 + I_2 \cos \phi_2 + \ldots}{I}$$

The quantities $\cos \phi_1$, $\sin \phi_1$, etc., can be obtained from the general formulae: $\cos \phi = $ resistance/impedance, and $\sin \phi = $ reactance/impedance, or $\sin \phi = [I_1 \sin \phi_1 + I_2 \sin \phi_2]/I$.

The equivalent impedance of the circuit is obtained by dividing the line current into the line voltage.

If the equivalent resistance and reactance of this impedance are required, they can be calculated by the formulae:

$$\frac{\text{Equivalent resistance}}{\text{of parallel circuits}} = \text{Impedance} \times \cos \phi$$

$$\frac{\text{Equivalent reactance}}{\text{of parallel circuits}} = \sqrt{[(\text{impedance})^2 - (\text{resistance})^2]}$$

$$= \text{impedance} \times \sin \phi$$

Current in a series-parallel circuit. The first step is to calculate the joint impedance of the parallel portion of the circuit. (Figure 1.14.) The easiest way of doing this is to calculate the branch currents, the joint impedance, and the equivalent resistance and reactance exactly as for a simple parallel circuit. The

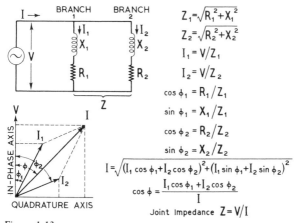

$$Z_1 = \sqrt{R_1^2 + X_1^2}$$
$$Z_2 = \sqrt{R_2^2 + X_2^2}$$
$$I_1 = V/Z_1$$
$$I_2 = V/Z_2$$
$$\cos \phi_1 = R_1/Z_1$$
$$\sin \phi_1 = X_1/Z_1$$
$$\cos \phi_2 = R_2/Z_2$$
$$\sin \phi_2 = X_2/Z_2$$
$$I = \sqrt{(I_1 \cos \phi_1 + I_2 \cos \phi_2)^2 + (I_1 \sin \phi_1 + I_2 \sin \phi_2)^2}$$
$$\cos \phi = \frac{I_1 \cos \phi_1 + I_2 \cos \phi_2}{I}$$

Joint Impedance $Z = V/I$

Figure 1.13
Parallel circuits

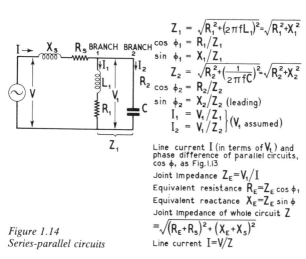

$$Z_1 = \sqrt{R_1^2 + (2\pi f L_1)^2} = \sqrt{R_1^2 + X_1^2}$$
$$\cos \phi_1 = R_1/Z_1$$
$$\sin \phi_1 = X_1/Z_1$$
$$Z_2 = \sqrt{R_2^2 + \left(\frac{1}{2\pi f C}\right)^2} = \sqrt{R_2^2 + X_2^2}$$
$$\cos \phi_2 = R_2/Z_2$$
$$\sin \phi_2 = X_2/Z_2 \text{ (leading)}$$
$$\left.\begin{array}{l} I_1 = V_1/Z_1 \\ I_2 = V_1/Z_2 \end{array}\right\} (V_1 \text{ assumed})$$

Line current I (in terms of V_1) and phase difference of parallel circuits, $\cos \phi$, as Fig.1.13

Joint Impedance $Z_E = V_1/I$
Equivalent resistance $R_E = Z_E \cos \phi_1$
Equivalent reactance $X_E = Z_E \sin \phi$
Joint Impedance of whole circuit Z
$$= \sqrt{(R_E + R_s)^2 + (X_E + X_s)^2}$$
Line current $I = V/Z$

Figure 1.14
Series-parallel circuits

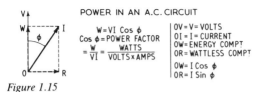

POWER IN AN A.C. CIRCUIT

$$W = VI \cos\phi$$
$$\cos\phi = \text{POWER FACTOR}$$
$$= \frac{W}{VI} = \frac{\text{WATTS}}{\text{VOLTS} \times \text{AMPS}}$$

OV = V = VOLTS
OI = I = CURRENT
OW = ENERGY COMPT
OR = WATTLESS COMPT

OW = I Cos ϕ
OR = I Sin ϕ

Figure 1.15

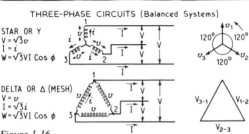

THREE-PHASE CIRCUITS (Balanced Systems)

STAR OR Y
$V = \sqrt{3}\,\upsilon$
$I = i$
$W = \sqrt{3}\,VI \cos\phi$

DELTA OR Δ (MESH)
$V = \upsilon$
$I = \sqrt{3}\,i$
$W = \sqrt{3}\,VI \cos\phi$

Figure 1.16

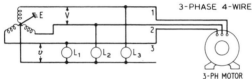

3-PHASE 4-WIRE

3-PH MOTOR

L_1 L_2 L_3 ARE SINGLE-PHASE LOADS AT VOLTAGE υ THREE-PHASE LOADS ARE TAKEN FROM LINES 1,2 AND 3 AT VOLTAGE V. NOTE:- $V = \sqrt{3}\,\upsilon$

Figure 1.17

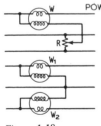

POWER IN 3-PHASE

USE OF ONE WATTMETER FOR
BALANCED LOAD.
NEUTRAL IS OBTAINED BY USE OF
RESISTANCE R
TOTAL POWER = 3 × W

TWO WATTMETER METHOD FOR
BALANCED OR UNBALANCED LOADS
TOTAL POWER = $W_1 + W_2$
POWER FACTOR IS OBTAINED FROM
$$\tan\phi = \frac{\sqrt{3}\,(W_1 - W_2)}{W_1 + W_2}$$

Figure 1.18

calculations can be made without a knowledge of the voltage across the parallel portion of the circuit (which is unknown at the present stage) by assuming a value, V_1.

Having obtained the joint impedance (Z_E) of the parallel portion of the circuit, this is added vectorially to the series impedance (Z_S) to obtain the joint impedance (Z) of the whole circuit, whence the current is readily obtained in the usual manner. Thus, the joint impedance of the parallel circuits (Z_E) must be split into resistance and reactance, i.e. $R_E = Z_E \cos \phi_E$ and $X_E = Z_E \sin \phi_E$, where ϕ_E is the phase difference. This resistance and reactance is added to the resistance and reactance of the series portion of the circuit, in order to calculate the joint impedance of the whole circuit.

Thus, the resistance term (R) of the joint impedance (Z) of the series-parallel circuit is equal to the sum of the resistance terms ($R_E + R_S$) of the separate impedances. Similarly, the reactance term (X) is equal to the sum of the reactance terms of the separate impedances, i.e. $X_E + X_S$. Hence the joint impedance of the series-parallel circuit is

$$Z = \sqrt{(R^2 + X^2)}$$
and line current $= V/Z$

Three-phase circuits. Three-phase currents are determined by considering each phase separately, and calculating the phase currents from the phase voltages and impedances in the same manner as for single-phase circuits. In practice, three-phase systems are usually symmetrical, the loads being balanced. In such cases the calculations are simple and straightforward. For the methods of calculations when the loads are unbalanced or the system is unsymmetrical reference should be made to the larger textbooks.

Having calculated the phase currents, the line currents are obtained from the following simple rules.

With a *star-connected system*:

line current $=$ phase current
line voltage $= 1.73 \times$ phase voltage

With a *delta-connected system*:

line current $= 1.73 \times$ phase current
line voltage $=$ phase voltage

Power in AC circuits. The power in a single-phase circuit is given by $W = VI \cos \phi$, where W is the power in watts, V the voltage (rms) and I the current (rms). $\cos \phi$ represents the power-factor of the circuit, so that

$$\text{power-factor} = \cos \phi = \frac{W}{VI} = \frac{\text{watts}}{\text{volt-amperes}}$$

Referring to Figure 1.15, I represents a current lagging by angle ϕ. This current can be split into two components, OW, the energy component, and OR, the wattless component. Only the energy component has any power value, so that the power is given by OV $\times$ OW = OV $\times$ OI $\cos \phi$ = $VI \cos \phi$.

Three-phase working. The three windings of a three-phase alternator or transformer can be connected in two ways, as shown in Figure 1.16. The relations between the phase voltages and currents and the line voltages and currents are indicated in this diagram. It should be noted that with the star or Y connection a neutral point is available, whereas with the delta or mesh connection this is not so. Generators are generally star wound and the neutral point used for earthing. Motors can be either star or delta, but for low voltage small-size motors a delta connection is usually used to reduce the size of the windings.

Power in a three-phase circuit. The total power in a three-phase circuit is the sum of the power in the three phases. Taking the star system in Figure 1.17 and assuming a balanced system (i.e. one in which the three voltages and currents are all equal and symmetrical), the total power must by 3 $\times$ power per phase. Therefore $W = 3vi \cos \phi$. Substituting the line values for phase volts and phase current, we get

$$W = 3(vi) \cos \phi = 3 \left(\frac{V}{\sqrt{3}} \times I \right) \cos \phi = \sqrt{3} \, VI \cos \phi$$

It will be found that the same expression gives the power in a delta-connected system, and so for any balanced system the power is given by $W = \sqrt{3} \, VI \cos \phi$, where V and I are the line volts and line current and $\cos \phi$ represents the power factor. For unbalanced or unsymmetrical systems the above expression does not hold good.

(Most three-phase apparatus such as motors can be assumed to form a balanced load, and calculations for current, etc., can be based on this assumption, using the above expression.)

22

Power in a three-phase circuit can be measured in several ways. For permanent switchboard work a three-phase wattmeter unit is used in which there are usually two elements, so that the meter will indicate both balanced and unbalanced loads. For temporary investigations either of the methods shown in Figure 1.18 can be used. The total power = $3W$ where W is the reading on the single meter.

For an unbalanced load two units must be used, and these are connected as indicated in Figure 1.18. In addition to giving the total power by adding the readings on the two meters, the power factor can be obtained. It is important to note, however, that the reading of one meter will be reversed if the power factor of the system is less than 0.5. In this case the leads of one of the meters may have to be reversed in order to get a positive reading. For power factors of less than 0.5 the readings must be *subtracted* instead of added.

The power factor of the system can be obtained from

$$\tan \phi = \frac{\sqrt{3}(W_1 - W_2)}{(W_1 + W_2)}$$

which gives the tangent of the angle of lag, and the cosine can be obtained from the tables.

Power in six-phase. In a six-phase system, such as is often used for rotary converters and other rectifiers, the power of the system (assumed balanced) is given by

$$W = 6VI \cos \phi$$

where V is the phase voltage and I the phase current.

In terms of line voltage V_L and line current I_L the power equation becomes

$$\frac{3}{\sqrt{2}} V_L I_L \cos \phi$$

In both cases $\cos \phi$ is the phase angle between the phase voltage and phase current.

Three-phase 4-wire. This system (Figure 1.16) is now used in most districts in the UK for distribution. There are three 'lines' and a neutral. The voltage between any one 'line' and neutral is usually 240 V and voltage between the 'lines' is $\sqrt{3}$ times the voltage to neutral. This gives a three-phase voltage of 415 V for motors, etc. Single-phase loads are therefore taken from all 'lines' to neutral and three-phase loads from the three lines marked 1, 2 and 3.

In the distribution cable the neutral may be either equal to the 'lines' or half-size. Modern systems generally use a full-size neutral, particularly where fluorescent lighting loads predominate.

2

Properties of materials

MAGNETIC MATERIALS

Permanent magnets (cast).   Great advances have been made in the development of materials suitable for the production of permanent magnets. The earliest materials were tungsten and chromium steel, followed by the series of cobalt steels.

Alni was the first of the aluminium-nickel-iron alloys to be discovered and with the addition of cobalt, titanium and niobium, the Alnico series of magnets was developed, the properties of which varied according to composition. These are hard and brittle, and can only be shaped by grinding, although a certain amount of drilling is possible on certain compositions after special heat treatment.

The Permanent Magnet Association (disbanded March 1975) discovered that certain alloys when heat-treated in a strong magnetic field became anisotropic. That is they develop high properties in the direction of the field at the expense of properties in other directions. This discovery led to the powerful Alcomax and Hycomax series of magnets. By using special casting techniques to give a grain oriented structure, even better properties are obtained if the field applied during heat treatment is parallel to the columnar crystals in the magnet.

Permanent magnets (sintered).   The techniques of powder metallurgy have been applied to both the isotropic and anisotropic Alnico types and it is possible to produce sintered permanent magnets which have approximately 10 per cent poorer remanence and energy than cast magnets. More precise shapes are possible when using this method of production and it is economical for the production of large quantities of small magnets. Sintering techniques are also used to manufacture the oxide permanent magnets based on barium or strontium hexaferrite. These magnets which may be isotropic or anisotropic, have higher coercive force but lower remanence than the alloy magnets described above. They have the physical properites of ceramics, and inferior temperature stability, but their low cost makes them ideal for certain applications. Barium ferrite bonded in rubber or plastics is available as extruded strip or rolled sheet.

The newest and most powerful permanent magnets discovered to date, based on an intermetallic compound of cobalt and samarium, are also made by powder metallurgy techniques.

Table 2.1   Properties of permanent magnets*

Material	Reman-ence T	Coercive force kAm^{-1}	BH_{max} kJm^{-3}	Sp. Gr.	Description
ISOTROPIC					
Tungsten steel 6%W	1.05	5.2	2.4	8.1	Rolled or forged steel
Chromium steel 6%Cr	0.98	5.2	2.4	7.8	Rolled or forged steel
Cobalt steel 3%Co	0.72	10.4	2.8	7.7	Rolled or forged steel
Cobalt steel 6%Co	0.75	11.6	3.5	7.8	Rolled or forged steel
Cobalt steel 9%Co	0.78	12.8	4.0	7.8	Rolled or forged steel
Cobalt steel 15%Co	0.82	14.4	5.0	7.9	Rolled or forged steel
Cobalt steel 35%Co	0.90	20	7.6	8.2	Rolled or forged steel
Alni	0.55	38.5	10	6.9	Cast Fe-Ni-Al
Alnico	0.75	58	13.5	7.3	Cast Fe-Ni-Al
Feroba 1 (sintered)	0.21	136	6.4	4.8	Barium ferrite
Bonded Feroba	0.17	128	5.6	3.6	Flexible strip or sheet
ANISOTROPIC					
Alcomax II	1.20	46	41	7.35	Cast Fe-Co-Ni-Al
Alcomax III	1.30	52	44	7.35	Cast Fe-Co-Ni-Al-Nb
Alcomax IV	1.15	62	36	7.35	Cast Fe-Co-Ni-Al-Nb
Columax	1.35	59	60	7.35	Grain oriented Alcomax III
Hycomax II	0.75	96	32	7.3	Cast Fe-Co-Ni-Al-Nb-Ti
Hycomax III	0.92	132	44	7.3	Cast Fe-Co-Ni-Al-Ti
Hycomax IV	0.78	160	46	7.3	Cast Fe-Co-Ni-Al-Ti
Columnar Hycomax III	1.05	128	72	7.3	Grain oriented
Feroba II	0.35	144	26.4	5.0	Barium ferrite
Feroba III	0.25	200	20	4.7	Barium ferrite
Sintered Sm Co₅	0.80	600	128	8.1	Cobalt-samarium

* *Permanent Magnets*, by Malcolm McCaig, Pentech Press, 1977.

Nickel-iron alloys.   Nickel-iron alloy containing about 25 per cent of nickel is practically non-magnetic, but with increased nickel content and suitable treatment some remarkably high permeability materials have been obtained. Some of the more popular alloys and their magnetic properties are shown in Tables 2.2(a) and 2.2(b).

From these tables it will be seen that there are two groups falling within the range 36-50 per cent. The alloys with the higher nickel content have higher initial and maximum permeabilities but lower saturation inductions, remanence and coercivity.

Table 2.2(a)   Properties of high-permeability
nickel-iron alloys (75-80% Ni-Fe alloys)

Property	Mumetal*	Nilomag*	Permalloy C*
Initial permeability	60 000	50 000	50 000
Maximum permeability	240 000	250 000	250 000
Saturation induction			
B_{sat} Tesla	0.77	0.70	0.80
Remanence			
B_{rem} Tesla	0.45	0.40	0.35
Coercivity H_c(A/m)	1.00	1.60	2.4

*Grades of higher magnetic quality, Mumetal plus Supermetal, Permalloy 'Super C' and Nilomag 771 are available

Table 2.2(b)   Properties of high-permeability nickel-iron alloys (36-50% Ni-Fe alloys)

Property	Radio-metal* 50	Permalloy † B	Nilo alloy 45	Radio-metal 36	Nilo alloy 36
Inital permeability	6 000	5 000	6 000	3 000	4 000
Maximum permeability	30 000	30 000	30 000	20 000	18 000
Saturation induction					
B_{sat} Tesla	1.6	1.6	1.2	1.2	0.8
Remanence B_{rem} Tesla	1.0	0.4	1.1	0.5	0.4
Coercivity H_c (A/m)	8.0	12.0	16.0	10.0	10.0

*Super, Hyrno and Hyrem Radiometal are derived from Radiometal 50 and offer improved permeability, electrical resistivity and remanence respectively
†Permalloys D and F offer higher electrical and remanence respectively

Radiometal and Mumetal are trade names of Telcon Metals Ltd; Nilo and Nilomag are trade names of Henry Wiggin Ltd; Permalloy is a trade name of Standard Telephones and Cables Ltd.

Typical applications for these nickel-iron alloys are detailed in Table 2.3. From this table it will be seen that the materials are particularly suitable for high frequency applications.

Applications	% Nickel		
	75-80	45-50	36
Transformer			
Pulse	x	x	x
Audio	x	x	
Microphone	x		
Current	x		
Output		x	
Small power		x	
High frequency		x	x
Magnetic amplifiers	x	x	
Magnetic screening	x	x	
Tape recorder heads		x	
Relays			
Cores and armatures		x	x
Small motors, synchros, rotors and stators		x	
Inductors, chokes (h.f.) and filter circuits	x		

COPPER AND ITS ALLOYS

The electrical resistance of copper, as of all other pure metals, varies with the temperature. This variation is sufficient to reduce the conductivity of high conductivity copper at 100°C to about 76% of its value at 20°C.

The resistance $\quad R_t' = R_t[1 + \alpha_t(t' - t)]$

where α_t is the constant mass temperature coefficient of resistance of copper at the reference t°C. For a reference temperature of 0°C the formula becomes

$$R_t = R_0(1 + \alpha_0 t)$$

Although resistance may be regarded for all practical purposes as a linear function of temperature, the value of the temperature coefficient is not constant but is dependent upon, and varies with, the reference temperature according to the law

$$\alpha_t = \frac{1}{\dfrac{1}{\alpha_0} + t} = \frac{1}{234.45 + t}$$

Thus the constant mass temperature coefficient of copper referred to a basic temperature of 0°C is

$$\alpha_0 = \frac{1}{234.45} = 0.004265 \text{ per } °\text{C}$$

At 20°C the value of the constant mass temperature coefficient of resistance is

$$\alpha_{20} = \frac{1}{234.45 + 20} = 0.00393 \text{ per } °\text{C}$$

which is the value adopted by the I.E.C.

Multiplier constants and their reciprocals, correlating the resistance of copper at a standard temperature, with the resistance at other temperatures, may be obtained from tables which are included in BS 125, 1432–1434, 4109.

Five alloys discussed below also find wide application in the electrical industry where high electrical conductivity is required. These are cadmium copper, chromium copper, silver copper, tellurium copper and sulphur copper. They are obtainable in wrought forms and also, particularly for chromium copper, tellurium copper and sulphur copper, as castings and forgings. The electrical resistivity varies from 1.71 microhm cm for silver copper in the annealed state at 20°C to 4.9 microhm cm for solution heat-treated chromium copper at the same temperature.

The main output of each alloy is determined by its major applications. For instance cadmium copper is produced as heavy gauge wire of special sections while silver copper is made generally in the form of drawn sections and strip. Much chromium copper is produced as bar and also as castings and forgings, though strip and wire forms are available.

Quantities of the five elements required to confer the differing properties on these alloys are quite small, the normal commercial ranges being: cadmium copper 0.7-1.0% cadmium; chromium copper 0.4-0.8% chromium; silver copper 0.03-0.1% silver; tellurium copper 0.3-0.7% tellurium; and sulphur copper 0.3-0.6% sulphur.

Cadmium copper, chromium copper and sulphur copper are deoxidised alloys containing small controlled amounts of deoxidant. Silver copper, like high-conductivity copper, can be 'tough pitch' (oxygen-containing) or oxygen-free, while tellurium copper may be either tough-pitch or deoxidised. 'Tough pitch' coppers

29

and alloys become embrittled at elevated temperatures in a reducing atmosphere. Thus, when such conditions are likely to be encountered, oxygen-free or deoxidised materials should be used. Advice can be sought from the Copper Development Association which assisted in the compilation of these notes.

Cadmium copper. This material is characterised by having greater strength under both static and alternating stresses and better resistance to wear than ordinary copper. As such it is particularly suitable for the contact wires of electric railways, tramways, trolley-buses, gantry cranes and similar equipment. It is also employed for Post Office line wires and overhead transmission lines of long span.

Because cadmium copper retains the hardness and strength imparted by cold work at temperatures well above those at which high-conductivity copper would soften it has another field of application. Examples are electrode holders for resistance welding machines and arc furnaces, and electrodes for spot and seam welding of steel. Cadmium copper has also been employed for the commutator bars of certain types of electric motors.

Because of its comparatively high elastic limit in the work-hardened condition, cadmium copper is also used to a limited extent for small springs required to carry current. In the form of thin hard-rolled strip an important use is for reinforcing the lead sheaths of cables which operate under internal pressure. Castings of cadmium copper, though rare, do have certain applications for switchgear components and the secondaries of transformers for welding machines.

On exposure to atmosphere the material acquires the normal protective patina associated with copper. Cadmium copper can be soft soldered, silver soldered and brazed in the same manner as ordinary copper. Being a deoxidised material there is no risk of embrittlement by reducing gases during such processes.

Chromium copper. Chromium copper is particularly suitable for applications in which considerably higher strengths than that of copper are required. For example for both spot and seam types of welding electrodes. Strip, and, to a lesser extent, wire are used for light springs destined to carry current. Commutator segments that are required to operate at temperatures above those normally encountered in rotating machines are another application. In its heat-treated state, the material can be used at temperatures up to about 350°C without risk of deterioration of properties.

30

In the solution heat-treated condition chromium copper is soft and can be machined. It is not difficult to cut in the hardened state but is not free-machining like leaded brass or tellurium copper. Chromium copper is similar to ordinary copper in respect of oxidation and scaling at elevated temperatures. Joining methods similar to cadmium copper outlined above are applicable. As in the case of cadmium copper special fluxes are required under certain conditions, and these should contain fluorides. Chromium copper can be welded using modern gas-shielded arc-welding technology.

Silver copper. Silver copper has an electrical conductivity equal to that of ordinary high-conductivity copper, but in addition, it possesses two properties which are of practical importance. Its softening temperature, after hardening by cold work, is considerably higher than that of ordinary copper, and its resistance to creep at moderately elevated temperatures is enhanced.

The principal uses of this material are in connection with electrical machines which either run at higher than normal temperatures or are exposed to them during manufacture. Soft soldering or stoving of insulating materials are examples of the latter.

Silver copper is obtainable in the form of hard drawn or rolled rods and sections, especially those designed for commutator segments, rotor bars and similar applications. It is also available as hollow conductors and strip. It is rarely called for in the annealed condition since its outstanding property is associated with retention of work hardness at elevated temperatures.

Silver copper can be soft soldered, silver soldered, brazed or welded without difficulty but the temperatures involved in all these processes, except soft soldering, are sufficient to anneal the material if in the cold-worked condition. Because the tough pitch material contains oxygen in the form of dispersed particles of cuprous oxide, it is important to avoid heating to brazing and welding temperatures in a reducing atmosphere.

While silver copper cannot be regarded as a free-cutting material, it is not difficult to machine. This is specially true when it is in the work-hardened condition, the state in which it is usually supplied. It is similar to ordinary copper in its resistance to corrosion. If corrosive fluxes are employed for soldering, the residues should be carefully washed away after soldering is completed.

Tellurium copper. Special features of this material are ease of machining combined with high electrical conductivity, retention of work hardening at moderately elevated temperatures and good resistance to corrosion. Tellurium copper is unsuitable for welding with most procedures, but gas-shielded arc welding and resistance welding can be effected with care. A typical application of this material is for magnetron bodies, which in many cases are machined from solid blocks of the material.

Table 2.4 Physical properties of copper alloys

Property	Cadmium copper	Chromium copper	Silver copper	Tellurium copper	Sulphur copper
Density at 20°C (10^3 kg m^{-3})	8.9	8.90	8.89	8.9	8.9
Coefficient of linear expansion (20–100°C) (10^{-6} K^{-1})	17	17	17.7	17	17
Modulus of elasticity† (10^9 N m^{-2})	132	108	118	118	118
Specific heat at 20°C (kJ kg^{-1} K^{-1})	0.38	0.38	0.39	0.39	0.39
Electrical conductivity at 20°C (10^6 S m^{-1})					
annealed	46–53	–	57.4–58.6	56.8*	55.1
solution heat treated	–	20	–	–	–
precipitation hardened	–	44–49	–	55.7*	–
Resistivity at 20°C (10^{-8} ohm m)					
annealed	2.2–1.9	–	1.74–1.71	1.76*	1.81
solution heat treated	–	4.9	–	–	–
precipitation hardened	–	2.3–2.0	–	–	–
cold worked	2.3–2.0	–	1.78	1.80	1.85

*Oxygen-bearing (tough pitch) tellurium copper
†Solution heat treated or annealed

Tellurium copper can be soft soldered, silver soldered and brazed without difficulty. For tough pitch, tellurium copper brazing should be carried out in an inert atmosphere (or slightly oxidising) since reducing atmospheres are conducive to embrittlement. Deoxidised tellurium copper is not subject to embrittlement.

Sulphur copper. Like tellurium copper, sulphur copper is a high-conductivity free-machining alloy, with greater resistance to softening than high-conductivity copper at moderately elevated temperatures, and with good resistance to corrosion. It is equivalent in machinability to tellurium copper, but without the tendency shown by the latter to form coarse stringers in the structure which can affect accuracy and finish of fine machining operations.

Sulphur copper finds application for all machined parts requiring high electrical conductivity, such as contacts, connectors and other electrical components. Joining characteristics are similar to those of tellurium copper.

Sulphur copper is deoxidised with a controlled amount of phosphorus and therefore does not suffer from hydrogen embrittlement in normal torch brazing operations; long exposure to reducing atmospheres can result in some loss of sulphur and consequent embrittlement.

ALUMINIUM AND ITS ALLOYS

For many years aluminium has been used as a conductor material in most branches of electrical engineering. In addition to the pure metal, several aluminium alloys are also good conductors, combining structural strength with an acceptable conductivity. The material is lighter than copper and therefore easier to handle; it is also cheaper. Another advantage is that its price is not subject to wide fluctuations as is copper.

There are two specifications for aluminium, one covering pure metal grade 1E and the other, a heat treatable alloy 91E. Grade 1E is available in a number of forms i.e. extruded tubes and sections (E1E); solid conductor (C1E); wire (G1E); and rolled strip (D1E). The heat treatable alloy, which has moderate strength and a conductivity approaching that of pure aluminium, is available in tubes and sections (E91E).

Relevant British standards covering aluminium for electrical purposes are:

BS 215 Part 1: Aluminium stranded conductors for overhead power transmission purposes. Part 2: Aluminium conductors, steel-reinforced for overhead power transmission purposes;

BS 2627. Wrought aluminium for electrical purposes – wire;

BS 2897. Wrought aluminium for electrical purposes – strip with drawn or rolled edges;

BS 2898. Wrought aluminium for electrical purposes – bars, extruded round tubes and sections;

BS 3242. Aluminium alloy stranded conductors for overhead power transmission;

BS 3988. Wrought aluminium for electrical purposes – solid conductors for insulated cables.

BS 6360. Specifications for conductors in insulated cables and cords.

Busbars.   Aluminium has been used for busbars for the past sixty years but only recently has it been accepted for general use. Full tables of current ratings are available for the different sizes and sections that are manufactured. The supply industry has adopted aluminium busbars as standard for its 400 kV substations. It is also widely used in switchgear, plating shops, rising mains and in the UK aluminium smelter plants. Sometimes the busbars are tin-plated for application where joints have to be dismantled and re-made frequently.

Cable.   Aluminium is extensively employed as the conductor for l.v. distribution cables and for high voltages up to 11 kV. House wiring cables above $2\frac{1}{2}$ mm^2 are also available with aluminium conductor.

Overhead lines.   The a.c.s.r. (aluminium conductor steel reinforced) conductor is the standard adopted for overhead transmission lines in most countries throughout the world. In the USA a.c.a.r. (aluminium conductor aluminium alloy wire reinforced) is rapidly gaining ground. The advantages of this new conductor are freedom from bimetallic corrosion and improved conductance for a given cross section.

Motors.   Cage rotors for induction motors often employ aluminium bars. Casings are also made from the material as are fans used for motor cooling purposes.

Foil windings.   Foil coil windings are suitable for transformers, reactors and solenoids. Foil thicknesses range from 0.040 mm to 1.20 mm in 34 steps. A better space factor than for a wire wound copper coil is obtained, the aluminium conductor occupying some 90% of the space as against 60% for copper. Heating and cooling is aided by the better space factor and the small amount of insulation needed for foil wound coils. Rapid radial heat transfer ensures an even temperature gradient.

Heating elements.   Aluminium foil heating elements have been developed but are not widely used at present. Applications include foil film wallpaper, curing concrete and possibly soil warming.

Table 2.5   Constants and physical properties of very high purity aluminium

Atomic number	13
Atomic volume	10 cm^3/g-atom
Atomic weight	26.98
Valency	3
Crystal structure	fcc
Interatomic distance (co-ordination number 12)	2.68 kX
Heat of combustion	200 k cal/g-atom
Latent heat of fusion	94.6 cal/g
Melting point	660.2 °C
Boiling point	2480 °C
Vapour pressure at 1200°C	1 × 10^{-2} mmHg
Mean specific heat (0 − 100°C)	0.219 cal/g°C
Thermal conductivity (0 − 100°C)	0.57 cal/cm s°C
Temperature coefficient of linear expansion (0 − 100°C)	23.5 × 10^{-6} per°C
Electrical resistivity at 20°C	2.69 microhm cm
Temperature coefficient of resistance (0 − 100°C)	4.2 × 10^{-3} per°C
Electrochemical equivalent	3.348 × 10^{-1} g/Ah
Density at 20°C	2.6898 g/cm^3
Modulus of elasticity	68.3 kN/mm^2
Modulus of torsion	25.5 kN/mm^2
Poisson's ratio	0.34

Heat sinks.   High thermal conductivity of aluminium and ease of extruding or casting into solid or hollow shapes with integral fins makes the material ideal for heat sinks. Semiconductor devices and transformer tanks illustrate the wide diversity of applications in this field. Its light weight makes it ideal for pole-mounted transformer tanks and it has the added advantage that the material does not react with transformer oil to form a sludge.

INSULATING MATERIALS

The publication in 1986 of BS 2757 has introduced a different concept of insulating materials than that outlined in the same standard issued in 1956. Alteration of the title to *Determining the Thermal Classification of Electrical Insulation* without reference to electrical machinery and apparatus which appeared in the title of the edition of 1956 is indicative of this.

Table 2.6 Class 2 stranded conductors for single-core and multicore cables (from BS 6360)

1	2	3	4	5	6	7	8	9	10
	Minimum number of wires in the conductor						Maximum resistance of conductor at 20°C		
	Circular conductor		Circular compacted conductor		Shaped conductor		Annealed copper conductor*		Aluminium conductor plain or metal-clad wires
Nominal cross-sectional area	Cu	Al	Cu	Al	Cu	Al	Plain wires	Metal-coated wires	
mm²							Ω/km	Ω/km	Ω/km
0.5	7	–	–	–	–	–	36.0	36.7	–
0.75	7	–	–	–	–	–	24.5	24.8	–
1	7	–	–	–	–	–	18.1	18.2	–
1.5	7	–	6	–	–	–	12.1	12.2	–
2.5	7	–	6	–	–	–	7.41	7.56	–
4	7	7	6	–	–	–	4.61	4.70	7.41
6	7	7	6	–	–	–	3.08	3.11	4.61
10	7	7	6	–	–	–	1.83	1.84	3.08
16	7	7	6	6	–	–	1.15	1.16	1.91
25	7	7	6	6	6	6	0.727	0.734	1.20
35	7	19	6	6	6	6	0.524	0.529	0.868
50	19	19	6	6	6	6	0.387	0.391	0.641
70	19	19	12	12	12	12	0.268	0.270	0.443
95	19	19	15	15	15	15	0.193	0.195	0.320
120	37	37	18	15	18	15	0.153	0.154	0.253

150	37	37	18	15	18	15	0.124	0.126	0.206
185	37	37	30	30	30	30	0.0991	0.100	0.164
240	61	61	34	30	34	30	0.0754	0.0762	0.125
300	61	61	34	30	34	30	0.0601	0.0607	0.100
400	61	61	53	53	53	53	0.0470	0.0475	0.0778
500	61	61	53	53	53	53	0.0366	0.0369	0.0605
630	91	91	53	53	53	53	0.0283	0.0286	0.0469
800	91	91	53	53	53	–	0.0221	0.0224	0.0367
960 (4 × 240)	Number of wires not specified						0.0189	0.0189	0.0313
1000	91	91	53	53	–	–	0.0176	0.0177	0.0291
1200	Number of wires not specified			53			0.0151	0.0151	0.0247
1600	–	–	–	–	–	–	0.0113	0.0113	0.0186
2000	–	–	–	–	–	–	0.0090	0.0090	0.0149

* To obtain the maximum resistance of hard-drawn conductors the values in columns 8 and 9 should be divided by 0.97.

Table 2.7 Standard aluminium conductors, steel reinforced (from BS 215: Part 2)

Nominal aluminium area (mm²)	Stranding and wire diameter Aluminium (mm)	Steel (mm)	Sectional area of aluminium (mm²)	Total sectional area (mm²)	Approx. overall diameter (mm)	Approx. mass per km (kg)	Calculated d.c. resistance at 20°C per km (Ω)	Calculated breaking load (kN)
25	6/2.36	1/2.36	26.24	30.62	7.08	106	1.093	9.61
30	6/2.59	1/2.59	31.61	36.88	7.77	128	0.9077	11.45
40	6/3.00	1/3.00	42.41	49.48	9.00	172	0.6766	15.20
50	6/3.35	1/3.35	52.88	61.70	10.05	214	0.5426	18.35
70	12/2.79	7/2.79	73.37	116.2	13.95	538	0.3936	61.20
100	6/4.72	7/1.57	105.0	118.5	14.15	394	0.2733	32.70
150	30/2.59	7/2.59	158.1	194.9	18.13	726	0.1828	69.20
150	18/3.35	1/3.35	158.7	167.5	16.75	506	0.1815	35.70
175	30/2.79	7/2.79	183.4	226.2	19.53	842	0.1576	79.80
175	18/3.61	1/3.61	184.3	194.5	18.05	587	0.1563	41.10
200	30/3.00	7/3.00	212.1	261.5	21.00	974	0.1363	92.25
200	18/3.86	1/3.86	210.6	222.3	19.30	671	0.1367	46.55
400	54/3.18	7/3.18	428.9	484.5	28.62	1 621	0.06740	131.9

Thermal classes and the temperatures assigned to them are as follows:

Thermal class	Temperature (°C)
Y	90
A	105
E	120
B	130
F	155
H	180
200	200
220	220
250	250

Temperatures over 250°C should increase by 25°C intervals and classes designated accordingly. Use of letters is not mandatory but the relationship between letters and temperatures should be adhered to.

When a thermal class describes an electrotechnical product it normally represents the maximum temperature appropriate to that product under rated load and other conditions. Thus the insulation subjected to this maximum temperature needs to have a thermal capability at least equal to the temperature associated with the

Table 2.8 Class 1 solid conductors for single-core and multicore cables (from BS 6360)

1	2	3	4
Nominal cross-sectional area	Maximum resistance of conductor at 20°C		
	Circular, annealed copper conductors*		Aluminium conductors, circular or shaped, plain or metal-clad
	Plain	Metal-coated	
mm²	Ω/km	Ω/km	Ω/km
0.5	36.0	36.7	–
0.75	24.5	24.8	–
1	18.1	18.2	–
1.5	12.1	12.2	18.1†
2.5	7.41	7.56	12.1†
4	4.61	4.70	7.41†
6	3.08	3.11	4.61†
10	1.83	1.84	3.08†
16	1.15	1.16	1.91†
25	0.727	–	1.20
35	0.524	–	0.868
50	0.387	–	0.641
70	0.268	–	0.443
95	0.193	–	0.320
120	0.153	–	0.253
150	0.124	–	0.206
185	–	–	0.164
240	–	–	0.125
300	–	–	0.100
380 (4 × 95)	–	–	0.0800
480 (4 × 120)	–	–	0.0633
600 (4 × 150)	–	–	0.0515
740 (4 × 185)	–	–	0.0410
960 (4 × 240)	–	–	0.0313
1200 (4 × 300)	–	–	0.0250

* To obtain the maximum resistance of hard-drawn conductors the values in columns 2 and 3 should be divided by 0.97.
† Aluminium conductors 1.5 mm² to 16 mm² circular only.

thermal class of the product. However, the description of an electrotechnical product as being of a particular thermal class does not mean, and must not be taken to imply, that each insulating material used in its construction is of the same thermal capacity. It is also important to note that the temperatures in the table are actual

Table 2.9 Class 5 flexible copper conductors for single-core and multicore cables (from BS 6360)

1	2	3	4
Nominal cross-sectional area	Maximum diameter of wires in conductor	Maximum resistance of conductor at 20°C	
		Plain wires	Metal-coated wires
mm²	mm	Ω/km	Ω/km
0.22	0.21	92.0	92.4
0.5	0.21	39.0	40.1
0.75	0.21	26.0	26.7
1	0.21	19.5	20.0
1.25	0.21	15.6	16.1
1.35	0.31	14.6	15.0
1.5	0.26	13.3	13.7
2.5	0.26	7.98	8.21
4	0.31	4.95	5.09
6	0.31	3.30	3.39
10	0.41	1.91	1.95
16	0.41	1.21	1.24
25	0.41	0.780	0.795
35	0.41	0.554	0.565
50	0.41	0.386	0.393
70	0.51	0.272	0.277
95	0.51	0.206	0.210
120	0.51	0.161	0.164
150	0.51	0.129	0.132
185	0.51	0.106	0.108
240	0.51	0.0801	0.0817
300	0.51	0.0641	0.0654
400	0.51	0.0486	0.0495
500	0.61	0.0384	0.0391
630	0.61	0.0287	0.0292

temperatures of the insulation and not the temperature rises of the product itself.

The 1956 edition of BS 2757 gave typical examples of insulating materials and their classifications as Group Y, A, E etc. That concept no longer exists but Table 1 of BS 5691 Part 2 lists materials and the tests which may be appropriate for determining their thermal endurance properties. This table lists three basic classes of material which are then further subdivided. The three classes are: (a) solid

Table 2.10   Class 6 flexible copper conductors for single-core and
multicore cables (from BS 6360)

1	2	3	4
Nominal cross-sectional area	Maximum diameter of wires in conductor	Maximum resistance of conductor at 20°C	
		Plain wires	Metal-coated wires
mm²	mm	Ω/km	Ω/km
0.5	0.16	39.0	40.1
0.75	0.16	26.0	26.7
1	0.16	19.5	20.0
1.5	0.16	13.3	13.7
2.5	0.16	7.98	8.21
4	0.16	4.95	5.09
6	0.21	3.30	3.39
10	0.21	1.91	1.95
16	0.21	1.21	1.24
25	0.21	0.780	0.795
35	0.21	0.554	0.565
50	0.31	0.386	0.393
70	0.31	0.272	0.277
95	0.31	0.206	0.210
120	0.31	0.161	0.164
150	0.31	0.129	0.132
185	0.41	0.106	0.108
240	0.41	0.0801.	0.0817
300	0.41	0.0641	0.0654

insulation of all forms not undergoing a transformation during application; (b) solid sheet insulation for winding or stacking, obtained by bonding superimposed layers; and (c) insulation which is solid in its final state but applied in the form of a liquid or paste, for filling, varnishing, coating or bonding.

Examples under class (a) are inorganic sheet insulation like mica, laminated sheet insulation, ceramics, glasses and quartz, elastomers, thermosetting and thermoplastic moulded insulation.

Examples under class (b) are solid sheet insulation bonded together by pressure-sensitive adhesive, heat, simple fusion and fusion combined with chemical reaction. Again mica products fall into this category as do adhesive coated films, papers, fabrics and laminates.

In the final class (c) the insulating material may be formed by physical transformation such as congealing, evaporation or a solvent or gelation. Fusible insulation materials with and without fillers, plastisols and organosols are examples. Another method is to solidify the insulation by chemical reactions such as polymerisation, poly-condensation or polyaddition. Thermosetting resins and certain paste materials are examples. Table II in the same standard lists available tests, the methods of carrying them out (by reference to an IEC or ISO standard), the specimen and end-point criterion.

New definitions. New definitions are now included in BS 2757 but the reader is also referred to IEC 216 and its Parts 1, 2, 3 and 4.

Temperature index (TI). The number corresponding to the temperature in degrees Celsius derived from the thermal endurance relationship at a given time, normally 20,000 h.

Relative temperature index (RTI). The temperature index of a test material obtained from the time which corresponds to the known temperature index of a reference material, when both materials are subjected to the same ageing and diagnostics procedures in a comparative test.

Halving interval (HIC). The number corresponding to the tem-perature interval in degrees Celsius which expresses the halving of the time to the end point taken at the temperature of the TI or the RTI.

Properties. The following notes give briefly the chief points to be borne in mind when considering the suitability of any material for a particular duty.

Relative density is of importance for varnishes, oils and other liquids. The density of solid insulations varies widely, e.g. from 0.6 for certain papers to 3.0 for mica. In a few cases it indicates the relative quality of a material, e.g. vulcanised fibre and pressboard.

Moisture absorption usually causes serious depreciation of electrical properties, particularly in oils and fibrous materials. Swelling, warping, corrosion and other effects often result from absorption of moisture. under severe conditions of humidity, such as occur in mines and in tropical climates, moisture sometimes causes serious deterioration.

Thermal effects very often seriously influence the choice and application of insulating materials, the principal features being: melting-point (e.g. of waxes); softening or plastic yield temperature; ageing due to heat, and the maximum temperature which a material will withstand without serious deterioration of essential properties; flash point or ignitibility; resistance to electric arcs; liability to carbonize (or 'track'); ability to self-extinguish if ignited; specific heat; thermal resistivity; and certain other thermal properties such as coefficient of expansion and freezing-point.

Mechanical properties. The usual mechanical properties of solid materials are of varying significance in the case of those required for insulating purposes, *tensile* strength, *transverse* strength, *shearing* strength and *compressive* strength often being specified. Owing, however, to the relative degree of inelasticity of most solid insulations, and the fact that many are quite brittle, it is frequently necessary to pay attention to such features as *compressibility, deformation under bending* stresses, *impact* strength and *extensibility, tearing* strength, *machinability* and ability to fold without damage.

Resistivity and insulation resistance. In the case of insulating material it is generally manifest in two forms (*a*) volume resistivity (or specific resistance) and (*b*) surface resistivity.

Electric strength (or dielectric strength) is the property of an insulating material which enables it to withstand electric stress without injury. It is usually expressed in terms of the minimum electric stress (i.e. potential difference per unit distance) which will cause failure or 'breakdown' of the dielectric under certain specified conditions.

Surface breakdown and flash-over. When a high-voltage stress is applied to conductors separated only by air and the stress is increased, breakdown of the intermediate air will take place when a certain stress is attained, being accompanied by the passage of a spark from one conductor to the other.

Permittivity (specific inductive capacity). Permittivity is defined as the ratio of the electric flux density produced in the material to that produced in free space by the same electric force, and is expressed as the ratio of the capacitance of a capacitor in which the material is the dielectric, to the capacitance of the same capacitor with air as the dielectric.

Table 2.12 Representative properties of typical insulating materials

Insulant	n^*	ϵ_r	tan δ 50 Hz	tan δ 1 MHz
Vacuum	∞	1.0	0	0
Air	∞	1.0006	0	0
Mineral insulating oil	11–13	2–2.5	0.0002	–
Chlorinated polyphenols	10–12	4.5–5	0.003	–
Paraffin wax	14	2.2	–	0.0001
Shellac	13	2.3–3.8	0.008	–
Bitumen	12	2.6	0.008	–
Pressboard	8	3.1	0.013	–
Ebonite	14	2.8	0.01	0.009
Hard rubber (loaded)	12–16	4	0.016	0.01
Paper, dry	10	1.9–2.9	0.005	–
Paper, oiled	–	2.8–4	0.005	–
Cloth, varnished cotton	13	5	0.2	–
Cloth, silk	13	3.2–4.5	–	–
Ethyl cellulose	11	2.5–3.7	0.02	0.02
Cellulose acetate film	13	4–5.5	0.023	–
S.R.B.P.	11–12	4–6	0.02	0.04
S.R.B. cotton	7–10	5–11	0.03	0.06
S.R.B. wood	10	4.5–5.4	–	0.05
Polystyrene	15	2.6	0.0002	0.0002
Polyethylene	15	2.3	0.0001	0.0001
Methyl methacrylate	13	2.8	0.06	0.02
Phenol formaldehyde wood-filled	9–10	4–9	0.1	0.09
Phenol formaldehyde mineral-filled	10–12	5	0.015	0.01
Polystyrene mineral-filled	–	3.2	–	0.0015
Polyvinyl chloride	11	5–7	0.1	–
Porcelain	10–12	5–7	–	0.008
Steatite	12–13	4–6.6	0.0012	0.001
Mycalex, sheet, rod	12	7	–	0.002
Mica, Muscovite	11–15	4.5–7	0.0003	0.0002
Glass, plate	11	6–7	–	0.004
Quartz, fused	16	3.9	–	0.0002

*Volume resistivity: $\rho = 10^n$ohm-m. The value of n is tabulated. Information in the above table is taken from the 13th Edition of Electrical Engineer's Reference Book published by Butterworths.

Liquid dielectrics. Liquid dielectrics used for insulation purposes are refined mineral oils and synthetic oils such as chlorinated diphenyl. Many other oils have quite good insulating properties, but for various reasons are not suitable for general use directly as insulation, although they are invaluable for such purposes as treating absorbent solid materials and as ingredients of varnishes, paints and enamels.

Silicone fluids. A further type of liquid dielectric suitable for insulation purposes is silicone fluid. This is used in small aircraft transformers, power and distribution transformers, and as an impregnant for capacitors.

The principal uses of liquid dielectrics are:

(*a*) As a filling and cooling medium for transformers, capacitors and rheostats.

(*b*) As an insulating and arc-quenching medium in switchgear, such as circuit breakers.

(*c*) As an impregnant of absorbent insulations, e.g. wood, slate, paper, and pressboard, used mainly in transformers, switchgear, capacitors and cables.

The more important properties of these liquids are, therefore, (i) electric strength, (ii) viscosity, (iii) chemical stability, and (iv) flash point.

Insulating oils. These are highly refined, hydrocarbon mineral oils obtained from selected crude petroleum, and have a relative density ranging from 0.85 to 0.88.

The oils of this type, used mainly for transformers and switchgear (such as oil circuit breakers), are dealt with in BS 148. A number of special mineral oils are employed for impregnated paper capacitors and cables, and others – usually of higher viscosity and flash point – for rheostats and for filling busbar chambers in switchgear.

The electrical properties of the oils are usually good, the resistivity being of a high order (e.g. 2.5 to 40×10^6 MΩ cm) and the power factor about 0.0002 at 20°C, 0.001 at 60°C and approximately 0.008 at 100°C. The permittivity varies from 2.0 to 2.5.

Synthetic insulating liquids. Synthetic liquids are now used as alternatives to mineral oils for insulation purposes, chiefly in transformers and capacitors, particularly where the fire risk associated with oil is unacceptable. Fire resistant replacement fluids were developed and for many years Askarels dominated the market. They are a mixture of polychlorinated biphenyls, marketed under a variety of trade names like Arocolor, Pyroclor, Inerteen and Pyranol.

During the last few years it has become widely realised that Askarels are toxic and are also resistant to biological and chemical degradation. It is the long term chronic effects that are most important particularly if they get into the food chain and are ingested. Now a policy is being adopted by responsible electricity authorities and industrial concerns with Askarel-filled transformers on their premises, to replace the fluid by an alternative – the operation being known as retrofilling.

There are a number of replacement fluids available having excellent qualities as a transformer coolant and insulating medium. They include silicone, synthetic esters, and high molecular weight hydrocarbons. The latest alternative is Formal NF which is non-flammable, and said to possess very good heat transfer properties, and acceptable toxicity under all conditions of operation and not to produce explosive gases. It has been produced by the former Electricity Council in conjunction with ISC Chemicals.

Gas insulation. A further means of insulation is by gases. Two such gases already in common use are nitrogen and sulphur hexafluoride SF_6. Nitrogen in particular is used as an insulation medium in some sealed transformers, while SF_6 is finding increasing use in switchgear both as an insulant and an arc-extinguishing media.

Vacuum insulation. Vacuum is now being employed in switchgear and motor control gear, mainly for arc-extinguishing applications but also partly for insulation.

Power factor and dielectric losses. When an alternating stress is applied to, say, the plates of a capacitor in which the dielectric is 'perfect,' e.g. dry air or a vacuum, the current passed is a pure capacitance current and leads the voltage by a phase angle of 90°. In the case of practically all other dielectrics conduction and other effects (such as dielectric hysteresis*) cause a certain amount of

---

*Dielectric hysteresis is a phenomenon by which energy is expended, and heat produced, as the result of the reversal of electrostatic stress in a dielectric subjected to alternating electric stress.

energy to be dissipated in the dielectric, which results in the current leading the voltage by a phase angle less than 90°. The value of the angle which is complementary with the phase angle is, therefore, a measure of the losses occurring in the material when under alternating electric stress.

Figure 2.1   Vector diagram for a material having dielectric loss

In the vector diagram, Figure 2.1, the phase angle is ϕ and the complementary angle δ is known as the loss angle. As this angle is usually quite small, the power factor (cos ϕ) can be taken as equal to tan δ (for values of cos ϕ up to, say, 0.1).

The energy loss (in watts) is $V^2 C\omega \tan \delta$, where V is the applied voltage, C the capacity in farads, and $\omega = 2\pi f$, where f is the frequency in hertz.

This loss, known as the dielectric loss, is seen to depend upon the capacity, which, for given dimensions of dielectric and electrodes, is determined by the permittivity of the insulating material. The properties of the dielectric which determine the amount of dielectric losses are, therefore, power factor (tan δ) and permittivity. It is consequently quite usual practice to quote figures for the product of these two, i.e. $k \times \tan \delta$, for comparing insulating materials in this respect. It will also be noted that the losses vary as the square of the voltage.

Power factor varies, sometimes considerably, with frequency, also with temperature, values of tan δ usually increasing with rise of temperature, particularly when moisture is present, in which case the permittivity also rises with the temperature, so that total dielectric losses are often liable to a considerable increase as the temperature rises. This is very often the basic cause of electrical breakdown in insulation under a.c. stress, especially if it is thick, as the losses cause internal temperature rise with consequent increase in the dielectric power factor and permittivity, this becoming cumulative and resulting in thermal instability and, finally, breakdown, if the heat developed in the interior cannot get away faster than it is generated.

47

These properties are, of course, of special importance in the case of radio and similar uses where high frequencies are involved.

SUPERCONDUCTIVITY

The ideal superconducting state exhibited by certain materials is characterised by two fundamental properties (a) the disappearance of resistance when the temperature is reduced to a critical value and (b) the expulsion of any magnetic flux which may be in the material when the critical, or transition temperature, is reached. The discovery of superconductivity was made at the University of Leiden in 1911 by Professor Onnes when he was examining the relationship between the resistance and temperature of mercury. In the years which followed many other elements were found to exhibit superconductivity and theories were developed to explain the phenomenon. The transition temperatures were typically about $10K$ ($-263°C$) which, in practice, meant that they had to be cooled with liquid helium at $4K$. In general these materials were of little more than academic value because they could only support a low current density in a low magnetic field without losing their superconducting properties.

In the 1950s a new class of superconductors was discovered which were alloys or compounds and which would operate with very high current densities – typically 10^5 A/cm^2 and high magnetic flux densities – typically 8 Tesla. The most important materials in this class were the alloy Nb Ti and the compound Nb$_3$ Sn. The consequence of these discoveries was the initiation of a significant activity worldwide on their application to many types of power equipment and magnets for research purposes. Other applications investigated were computers and very sensitive instruments.

In the UK the former CEGB (now privatised) commenced studies on the superconducting power cables and magnetohydrodynamic (MHD) power generation and an electrical machines programme was commenced by NEI International Research and Development Co Ltd. This company designed and built the world's first superconducting motor in 1966 (now in the Science Museum) and followed this with a series of other d.c. motors and generators up to the early 1980s; they also played a leading role in the development of superconducting a.c. generators for central power stations and a fault current limiter for use in power transmission/distribution networks. The driving force for these developments was a reduction of electrical losses, in some cases the elimination of magnetic iron, and reductions in the size and weight of plant. Many other countries had similar programmes and there was a good measure of international collaboration through conferences. Milestones were the design, construction and test by NEI-IRD of a 2.44 MW low-speed superconducting motor in 1969; a 87 kVA a.c. generator by the Massachusetts Institute of Technology in the late 1960s; a very large superconducting Bubble Chamber magnet at CERN in Switzerland in the late 1960s and a superconducting levitated train by Siemens in the mid 1970s. Many other countries have made significant achievements since these dates.

One of the problems which arose during the 20 years or so of these developments was the high cost of designing to meet the engineering requirements at liquid helium temperature and, by the early 1980s, many programmes had been terminated and others were proceeding very slowly.

In late 1986, Bednorz and Mueller working in Zurich discovered that a ceramic material LaBaCuO was superconducting at 35K (they were awarded a Nobel prize) and in 1987, Professor Chu at the University of Houston in Texas discovered that YBaCuO was superconducting at 92K and since that time the transition temperature of other materials has crept up to over 105K. The enormous significance of these discoveries is that the materials will work in liquid nitrogen instead of liquid helium. The consequence has been an unprecedented upsurge of activity in every country in the world with a technology base. Much of this work is directed at seeking new superconductors with higher temperatures transition and to establishing production routes for the materials. Some of the major problems are that the new materials are brittle and, unless they are in the form of very thin films, the current density is rather low $10^3 - 10^4$ A/cm^2; indications are that they will operate at very high flux densities in excess of 50T. However, good progress is being made with the development of materials and attention is being turned to applications. There are very significant advantages in using liquid nitrogen instead of liquid helium – for example the efficiency of refrigeration is nearly 50 times better. Many of the organisations in different countries who were working with the earlier lower temperature materials are now re-examining their designs but based upon the use of liquid nitrogen. The impact upon industry is expected to be as important as the silicon chip and many new applications will probably be identified which will open up completely new markets.

3

Plastics and rubber in electrical engineering

PROPERTIES OF MOULDING MATERIALS

Plastics have become established as very important materials for the electrical engineer especially as insulation but also for structural parts and in some cases as replacements for metals. The term 'plastics' is an omnibus one covering a great number of substantially synthetic materials which have rapidly increasing fields of application.

Plastics can be conveniently divided into two different groups, known as thermosetting and thermoplastics materials. The two groups behave differently when external heat is applied. With thermosetting materials the application of heat initially causes softening and during this period the material can be formed or moulded. Continuation of heating however results in chemical changes in the material generally resulting in rigid cross-linked molecules. These are not appreciably affected by further heating at moderate temperatures. Overheating however may cause thermal decomposition. In addition to heat, thermosetting resins may be hardened (or 'cured') by catalysts, radiation, etc.

When thermoplastics materials are heated, they soften and become less stiff: they may eventually reach a stage at which they become a viscous liquid. On cooling, such a material stiffens and returns to its former state. This cycle of softening and hardening may in theory be repeated indefinitely. Overheating can however result in irreversible decomposition.

A major difference between thermosetting and thermoplastics materials is that the former are seldom used without the addition of various reinforcing or filling materials such as organic or inorganic fibres or powders. Thermoplastics are more often used in unfilled form, but fibres, fillers and other additives can be added when special properties are required.

50

In general, many thermoplastics come into the low loss, low permittivity category whereas thermosetting materials often have higher loss and permittivity values.

THERMOSETTING MATERIALS

The number of thermosetting materials available to industry is smaller than for the thermoplastics, and mention will be made only of the more important groups used in electrical engineering. These are the phenolic, aminoplastic, polyester, epoxy, silicone, polyimide and polyaralkylether/phenol resins. More details about each of these materials is given later in this section.

As already noted, thermosetting resins are seldom used untreated and the following basic processing methods may be considered as typical.

Laminating. The impregnation of fibrous sheet materials such as glass and cotton materials, cellulose paper, synthetic fibre and mica with resins and forming into sheets, tubes and other shapes by the action of heat and pressure in a press or autoclave.

Compression moulding. Complex shaped components produced by curing filled and reinforced compounds under heat and pressure in a matched metal mould cavity.

Transfer moulding. Similar to compression moulding but involves transferring fluxed moulding material from a heated transfer pot by means of a plunger through a runner system.

Casting. Used for liquid resin based compounds such as polyester and epoxy, to form complex shapes. Large insulators and encapsulated components are examples.

Properties. Most thermosetting materials are used as composites and the resultant characteristics are naturally highly dependent on the properties of the constituents. For example, the use of glass fabric to reinforce a particular resin system will produce a material with a higher modulus, a better impact strength, a better resistance to high temperature than a similar material using a cellulose paper as reinforcement. The variations in properties

51

which can be produced by using combinations of the various resins and reinforcing materials is very wide and it is possible only to give a few typical examples.

Phenolic resin (phenol formaldehyde) – PF.   This is the bakelite type material so well known in industry. The resin is made by reacting phenolic or cresylic materials with formaldehyde at temperatures around 90–100°C either with or without a catalyst. The processing is done in a digester equipped for refluxing, usually with arrangements for the removal of water formed during the reaction. Ammonia, soda or other alkaline catalysts are generally employed although sometimes acid catalysts are used. The final polymerising or 'curing' time of the resin, which vitally affects its utility, is varied as desired by the manufacturing process. Some resins cure in a few seconds at temperatures around 150°C whereas others may require an hour or more. The resins are sometimes in a semi-liquid form but more usually they are solids with softening temperatures ranging from 60–100°C.

Like all thermosetting materials, PF may be compounded with fillers, pigments and other ancillary materials to form moulding compounds which can be processed by several techniques. Perhaps the most widely used method involves curing under heat and pressure in metal moulds to produce the finished article. The fillers may be cheap materials used partly to economise in resin but also to improve performance and often to reduce difficulties due to effects such as mould shrinkage, coefficient of thermal expansion, etc. Examples of some typical fillers are chopped cotton cloth to produce greater strength, and graphite to produce a material that is an electrical conductor with good wearing properties. Wood flour is often used as a general purpose filler. The PF resins are relatively cheap.

As already mentioned this resin can be used to impregnate fabrics, wood veneers or sheets of paper of various types. When these are hot-pressed high strength laminates are produced. The liquid resins suitably modified can be used as adhesives and insulating varnishes.

Depending on the fillers used, PF materials may be considered as being suitable for long term operation at temperatures in the 120°C to 140°C region although in some forms they may be suitable for even higher temperatures.

The main disadvantages of PF materials are the restriction to dark brownish colours and, from an electrical point of view, the poor resistance to tracking – when surface contaminants are present. Nevertheless, their good all round electrical performance and low cost make them useful for a wide range of applications. They are employed extensively in appliances and in some electrical accessories.

Aminoplastic resins (ureaformaldehyde – UF and melamine-formaldehyde – MF). These two resins are aminoplastics and are more expensive than phenolics. Resins are clear and uncoloured while compounds are white or pastel coloured, and are highly resistant to surface tracking effects. They are particularly suitable for domestic applications but should not be used where moisture might be present.

The resins are produced by reaction of urea of melamine with formaldehyde. If the condensation process for UF resins is only partially carried out, useful water-soluble adhesives are obtained. They can be hardened after application to joints by means of the addition of suitable curing agents. Hot-setting MF resins with good properties are also available.

By addition of various fillers UF and MF resins may be made into compounds which can be moulded to produce finished articles. In addition, melamine resins are employed to produce fabric laminated sheet and tubes. In decorative paper-based sheet laminates for heat resistant surfaces the core of the material is generally formed from plies of cellulose paper treated with PF resin but the decorative surfaces comprise a layer of paper impregnated with MF resin.

Depending on the fillers used, UF and MF resin-bonded materials may be considered as being suitable for long term operation at 110–130°C.

As a result of their good electrical properties and excellent flammability resistance, UF materials are suitable for domestic wiring devices.

Alkyd and polyester resins (UP). The group known as alkyds is mainly used in paints and varnishes but slightly different materials are used for mouldings. These alkyd resins are the condensation products of polybasic acids (e.g. phthalic and maleic acids) with polyhydric alcohols (e.g. glycol and glycerol). They are substantially non-tracking and some newer types, when used in conjunction with fillers, have very high heat resistance.

The unsaturated polyester resins (UP) are usually solutions of unsaturated alkyd resins in reactive monomers, of which styrene is the most commonly used. By the addition of suitable catalysts these resins may be cured at ambient temperatures with zero pressure (contact moulding) to produce large strong structures at comparatively low capital costs. Alternatively, preimpregnated glass fibre reinforced compounds, known as dough moulding compounds (DMC), may be rapidly cured under high temperatures and pressures to form durable and dimensionally accurate mouldings and insulating sheets with good mechanical properties and thermal stability. Such mouldings are used in contactors.

As well as sheets for insulation, glass fibre polyester or reinforced plastics (RP or GRP) mouldings are used for covers and guards, line-operating poles, insulating ladder and many other applications where large, strong, complex insulated components are required.

Silicones. In addition to hard, cured resins, silicone elastomers (rubbers) are also available. These materials have outstanding heat and chemical resistance and their resistance to electrical discharges is excellent. Silicone elastomers can be applied to glass fabrics and woven sleevings to produce flexible insulants suitable for high temperature use. Filled moulding compositions, encapsulating and dielectric liquids based on silicone resins for use at elevated temperatures are also available.

Polyimide resins (PI). A fairly recent development has been the organic polyimide resins which give good performance at temperatures in the 250–300°C region. For this reason, they are generally used with glass or other high temperature fibrous reinforcements. Curing is by heat and pressure and care is needed during the processing operation if good properties are to be obtained. This is because volatile products are evolved during the process although it is hoped that this difficulty will be overcome in time. Electrical properties are also excellent.

Polyaralkylether/phenol resin. Recently developed by the Friedel Crafts route is a resin based on a condensation reaction between polyaralkylether and phenol. The material is similar in mechanical and electrical properties to some of the epoxy resins but having temperature capabilities more like those of silicone resins. At present the cost is approximately between the two. Long term operation at 220–250°C is possible.

These resins are generally used with glass fabric reinforcements to produce laminates or tubes and with fibrous asbestos fillers to produce mouldings.

Epoxy resins. Epoxy resins are produced by a reaction between epichlorohydrin and diphenylalpropane in an alkaline medium. The electrical properties of this thermosetting material are outstanding with resistance to alkalis and non-oxidising acids good to moderate. Water absorption is very low and stable temperature range is between about $-40°C$ to $+90°C$.

Epoxy resins are widely used by the electrical industry for insulators, encapsulating media for distribution and instrument transformers and, when used with glass reinforcement for printed circuits.

THERMOPLASTICS MATERIALS

The main thermoplastics materials used in electrical engineering applications are discussed below with an indication of their uses. Later their electrical properties are treated in some detail.

Polyethylene (PE). This tough resilient material was first used as an insulant for high frequency low voltage cables in radar and its low-loss properties are also exploited in high performance submarine cables, and as an insulator and sheathing for telephone cables.

Polytetrafluorethylene (PTFE). PTFE is a relatively soft flexible material which is chemically inert, can withstand continuous temperatures in the range $\pm 250°C$, has excellent insulating and non-tracking properties over a wide temperature range. It is used as a dielectric and insulator in high temperature cables, as spacers and connectors in high frequency cables, as a hermetic seal for capacitors and transformers and in valve holders.

Polyvinylchloride (PVC). Unplasticised PVC is hard, stiff and tough, and has good weathering, chemical and abrasion resistance. It is employed for conduit and junction boxes. Incorporation of plasticisers can produce PVC compounds with a wide range of flexibilities, for which the main electrical use is in low frequency cable insulation and sheathing and for moulded insulators.

Table 3.1 Typical properties of some moulding materials*

Type of thermosetting resin	Phenol formaldehyde	Urea formaldehyde	Melamine formaldehyde	Epoxy	Unsaturated polyester
Type of filler	Wood flour/cotton flock	Alpha cellulose	Alpha cellulose	Mineral	Mineral
Density, kg/m³	1320–1450	1470–1520	1470–1520	1700–2000	1600–1800
Heat distortion temperature, °C	125–170	130–140	200	105–150	90–120
Water absorption after 24h at room temperature, %	0.3–1.0	0.4–0.8	0.1–0.6	0.05–0.2	0.1–0.5
Volume resistivity at room temp. Ωm	10^7–10^{11}	10^{10}–10^{11}	10^{10}–10^{12}	$>10^{13}$	10^{11}
Electric strength at room temp. kV rms/mm	8–17	12–16	12–16	18–21	15–21
Loss tangent at 1MHz and room temp, tan δ	0.05–0.10	0.025–0.035	0.025–0.050	0.01–0.03	0.015–0.030
Permittivity at 1MHz and room temp	4.5–6.0	6.0–7.5	7.0–8.0	3.0–3.8	3.5–4.2
Flexural strength at room temp. MN/m²	55–85	70–110	70–110	80–110	60–100
Tensile strength at room temp. MN/m²	45–60	40–90	50–100	60–85	20–35
Elastic modulus at room temp. GN/m²	5–8	7–9	7–9	8–10	7–9
Thermal conductivity perpendicular to surface. W/m°C	0.17–0.21	0.3–0.4	0.25–0.4	0.5–0.7	0.4–0.6

*It should be noted that the values given in this table have been collected from a wide variety of sources and the test methods and specimen dimensions may therefore be such as to make direct comparisons impossible. Wide limits have been given because of the wide types and grades of materials available.

Table 3.2 Typical properties of some laminated materials†

Type of thermosetting resin	Phenol formaldehyde				Melamine formaldehyde	Unsaturated polyester	Epoxy	Silicone	Polyimide	Polyaralkyd-etheriphenols
Reinforcement	Cellulose paper	Cotton fabric	Wood veneer*	Asbestos paper	Glass fabric	Glass fabric	Glass fabric	Glass fabric	Glass fabric	Glass fabric
Density, kg/m³	1340	1330	1300	1700	1700	1750	1770	1650	1850	1770
Water absorption after 24h in water, %	0.5–3.0	0.5–3	1.5–3	0.1–0.5	0.5–1	0.2–0.5	0.1–0.3	0.1–0.2	0.1–0.2	0.1–0.2
Loss tangent at 1MHz, tan δ	0.02–0.04	0.04–0.06	0.05	N/a	0.02–0.04	0.01–0.03	0.005–0.02	0.001–0.003	0.01–0.02	0.01–0.03
Permittivity at 1MHz	5–6	5–6	4–5	N/a	6.5–7.5	4–5	4.5–5.5	3.5–4.5	4–4.5	4.8
Electric strength normal to laminate in oil at 90°C kVrms/mm	12–24	12–18	3–5	1–5	10–14	10–18	16–18	10–14	16–20	28–34
Electric strength along laminate in oil at 90°C kVrms/mm	40–50	25–35	25–30	5–15	25–35	30–50	35–45	30–40	60–80	
Flexural strength, MN/m²	140–210	105–140	105–210	140–210	200–310	280–350	350–450	105–175	350–520	520
Tensile strength, MN/m²	85–110	70–105	85–170	105–140	175–240	210–240	240–310	105–175	350–450	350–450
Elastic modulus, GN/m²	6–11	5–8	14–17	7–9	12–15	7–12	12–14	4–9	20–28	35
Coeff. of thermal expansion in plane of sheet per °C × 10⁶	10–15	17–25	8–15	15–25	10–15	10–15	10–15	10–12	5–10	N/a

* Some of the values are very dependent on the direction of the grain. N/a = No data available.
† Some of the values given in the table have been collected from a wide variety of sources and specimen dimensions therefore may be such as to make direct comparison impossible. Because of the wide variations in types and grades available wide limits have been given.

57

Polypropylene (PP). Polypropylene combines strength, fatigue resistance, stiffness and temperature withstand and excellent chemical resistance. Its good electrical properties are exploited in high frequency low loss cable insulation. Biaxially oriented polypropylene film is used to make power capacitors of either film-foil or film-paper-foil construction, and also high energy rapid discharge capacitors.

Thermoplastics polyester (PTP). As a biaxially oriented film, polyethylene terephthalate has high dielectric strength, high volume resistivity, flexibility, toughness, excellent mechanical strength and a high working temperature. Major electrical applications include motor insulation, cable wrapping, insulation in transformers, coils, and relays, printed circuit flat cables, and in capacitors as metallised film.

Other plastics. In film form polycarbonate (PC), polyphenylene oxide (PPO) and polysulphone are used as capacitor dielectrics. Many thermoplastics find applications in electrical engineering for mechanical rather than electrical reasons. For example housings, casings and containers are made from ABS (acrylonitrile-butadiene-styrene), PVC (polyvinylchloride), POM (acetal), PC, PPO PP and nylon. Outdoor illuminated signs make use of CAB (cellulose acetate butyrate), PMM (acrylic) and PVC while diffusers for fluorescent lighting fittings employ PMM and PS (polystyrene).

Electrical properties. Most thermoplastics are good electrical insulators, some outstandingly so. Often the choice of a plastics for a particular application depends primarily on factors other than electrical properties. For example mechanical properties such as creep, long-term strength, fatigue and impact behaviour (see BS 4618) are often the deciding factors. Corrosion resistance or thermal stability may also govern the choice. An increasingly important factor is flammability. Many thermosets, PVC, polycarbonate and nylon are inherently flame-retardant, whereas many other plastics materials, when unmodified, will support combustion.

Where electrical properties are important five characteristics are of interest: resistivity, permittivity and power factor, electrical breakdown, electrostatic behaviour and conductance.

Some plastics at room temperature (e.g. highly plasticised PVC), exhibit ohmic behaviour so that the current reaches a steady value. For most plastics however the resistivity depends on the time of electrification and Table 3.3 gives data on the apparent volume resistivity after various times of electrification. Surface resistance values depend on the state of the surface of the plastics, particularly on the presence of hydrophilic impurities which may be present or as additives in the plastics. Results depend greatly on the ambient conditions, particularly on the relative humidity.

The permittivity of many non-polar plastics, e.g. PE, is essentially constant with frequency and changes with temperature may be related to changes in density using the Clausius-Mosotti relationship. For popular materials these considerations do not apply.

Power factor (loss tangent and loss angle) data have maxima which depend on both frequency and temperature. The levels of dielectric loss can range from a few units of 10^{-6} in tan δ to values as high as 0.3. For low values of loss tangent it is usual to modify tan δ to the angular notation in microradians (1 micro-radian = 10^{-6} in tan δ units). Materials of high power factor such as PVC,

Table 3.3 Apparent volume resistivity at 20°C for differing times of electrification

Material	Time of electrification (sec)		
	10	100	1000
Low density polythene	10^{18}	10^{19}	$>10^{19}$
High density polythene	$>10^{16}$	$>10^{16}$	$>10^{16}$
Polypropylene	10^{17}	10^{18}	10^{18}
Flexible PVC	$>10^{14}$	$>10^{14}$	$>10^{14}$
Rigid PVC	2×10^{15}	3×10^{16}	10^{17}
Poly (methyl methacrylate)	2×10^{15}	2.5×10^{16}	10^{17}
PTFE	$>10^{18}$	$>10^{19}$	$>10^{19}$
Polyacetal	–	$>10^{14}$	–
Nylon 6	–	$>10^{12}$	–
Nylon 66	–	$>10^{12}$	–
Nylon 610	–	$>10^{13}$	–
Thermoplastics polyesters (oriented film)	10^{16}	10^{17}	10^{17}
Polyethersulphone (dried)	3×10^{16}	1.5×10^{17}	10^{18}
Polysulphone	4.1×10^{16}	5.2×10^{17}	3.2×10^{18}
Polycarbonate		9×10^{15}	

Table 3.4 **Frequency dependence of permittivity and power factor at 20°C and relative humidity of 50%**

Material	Permittivity					Power factor				
	100 Hz	1 kHz	10 kHz	100 kHz	1 MHz	100 Hz	1 kHz	10 kHz	100 kHz	1 MHz
Low density polythene	2.28	2.28	2.28	2.28	2.28	$<10^{-4}$	$<10^{-4}$	$<10^{-4}$	$<10^{-4}$	1.5×10^{-4}
High density polythene	2.35	2.35	2.35	2.35	2.35	$<10^{-4}$	$<10^{-4}$	$<10^{-4}$	$<10^{-4}$	$<10^{-4}$
Polypropylene	2.3	2.3	2.3	2.3	2.3	$<5 \times 10^{-4}$	$<5 \times 10^{-4}$	$<5 \times 10^{-4}$	$<5 \times 10^{-4}$	$<5 \times 10^{-4}$
Flexible PVC	5–6	5–6	4–5	3–4	2.3	$\sim 10^{-1}$	$\sim 10^{-1}$	$\sim 10^{-1}$	$\sim 10^{-1}$	$<10^{-1}$
Rigid PVC	3.5	3.4	3.3	3.3	3	10^{-2}	$\sim 2 \times 10^{-2}$	$\sim 2 \times 10^{-2}$	$\sim 2 \times 10^{-2}$	10^{-2}
Poly (methyl methacrylate) dried	3.2	3.0	2.8	2.8	2.7	5×10^{-2}	4×10^{-2}	3×10^{-2}	2×10^{-2}	1.5×10^{-2}
PTFE	2.1	2.1	2.1	2.1	2.1	$<5 \times 10^{-5}$	$<5 \times 10^{-5}$	$<5 \times 10^{-5}$	$<5 \times 10^{-5}$	$<10^{-4}$
PCTFE	2.7	2.65	2.56	2.5	2.48	1.9×10^{-2}	2.5×10^{-2}	2×10^{-2}	1.2×10^{-2}	8×10^{-3}
Polyacetal	3.7	3.7	3.7	3.7	–	2×10^{-3}	2.5×10^{-3}	2.8×10^{-3}	3.2×10^{-3}	–
Nylon 6	–	4	–	–	3.6	–	3×10^{-3}	–	–	–
Nylon 66	4.9	4.5	4.0	3.7	3.4	$\sim 5 \times 10^{-2}$	$\sim 6 \times 10^{-2}$	$\sim 6 \times 10^{-2}$	$\sim 6 \times 10^{-2}$	6×10^{-2}
Nylon 610	3.9	3.7	3.5	3.4	3.2	$\sim 3 \times 10^{-2}$	$\sim 4 \times 10^{-2}$	$\sim 4 \times 10^{-2}$	$\sim 4 \times 10^{-2}$	$\sim 5 \times 10^{-2}$
Polyesters (linear oriented film)	3	3	3	3	–	3×10^{-3}	7×10^{-3}	1.2×10^{-2}	1.5×10^{-2}	5×10^{-2}
Polycarbonates	2.96	2.95	2.95	2.95	–	1.2×10^{-3}	1×10^{-3}	2×10^{-2}	5×10^{-3}	–
Polyethersulphone. dried	3.6	3.6	3.6	–	–	1.5×10^{-3}	2.1×10^{-3}	3.1×10^{-3}	–	–
Polysulphone	3.16	3.16	3.16	–	–	2.2×10^{-3}	9.7×10^{-4}	1.6×10^{-3}	–	–

Table 3.5 Temperature dependence °C of permittivity and power factor at 1kHz

Material	Permittivity					Power factor				
	−50	0	50	100	150	−50	0	50	100	150
Low density polythene	2.28	2.28	2.28	2.28	2.28	$<10^{-4}$	$<10^{-4}$	$<10^{-4}$	$<10^{-4}$	—
Polypropylene	2.3	2.3	2.3	2.3	2.3	$<5 \times 10^{-4}$	$<5 \times 10^{-4}$	$<5 \times 10^{-4}$	$<5 \times 10^{-4}$	—
Flexible PVC	3	3–4	7	7–8	—	$\sim 2 \times 10^{-2}$	$\sim 8 \times 10^{-2}$	$\sim 8 \times 10^{-2}$	$\sim 2 \times 10^{-2}$	—
Rigid PVC	3.1	3.3	3.7	12	—	$\sim 2 \times 10^{-2}$	$\sim 2 \times 10^{-2}$	$\sim 2 \times 10^{-2}$	1×10^{-1}	—
Poly (methyl methacrylate) dried	2.7	2.9	3.3	4.2	—	1×10^{-2}	3×10^{-2}	7×10^{-2}	8×10^{-2}	—
PTFE	2.1	2.1	2.05	2.0	1.95	5×10^{-5}	$<5 \times 10^{-5}$	$<5 \times 10^{-5}$	$<5 \times 10^{-5}$	$<5 \times 10^{-5}$
PCTFE	2.43	2.55	2.75	2.83	2.89	5×10^{-3}	2×10^{-2}	1×10^{-2}	3×10^{-3}	2.5×10^{-3}
Polyacetal	3.5	3.7	3.75	3.8	—	2×10^{-2}	2.5×10^{-2}	1.5×10^{-3}	2.5×10^{-2}	—
Polyester (linear) oriented film	3	3	3	3	3.3	1.2×10^{-2}	1×10^{-2}	3×10^{-3}	5×10^{-3}	8×10^{-3}
Polycarbonate	3	3	3	3	3	6×10^{-3}	1.5×10^{-3}	1×10^{-3}	1×10^{-3}	2×10^{-3}
Polyethersulphone	—	3.7	3.57	3.56	3.54	—	—	8×10^{-4}	1.2×10^{-3}	7.4×10^{-4}

transform electrical energy into heat; at high frequencies the heat generated may lead to softening of the plastics and this is the basis of dielectric heating used commercially.

Permittivity and power factor data are best presented as contour maps of the relevant property on temperature-frequency axes, as recommended in BS 4618, but this is beyond the scope of this section. In Table 3.4 data are given for permittivity and power factor at 20°C at frequencies in the range 100 Hz to 1 MHz. Table 3.5 indicates the temperature dependence of these properties from −50°C to +150°C at 1 kHz.

When a dielectric/conductor combination is subject to high voltages in the absence of electrical discharges, the effect may be to induce a new set of thermal equilibrium conditions, or to produce thermal runaway behaviour resulting in electrical breakdown. However, with many plastics, failure results from the electrical (spark) discharges before the onset of thermal runaway, either by the production of conducting tracks through or on the surface of the material, or by erosion. This highlights the importance of the standard of surface finish in affecting track resistance. The designer should therefore aim for complete freedom from discharges in assemblies where long life is required.

One consequence of the high resistivities which many plastics have is the presence of electrostatic charge in or on an article. Often associated problems such as dust pick-up and build-up of charge on carpets can be resolved by applying a hygroscopic coating to the surface or by incorporating an anti-static agent in the plastics which will migrate to the polymer surface and reduce surface resistivity. Where a spark discharge could lead to a hazardous situation, anti-static grades of plastics should be used.

Electrically conducting plastics are usually produced by incorporating high concentrations of filamentary carbon black or certain pyrolised materials into an otherwise electrically insulating polymer matrix.

RUBBER IN ELECTRICAL ENGINEERING

Rubbers are organic compounds derived from the latex plant or by synthesis. They are characterised by some unique qualities, notably high elasticity. Natural or synthetic latex is the chief material of the rubber industry and vulcanised rubber in its many forms is the main end product. To be classed as a rubber rather than a plastic, a material must meet four requirements:

1. It must consist of very long chain molecules.
2. The long-chain molecules must be lightly cross-linked (vulcanised).
3. The glass transition temperature must be well below room temperature.
4. The melting point of the crystallites, if any, must also be below room temperature.

Vulcanised rubber meets all four requirements and finds applications in the cable industry and other electrical fields. The temperature limits in degrees Celsius for practical applications of the various rubbers is given in Table 3.6.

Table 3.6   Temperature limits in °C of some rubbers

Natural rubber	−50 to 60
Styrene butadiene rubber	−45 to 65
Butyl rubber	−30 to 80
Chloroprene rubber	−20 to 70
Nitrile rubber	−10 to 90
Polysulphide rubber	−50 to 100
Silicone rubber	−70 to 150
Polyurethane rubber	−55 to 90
Fluor rubber	−30 to 150
Ethylene propylene rubber	−50 to 90

The electrical properties of some rubbers are shown in Table 3.7. A glance at the table reveals that in general the material possesses good electrical properties with the exception of nitrile polychloropropene and polyurethane. The latter are not suitable for high frequency applications because of high dielectric losses. Ebonite is no longer used for electrical purposes as it has been replaced by plastics like polyethylene, but is included in the table for completeness. Gutta-percha, an isomeric form of natural rubber, which has been used in sea cables, has also been replaced by polyethylene.

The cable industry is a large user of rubbers for insulating and sheathing purposes. Ethylene propylene rubbers (EPR) and the hard version (HEPR) are specified in BS 6899 and BS 6469 respectively. The latter is suitable for use at temperatures up to 90°C.

Table 3.7   **Electrical properties of rubber**

Material	Volume resistance in 10^{-2} ohm–cm	Dielectric constant at 1 kHz	Tan δ at 1 kHz	Breakdown voltage in kV/cm (50 Hz after 2 hours)
Natural rubber	10^{15}–10^{17}	2.3–3.0	0.0025–0.0030	210
SBR	10^{15}	2.9	0.0030	260
Butyl rubber	10^{17}	2.1–2.4	0.0030	220
Nitrile rubber	10^{10}	13.5	0.055	165
Polychloroprene rubber	10^{11}	9.0	0.030	225
Polysulphide rubber	10^{12}	7.0–9.5	0.001–0.005	200
Silicone rubber	10^{11}–10^{17}	3.0–3.5	0.001–0.010	220
Fluor rubber	10^{18}	2.0	0.0002	250
Polyurethane rubber	10^{11}	7.0	0.017–0.09	165
Polyethylene rubber	10^{15}–10^{19}	2.3	0.0005	200
Ebonite	10^{14}–10^{16}	2.8–4.0	0.005	250

EPRs have good ozone resistance making them suitable for high voltages. Like natural rubber the material burns and is not particularly oil resistant so insulated cable cores have to be provided with some kind of sheathing.

Butyl rubber was used as a cable insulating medium but its rather limited mechanical strength has caused it to lose favour, although it is still used for some wiring cables. Nevertheless it can be operated in high ambient temperatures and is therefore used sometimes as the entry cable in a luminaire.

Silicone rubber is becoming increasingly popular as the insulating medium for fire resistant cables.

4

Valves and semiconductors

During the past decade great advances have been made in the field of 'light' current engineering particularly in the use of semiconductors. Whereas valves and transistors were the province of the electronics engineer, the development from these of the thyristor has caused a bridge to be built between the light and heavy current electrical engineer.

The main cause for this is that thyristors have reached a stage of being able to handle currents of several hundred amperes at voltages in excess of 2 kV, and the difficulties associated with operating them in series and parallel have been largely overcome. There is, therefore, a growing field of applications for the thyristor of which the power engineer must be aware.

In order to understand the operation of thyristors it is necessary to first discuss the principles of operation of the valve, including the function of the grid, for this was the forerunner of the present semiconductor rectifier systems. The transistor is closely allied to the thyristor, and so some space is devoted in this chapter to discussing its mode of operation.

As far as the power engineer is concerned the applications of these sophisticated devices are more important than the theory of their operation, and it is this side which will be covered most fully. It is necessary, however, to describe the construction and operation of both the transistor and thyristor to appreciate some of the difficulties associated with their applications.

VALVES AND SIMILAR DEVICES

Transistors and thyristors have reached a high state of development and have taken over some of the work formerly done by valve type devices. There is however still a wide range of applications for valves including thyratrons, magnetrons, klystrons and ignitrons. They find their greatest applications in the field of telecommunications.

Hydrogen thyratron. The hydrogen thyratron has become the most widely used switching device for medium and high power radar pulse modulators and other applications where accurately timed high power pulses are required. Thyratrons were first used in radar systems but they are now being employed in various circuits concerned with particle accelerators and high power energy diverter systems. The English Electric Valve Co's range includes both hydrogen and deuterium-filled triode and tetrode thyratrons with either glass or ceramic/metal envelopes. Multigap ceramic tubes are available for operation up to 160 kV.

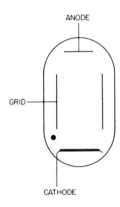

Figure 4.1   The structure of a simple negative grid thyratron (English Electric Valve Co Ltd)

Basically all thyratrons consist of a gas-filled envelope containing an anode, a control grid and a thermionic cathode, see Figure 4.1. Gas may be mercury vapour, hydrogen or an inert gas. The tube remains in a non-conductive state with a positive voltage on the anode if a sufficiently negative voltage is applied to the grid. Value of the grid voltage depends on the anode voltage and the geometry of the tube. If the grid voltage is made less negative the ability to hold off the positive anode voltage is reduced. This is the basis of operation of the negative grid thyratron.

The cathode emits electrons which, in the absence of grid control, are accelerated by the anode voltage and collide with the gas atoms present, to produce an ionised column of gas. A very low voltage, typically 50–100 V for hydrogen and deuterium, then exists across the discharge, through which a wide range of currents can be passed, their magnitudes depending on the external circuitry. During conduction the grid is sheathed with ions which effectively prevent any control being exercised. The tube returns to its non-conducting state only when the anode voltage is removed or reversed for a time sufficient to allow the charge density to decay to a low value. After this the grid regains control and when the anode voltage is re-applied the tube will not conduct. Thus the tube acts as an electronic switch, turned on by a positive-going charge of grid voltage and turned off only by removal or reversal of the anode voltage.

In some designs a disc baffle is fixed close to the cathode side of the grid and this usually modifies the tube's characteristics so that the potential of the grid may be taken positive without the tube conducting when the anode voltage is applied. This arrangement is called a positive grid thyratron. Baffles may also be used to protect the tube electrodes from deposition of cathode material which may cause mal-functioning. In practice this means that a positive pulse with respect to the cathode potential must be applied to the grid to initiate a grid cathode discharge.

Mercury-vapour filled thyratrons have been superseded by hydrogen-filled tubes because of the low limiting voltage that could be developed across the tube, approximately 30 V. Higher voltage across the mercury tube results in positive bombardment causing rapid cathode deterioration.

Most thyratrons can be mounted in any position, though with the larger tubes a base-down position is usually more convenient.

When the reduction of firing time variations is important in a thyratron circuit, the triode type can be replaced by a tetrode unit, which has two grids. These grids may be driven in a number of ways. The first grid may be continuously ionised while the second may act as a gate when pulse driven above its negative bias level. Another way is that of pulsing the grids successively with a delay of about one microsecond between the leading edges of the pulses. A third method is to pulse both grids from a single trigger source arranged to drive them separately.

Multi-gap thyratrons overcome the problems of operating thyratrons in series. When triggered, the multi-gap thyratron breaks down gap by gap in the normal gas discharge mode.

Hydrogen thyratons for pulse modulator series are available from English Electric Valve and the M-O Valve with a peak power output of 400 MW and a peak forward voltage of 160 kV, this tube being a 4-gap ceramic/metal tetrode.

Magnetrons. The cavity magnetron is a compact, efficient thermionic valve for generating microwaves, e.g. radar waves. Basically it is a diode, but it belongs to the family of valves known as 'crossed-field devices'.

Essentially it consists of a (usually) concentric heater, anode and cathode assembly and an embracing magnet (Figure 4.2). The magnetic field is at right angles to the electric field. High voltage pulses between the hot cathode and anode produce a spoked-wheel cloud of electrons which spirals round the cathode under the influence of the interacting electric and magnetic fields. The electron cloud only exists in a vacuum. The copper anode is so designed that a number of resonant cavities are formed around its inside diameter. As the 'tips' of the electron-cloud spokes brush past each very accurately dimensioned cavity the field interacts with the field existing in the cavity. Energy is absorbed and the oscillations within the cavity build up until the cavity resonates. The energy released is beamed through the output window and along a waveguide or cable to the aerial.

Pulsed type magnetrons are available in the frequency range from 1–80 GHz with peak power outputs from a few hundred watts to several thousand megawatts, and mean power a thousand times less than peak power. Tubes for continuous wave operation are used mainly for heating, having powers in the range of 25 kW–200 W at frequencies of 0.9 GHz and 2.45 GHz. Lower power types are available for beacon operation at frequencies of about 9 GHz.

The cathode of a magnetron must be operated at its correct temperature. If it is too low reduced emission can cause unstable operation which may damage the magnetron. Excessively high temperatures lead to rapid deterioration of the cathode. The combination of the magnetic and electric fields in the interaction space results in a back bombardment of electrons to the cathode. This causes dissipation of a proportion of the anode input power. In order to maintain the cathode at its optimum temperature under these conditions it is generally necessary to reduce the heater voltage.

Care must also be exercised when raising the cathode to its operating temperature, before applying the anode voltage. The cold resistance of the heater is typically less than one fifth of the hot resistance. The surge of current when switching on the heater must be controlled. Similar precautions must be taken when designing the modulator output circuit to prevent pulse energy being dissipated in the magnetron heater.

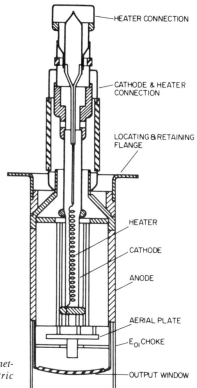

HEATER CONNECTION

CATHODE & HEATER CONNECTION

LOCATING & RETAINING FLANGE

HEATER

CATHODE

ANODE

AERIAL PLATE

E_{oi} CHOKE

OUTPUT WINDOW

Figure 4.2   The magnetron (English Electric Valve Co Ltd)

Operation of grid-controlled rectifier on A.C.   The most important applications of the grid-controlled rectifier, chiefly rely on its behaviour on alternating-current supplies.

When an alternating current is impressed on the anode circuit, the arc is extinguished once every half-cycle because the valve is fundamentally a rectifier. Hence the grid is given the opportunity of regaining control every half cycle. This means that when the arc is extinguished due to the reversal of the alternating voltage, a negative grid can prevent anode current starting when the anode becomes positive again on the next cycle.

Further, by supplying an alternating voltage of the same frequency but of variable phase to the grid, the average value of the anode current can be controlled. Figure 4.3 indicates how a

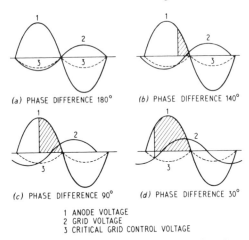

(a) PHASE DIFFERENCE 180°   (b) PHASE DIFFERENCE 140°

(c) PHASE DIFFERENCE 90°   (d) PHASE DIFFERENCE 30°

1 ANODE VOLTAGE
2 GRID VOLTAGE
3 CRITICAL GRID CONTROL VOLTAGE

Figure 4.3   Control of anode current by phase variation of grid voltage. The shaded areas show the current conducting periods

grid voltage of variable phase can delay the starting of the arc during the positive half-cycle of anode voltage, and thereby permit the mean value of the rectified current to be smoothly controlled from zero to the maximum value, corresponding to a phase change in the grid voltage of from 180° to 0°.

An alternative method of grid control having the same effect as phase control is to impose an adjustable d.c. bias voltage upon a constant phase a.c. grid voltage (amplitude control).

Rectification and power control.   It will be apparent from the above that by using the grid in conjunction with a phase-shifting device a d.c. load can be supplied from an a.c. source and at the same time the load current can be regulated from zero to maximum. Furthermore, if the phase shifting device is operated by the changes in the external conditions that are to be controlled, completely automatic control can be effected.

Figure 4.4 shows a typical circuit for rectification and power control, using two gas-filled triodes to obtain the advantages of full-wave rectification. The respective control grids are connected to the ends of the transformer secondary, T_2, so that at all times one grid is positive while the other is negative, in the same way as the corresponding anodes. The phase position of the applied grid voltage is readily varied by the phase-shifting device, consisting of a distributed field winding connected to the a.c. supply and a single-phase rotor connected to the primary of T_2. The phase of the induced rotor voltage is determined by the rotor position relative to the field, so that change of rotor position changes the phase of grid voltage in each tube and correspondingly varies the d.c. circuit current. This form of control has been used for many purposes, notably the speed control of motors.

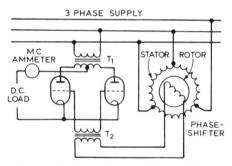

Figure 4.4   Circuit for supplying variable current to d.c. mains using gas triodes

Phase control circuits.   The phase-shifting device referred to above is a small induction regulator, constituting a very light controlling element. Alternative methods of applying phase control in terms of voltage are shown in Figure 4.5. At (a) the voltage across the inductance-resistance circuit is given by AB in the vector diagram. The same current flows through the inductance as

71

through the resistance and the voltage drops – AG across the resistance and GB across the inductance – are therefore approximately 90° out of phase with each other. The grid voltage is the voltage of point G relative to point O (which is connected to the cathode), that is, OG in the vector diagram. By varying R the phase of the voltage OG is varied with respect to the voltage OB, which is the anode voltage; increase of resistance, therefore, produces increase of anode current.

In Figure 4.5(b) the anode current decreases steadily as the resistance is increased, while in (c) the anode current is a maximum or zero according as the resistance is greater or less than a critical value.

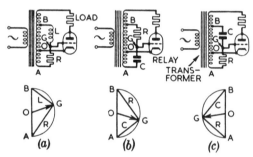

Figure 4.5   Methods for phase-control of grid-bias voltage

Where accuracy of timing is required, the a.c. grid voltage, instead of having a sinusoidal wave form, is arranged to have a peaky wave-form, and be superimposed on a steady d.c. bias voltage considerably more negative than the maximum critical value. The phase of the a.c. component is varied so as to cause the positive peak to occur at any point in the positive half-cycle of anode voltage at which anode current flow is required to commence.

Control of a.c. loads. The control of alternating-current load currents may be conveniently effected by means of a gas-filled triode used in conjunction with a *saturable reactor*. A single-phase circuit is shown in Figure 4.6. The saturable reactor has a laminated core of high permeability iron and carries two a.c. windings in series with the load on its outer limbs. The centre limb carries a d.c. winding supplied with direct current by the gas-filled triode. When the grid voltage phase difference, as controlled by the phase-shift device, is such that the tube is non-conducting, the a.c. winding has a high impedance due to the presence of the high permeability iron core. Under these conditions, the load current is negligible. As the grid-controlled rectifier output current is increased, the iron of the reactor gradually becomes saturated, reducing the permeability of the iron, and thus lowering the impedance of the reactor a.c. winding. At full output of the thyratron, the iron is completely saturated and substantially full-line voltage is impressed upon the load.

During the positive half-cycles the thyratron passes grid-controlled impulses of unidirectional current; during the negative half-cycles the thyratron can convey no current, but the reactor current is maintained by the discharge of the energy stored in its magnetic field through the rectifier (2) across its terminals. The rectifier may be of the selenium or gas-filled diode type.

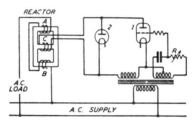

Figure 4.6 Circuit for control of a.c. load current using saturable reactor

Temperature measurement and control. In the control of a heating load where temperatures above 750°C are encountered, measurement of the temperature can be carried out by a photo-electric pyrometer. This gives instantaneous response to changes of temperature when viewing the hot body. The small photoelectric currents, which are a measure of the received radiation, are amplified by a vacuum valve amplifier which energizes the control grids of the vapour valves, thus increasing or decreasing the current to the load by saturable reactor control.

For low temperatures, a resistance thermometer connected in a bridge circuit is used with a vacuum valve amplifier for providing the signal voltage on the control grid. In electrically heated boilers an accuracy of control of plus or minus 1°C at 300°C has been obtained by this means.

Electronic control of machines and processes. Other physical qualities beside that of temperature referred to above may be employed as signals for driving the control grids, provided the quality is converted into terms of voltage. Thus speed, torque, acceleration, pressure, illumination, sound, mechanical movement, and the various electrical qualities are utilized for initiating control in gas-filled or mercury-vapour triode circuits. Sound, for example, can be detected by a carbon, inductive, or capacitor microphone, and R, L, or C used as the variable element in the grid circuit. Mechanical movement can be detected by change of R, L, or C in a carbon pile resistance, slider resistance, variable choke, or capacitor – or by means of a photo-cell. A tacho-generator is employed for registering changes in speed or acceleration, as for example in certain types of speed control of d.c. motors.

The application of electronic devices and particularly the gas-filled relay to the control of machines and processes is often desirable on the score of speed of operation, safety, economy, consistency of product, and lower maintenance costs.

Ignitrons. The ignitron is a high-current rectifier with a mercury pool cathode, usually in a water-cooled steel envelope. In its simplest form it consists of a cylindrical vacuum envelope with a heavy anode supported from the top by a glass insulator, dipping into the mercury pool at the bottom. For some applications tubes may be provided with additional ignitors, auxiliary anodes and internal baffles. Ignitrons are used in applications calling for high current levels such as resistance welding and high power rectification. There are also types intended for high current single-pulse operation such as discharging capacitor banks; these are used to pulse particle accelerator magnet coils, for electromagnetic forming of metals and similar applications.

The action of an ignitron is similar to a thyratron in that a control signal is needed to start conduction, which then continues until the current falls to zero. When the tube is operating as an a.c. rectifier it conducts during one half cycle of the supply frequency and must be ignited every alternate half cycle for as long as it is required to conduct. Figure 4.7 shows a cross section of an English Electric Valve rectifier ignitron.

The ignitor is a small rod of semiconducting material, with a pointed end dipping into the cathode pool. When a suitable current is passed through the ignitor-mercury junction, the ignitor, being positive, a cathode spot is formed on the surface of the mercury and free electrons are emitted. If the anode is sufficiently positive with respect to the cathode at this time, an arc will form between cathode and anode. Once the arc has struck the ignitor

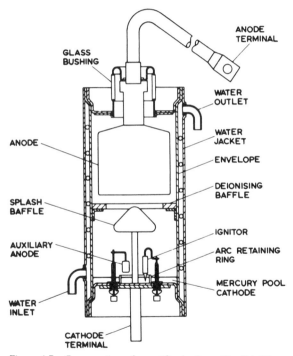

Figure 4.7   Cross-section of a rectifier ignitron (English Electric Valve Co Ltd)

has no further control and the tube continues to conduct until the voltage across it falls below the ionisation potential of the mercury vapour.

In a three-phase welding control circuit, the ignitron must de-ionise quickly in order to hold off the high inverse voltage which immediately follows the conduction cycle. This is accomplished by including a baffle (see Figure 4.7) which operates at cathode potential. No additional connections are required but the voltage drop across the tube is increased slightly. An auxiliary anode may be provided for power rectification at higher voltages. This is used to strike a small arc in a low-voltage circuit separate from the main load. This maintains the cathode spot at low load currents ensuring stable operation under these conditions. The large tubes designed for single pulse operation are also fitted with an auxiliary anode which may be used to prolong the ignition arc. Little or no baffling is used so as to keep the arc voltage drop as low as possible.

The cathode ray tube. The cathode ray tube consists of a highly evacuated conical-shaped glass container having at its narrow end or throat, a cathode capable of being heated, a control electrode and several anodes. Higher up the throat of the tube are two pairs of deflector plates arranged at right angles to each other. Figure 4.8.

The large end or base of the conical container is coated on the inside with a fluorescent compound which emits light when the cathode ray impinges on it.

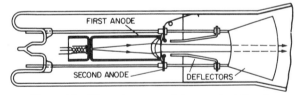

Figure 4.8   Electrode system of electrostatic C.R. tube

The action of the tube is as follows. When the cathode is heated, and the anodes connected to positive high voltages of increasing magnitude the electrons emitted from the cathode are accelerated and focussed into a beam which emerges from a small aperture in the final anode. This beam passes between the first pair of deflector plates, which are located so that when given an electrostatic charge they can attract the electrons say to the right or left. The beam then passes between the second pair of deflector plates arranged so that the beam can be attracted up or down. The beam, which consists of a stream of electrons moving at high speed, impinges on the fluorescent screen and causes a spot of light to appear on it.

If now a fluctuating voltage is applied to the first pair of deflector plates, the beam will move from side to side of the screen in synchronism with the voltage applied to the first deflectors. If another voltage is applied between the second deflector plates, the beam, whilst maintaining its side-to-side motion, will also receive a vertical displacement corresponding to the voltage applied to the second deflectors.

It will be seen that the spot of light on the screen will thus trace out a curve corresponding to the varying potentials applied to the two pairs of deflector plates.

In many applications the horizontal deflectors are given a steadily increasing voltage with instantaneous flyback after the beam or spot has reached the extreme edge of the screen.

The result of this is that the spot travels at a uniform speed from left to right in a given time period, e.g. $\frac{1}{50}$ of a second. If a 50 Hz alternating voltage is applied to the vertical deflector plates the vertical deflection of the spot at any instant will correspond to the instantaneous voltage of the supply. Thus the wave form of the supply voltage will be traced out on the fluorescent screen. As owing to the instantaneous flyback the above process is repeated 50 times per second, the wave-form of the applied alternating voltage will be seen on the screen as a stationary sine wave with or without harmonics, according to the wave form of the applied voltage.

When the cathode ray tube is applied to television, the horizontal deflector plates are used for causing the spot to sweep across the screen with flyback at the end of each sweep, whilst the vertical deflectors give the spot a small vertical displacement after each horizontal sweep, with vertical displacement after each horizontal sweep, with vertical flyback when the lower edge of the screen has been 'scanned'.

The incoming television signals are applied to the control electrode and have the effect of increasing or decreasing the intensity of the beam, according to the brightness of the picture spot, which is being transmitted at any instant.

Klystrons. The klystron is a thermionic device in which electrons from a heated cathode are accelerated to full anode potential and formed into a long parallel beam. This first traverses two grids or apertures between which is a hf voltage provided by a tuned circuit in the form of a resonant cavity. The resultant small periodic changes in speed eventually lead to conventional current modulation by a process known as bunching. Where this bunching is virtually complete another resonant cavity (catcher) is placed. If its impedance and tuning are correct it will extract from the beam much more power than was used to modulate it. The arrangement then acts as an amplifier. Efficency is moderate, output power can be very high, and gain can be almost indefinitely increased by

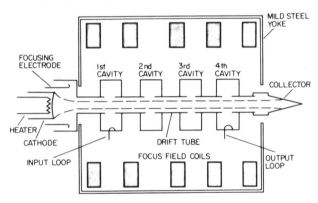

Figure 4.9   Schematic arrangements of a four-cavity amplifier complete with the coils and mild steel yoke used to provide the axial magnetic field (English Electric Valve Co Ltd)

putting more resonant cavities between the buncher and catcher. Oscillators up to the highest frequencies can be formed either by returning some of the collected power to the buncher, or by using a single resonant cavity and reflecting the beam back through it.

High performance klystrons require an axial magnetic field to focus and control the magnetic beam during its passage through the drift tubes, see Figure 4.9. The improved performances generally justifies the complication involved. Damage to a valve will occur if the beam is not focussed efficiently and it is essential to protect it against a failure of the focussing system, which would cause a rise in the electron current reaching the drift tubes. If a simpler installation with a reduced performance is acceptable, it may be possible to use a klystron which has no external means for focussing the beam. This eliminates the need for a magnetic field and the power supply required to maintain it. The precautions on the beam focussing are reduced simply to that of screening the electron beam from the influence of any stray magnetic fields present in the vicinity of the valve.

PHOTOELECTRIC DEVICES

Photo-cell relays. Basic component of a photo-cell relay is an integral light activated switch. It combines a silicon planar photodiode with integrated circuitry on a single substrate to provide a highly sensitive photoelectric device. Operation is such that when light of a selected intensity falls upon it, the device switches on and supplies current to an external load. When the light intensity falls below the critical level the load current is turned off. This critical level can be adjusted within wide limits.

The equipment comprises a projector containing a light emitting diode and optical system projecting a beam of light either directly or by reflection onto a photo-cell mounted in a receiver unit. The relay coil is energised when the light beam is made and de-energised when it is broken. Thus the relay contacts can provide a changeover operation which can be used to perform some external control function. The control unit can contain additional circuitry such as time delays or LED failure circuits to meet a wide variety of application requirements. Systems are available for operating over distances from 10–15 mm up to 50 m or more.

Applications include conveyor control, paper breakage alarm, carton sorting and counting, automatic spraying, machinery guarding, door opening, level controls, burglar alarms, edge alignment control and punched card reading.

Photoelectric switch units. Light-sensitive switches are used for the economical control of lighting. They consist of a photo-cell which monitors the intensity of the light and automatically switches the lighting on or off. Construction of a typical unit is shown in Figure 4.10 and this will switch a resistive load of 3 A at

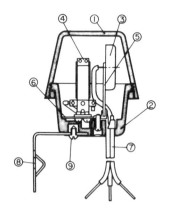

1. Housing, upper
2. Housing, lower
3. Cadmium-sulphide photocell
4. Relay, bi-metal
5. Insulation plate
6. Holder, relay
7. Lead wires
8. Mounting bracket
9. Mounting screw

Figure 4.10   Photain type PB0-2403   photoelectric switch unit

250 V a.c. The unit is based on a cadmium-sulphide cell and it incorporates a 2-minute time delay to prevent 'hunting'. Larger units are available with resistive switching capacities up to 10 A.

Silicon photo-electric cells.   These cells are designed to provide large output current even under low illumination intensities. Currents of several milliamperes are obtainable. Structure of a photo-electric cell is shown in Figure 4.11 in which it will be seen to consist of a thin *p* type layer on *n* type silicon. Due to its photo-voltaic effect there is no need for a bias power source. A linear output can be obtained by selecting a suitable load resistance for a wide range of illuminance. Like the silicon blue cell, described below, it has no directivity of receiving light, so there is no need to adjust the optical axis as is the case with photo-transistors.

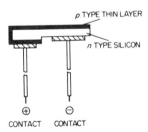

Figure 4.11   Photain photo-voltaic cell

Silicon blue cell. The Sharp's silicon blue cell manufactured by Photain Controls is claimed to be the world's first photo-electric diode possessing high sensitivity over the entire visible light spectrum. It is more reliable than the selenium or cadmium-sulphide photo-cells and has superior time response. No bias power is required, it has a lower noise level than the other two types and it is non-directional.

Applications include illumination meters, exposure meters, optical readouts of film sound tracks, colorimetry, flame spectrometry, photo spectrometry and colour or pattern recognition equipment.

SEMICONDUCTORS (Transistors, diodes, thyristors)

Midway between conducting materials (chiefly metals) and insulators such as glass, rubber and plastics, are certain substances, notably silicon and germanium, which act as insulators under certain conditions and as conductors under other conditions. These are known as semiconductors.

Germanium and silicon have been given great prominence, since the crystals of these elements have been used for the

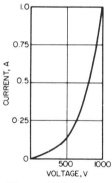

Figure 4.12 Current/voltage characteristic typical of a silicon-carbide resistor

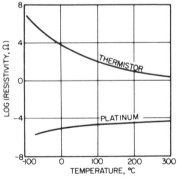

Figure 4.13 Resistivity/temperature relation of a typical thermistor compared with platinum

81

construction of rectifying and amplifying units, commonly known as *transistors*. It must not, however, be assumed that these are the only materials coming under the designation of semiconductors.

Mixtures of the oxides of various metals, e.g. manganese, nickel, cobalt and copper, also exhibit semiconductor characteristics. This group of semiconductors has a large negative temperature coefficient. At low temperatures they have a comparatively high resistance, but this decreases rapidly with rise of temperature. Materials having this characteristic are known as *thermistors*.

To give an idea of the resistance range obtainable it may be mentioned that a particular thermistor having a resistance of 10 000 ohms at a temperature of 0°C., has a resistance of only 10 ohms when the temperature is increased to 200°C.

A third type of semiconductor is the *non-linear resistor*; silicon-carbide is an example of this. It has been found that if a rod made of silicon-carbide is used as a resistance, the relation between current and voltage does not follow Ohm's Law, so that doubling the voltage may increase the current about 20-fold, and trebling it about 100-fold. Figure 4.12 shows a typical current/voltage characteristic. A firm of leading electrical manufacturers has perfected a process by which it is possible to make pure silicon-carbide by an unusual method; the product has a bright yellow colour and is a good electronic semiconductor of the 'excess electrons' type.

The fast-growing family of metal oxide protective devices must also be mentioned. These devices have a high coefficient of non-linearity, which makes them advantageous for all forms of surge suppression including high-power gapless diverters.

Transistors

Semiconductors in the germanium and silicon groups are now widely used in electronic circuits. Silicon crystals of great purity used in conjunction with a point-contact of tungsten wire have proved very satisfactory as rectifiers or detectors in radio reception. Figure 4.14 shows one form of silicon detector.

Silicon transistors are available with junction operating temperatures up to 175°C compared with 85°C allowable for germanium-type transistors. The silicon detector has limitations, however; certain types were found to be easily damaged by overload and investigations showed that germanium in a pure state had special

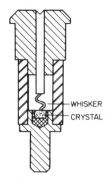

Figure 4.14   Silicon crystal detector

properties enabling it to be used as a rectifier and as a triode suitable for the amplication of radio signals. When a germanium crystal is used for rectifying purposes it may be either of the point-contact type or of the junction type.

Point-contact transistor.   If two fine wires closely spaced make contact with the upper surface of a germanium block, we then have a crystal triode. This device is called the point-contact transistor and is illustrated in Figure 4.15. The two wire electrodes, called the emitter and the collector, are spaced about 0.005–0.025 cm apart. The third electrode is the large area low resistance contact at the base. Either emitter and collector separately connected with the base electrode exhibit the properties of a high-back-voltage rectifier. Operation is illustrated by Figure 4.16.

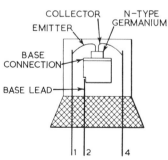

Figure 4.15   Germanium point-contact transistor

If a potential positive with respect to the base electrode is applied to the emitter and a negative bias to the collector, it is found that as the current to the emitter is varied by the voltage a corresponding and greater current change appears in the collector circuit, thus giving rise to an amplification effect. Amplifications up to 100 times (20 dB.) can thus be obtained.

Junction-type transistor. In practice, the point-contact transistor has several limitations and, in consequence, the junction-type transistor was developed. This consist of a suitably prepared thin section of silicon (or germanium) with a deficiency of electrons sandwiched between two large crystals with an excess of electrons. The centre section is known as the *p* or positive, and the two end sections are known as the *n* or negative sections. The centre section corresponds to the grid of a valve whilst the *n* sections, known as the emitter and collector respectively, correspond to the cathode and anode. This type of transistor is known as the *npn* transistor. If the centre section comprises *n*-type material and the two end sections are of *p*-type material, the resulting transistor is said to be of the *pnp* type.

Operation. In order to understand the operation of the *pnp* or *npn* junction transistor it is necessary to consider the conditions existing on both sides of the particular *pn* junction. In the *n*-type material there is a concentration of mobile electrons while on the *p*-type side there is a concentration of mobile holes. Due to a potential barrier existing at the junction it is not possible for the free electrons to move across to the *p*-type side or for the holes to move across to the *n*-type side.

If a battery is connected to the *pn* junction as shown in Figure 4.17(a) an increased bias is exerted on the junction causing the free electrons and the free holes to move away from the junction. When the battery connection is reversed, Figure 4.17(b), the external potential 'overcomes' the internal potential barrier and there is a movement of electrons and holes from the respective materials across the junction and round the outside circuit. This movement constitutes a current flow.

Thus it will be seen that the junction acts as a rectifying device, with conduction taking place in one direction only. However, it must be appreciated that if the reverse voltage is too large (Figure 4.18), a breakdown of the potential barrier will result and the rectifying qualities of the transistor will be destroyed.

84

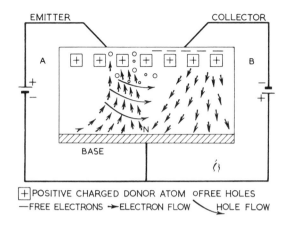

+ POSITIVE CHARGED DONOR ATOM o FREE HOLES
− FREE ELECTRONS → ELECTRON FLOW ＼ HOLE FLOW

Figure 4.16   Operations of the point-contact transistor (RCA Ltd)

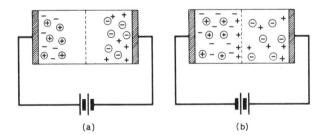

(a)   (b)

Figure 4.17   Diagrammatic representation of a p-n junction. In (a) the battery is connected in such a way as to increase the bias. In (b) the bias is removed by reversing the battery, which now assists current flow.

The characteristic curve for a junction rectifier is shown in Figure 4.18. From the curve it will be noticed that for a small applied voltage in the conducting or forward direction a comparatively large current flows. However, when a large reverse voltage is applied very little current flows until the breakdown point is reached. It is obvious therefore for correct operation the device must not be used in a circuit where the system voltage exceeds the breakdown voltage.

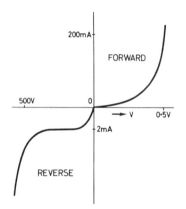

Figure 4.18   Characteristic curve for a junction transistor. Note the different scales for the voltages.

Theory of action.   The *pnp* junction transistor structure is shown schematically in Figure 4.19. The letters *e*, *b* and *c* indicate respectively the emitter, base and collector. Two *pn* junctions have been formed in the transistor and across each there will be a potential barrier as described earlier. The arrows in the diagram show the direction of the electric fields across the two junctions and are the directions in which positive charges would move under the influence of the electric field.

If the battery is connected as shown in Figure 4.19(a), the righthand *pn* junction is biased in the reversed or non-conducting direction, the length of the arrows serving to indicate the strength of the electric fields at the junctions. Under these conditions the collector current is very small, and in a good transistor is only a few microamperes.

When the battery potential is applied as shown in Figure 4.19(b), it will be seen that the field across the barrier, in the

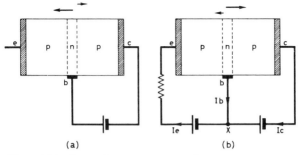

Figure 4.19 A p-n-p junction transistor. The arrow lengths denote the strengths of the fields at the junctions, as well as their direction of operation

forward or conducting direction is reduced, so that it is relatively easy for holes and electrons to move across the barrier. But the material of the transistor is such that there is a large concentration of holes in the *p*-region and relatively few electrons in the *n*-region so that the current flowing consists almost entirely of holes and the emitter is said to inject holes into the *n*-region.

These holes move across the emitter-base junction into the *n*-region, which has a width of about 0.005 mm, and find themselves near the base-collector junction across which the field is such as to drive the holes towards the collector. Thus the effect of applying a voltage to the emitter is to cause an increase in the collector current by an amount nearly equal to the emitter current. An increase in the emitter causes an increase in emitter and collector current. The amount of emitter current may be several milliamperes and so the collector current due to this is very much larger than the small collector current flowing when the emitter current is zero.

The mode of action of the *npn* junction transistor is similar to the above except that the applied voltages are reversed and the part played by the holes is taken over by the electrons.

Field effect transistors. Field effect transistors (FET) may be divided into two main categories, namely junction and insulated gate. Basically a junction FET is a slice of silicon whose conductance is controlled by an electric field acting perpendicularly to the current path. This electric field results from a reversed-bias *pn* junction and because of the importance of this transverse field the device is so named.

One of the main differences between a junction FET and a conventional transistor is that for the former current is carried by only one type of carrier, the majority carriers. In the case of the latter both majority and minority carriers are involved. Hence the FET is sometimes referred to as a unipolar transistor while the conventional type is called a bipolar transistor.

Another important difference is that the FET has a high input impedance, while ordinary transistors have a low input impedance. Because of this, junction FET's are voltage-operated as opposed to current-operated bipolar transistors.

Space does not allow a decription of the operation of FET's but two of the applications of the junction FET are: compound or hybrid source follower and as a voltage variable resistor.

The insulated-gate FET is available in two basic forms, the enhancement type and the depletion type. Like the junction FET they are available as *p*- or *n*-channel versions. The essential difference between the junction FET and the insulated-gate FET is that with the former the conductivity of the channel between the source and the drain is controlled by the transverse electric field created by the reverse-biased *pn* junction. With the insulated-gate FET the conductivity of the channel is controlled by a transverse electric field which is induced capacitively across an insulator or dielectric.

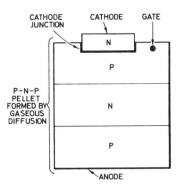

Figure 4.20 Diagrammatic representation of npnp thyristor

The insulated-gate FET can be used as an amplifier and it has an important advantage over the junction FET of a much higher input resistance typically of the order of $10^{12}\,\Omega$. Leakage current is negligible.

Because the gate current is extremely small, of the order of 10^{-15} A the device is suitable for use as an electrometer. It also has advantages when used as an electronic switch.

Diodes

It is about 30 years since the introduction of high-power semiconductor diodes and less than 20 years since high-power thyristors appeared on the market. In that time the silicon rectifier has supplanted virtually all other forms of d.c. power supplies, except storage batteries. GEC Transmission & Distribution Projects has installed over 6000 MW of electrochemical equipments and nearly 2000 MW of traction rectifiers. Variable-speed motor drives of all sizes are normally fed by thyristor equipments. Mercury arc rectifiers are not now used in these applications and have been supplanted by silicon thyristor equipments in the high-voltage d.c. transmission field.

Principles of operation. A silicon atom has a positive nucleus and four valence electrons outside its nucleus. In pure crystal form the valence electrons of adjoining atoms are locked in covalent bonds in a symmetrical lattice. Ideally, there are no free electrons for current carrying in this high resistance semiconductor material.

The effect of the addition of an impurity into the material, one part in 10^9 to 10^{10}, having five valence electrons is shown in Figure 4.21. The fifth electron of each impurity atom does not fit into the lattice and is available for current conduction. This type of doped crystal is called n type. The addition of an impurity having three valence electrons, also shown in Figure 4.21 causes a deficiency of electrons. The deficient bond, or hole, in the lattice may be filled by an electron from a neighbouring atom if a suitable potential is applied. The hole has then been transferred and this action has some of the characteristics of the movement of a positive particle. This material is called p type.

A rectifier element is constructed by forming a junction of p and n type materials in a single thin slice of silicon crystal. Figure 4.22 shows that a forward voltage gives a free current flow, by

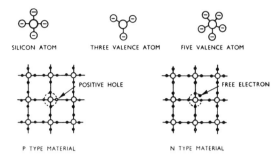

SILICON ATOM THREE VALENCE ATOM FIVE VALENCE ATOM

POSITIVE HOLE FREE ELECTRON

P TYPE MATERIAL N TYPE MATERIAL

Figure 4.21   Atomic structure of p and n materials

passage and re-combination of electrons and holes through the junction. If a reverse voltage is applied, the electrons and holes move away from the junction leaving a high resistance area devoid of current carriers; the reverse current therefore is very small. Figure 4.23 shows the shape of a diode characteristic with a low voltage drop at high forward current and a high reverse voltage withstand at minute reverse currents. The *pn* element is hard soldered or compression bonded between silver surfaces in a sealed container having a heavy copper base to ensure good contact with the cooling device, its mass providing a heat sink during short duration overcurrents.

Physical arrangements.   In its simplest form a silicon rectifier comprises an assembly of silicon diodes on heat sinks together with fuse and surge voltage protection components. These are all normally housed in a sheet enclosure and supplied by a separately mounted double-wound transformer. Additional items may include control cubicles, switchgear, voltage regulators and connections. A controlled rectifier will be similar, with thyristors in place of the diodes and with appropriate electronic firing gear and controls.

GEC Transmission & Distribution Projects makes a range of small naturally-cooled equipments which can be wall-mounted above the rectifier transformer. Ratings are typically up to 300 kW. Higher rated equipments have the rectifier assembly contained in a floor-mounted cubicle with the associated transformer alongside. Cooling may be air natural or in some cases air

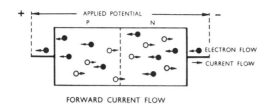

FORWARD CURRENT FLOW

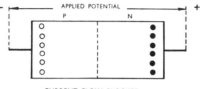

CURRENT FLOW BLOCKED

Figure 4.22   Application of forward and reverse voltage across silicon rectifier element

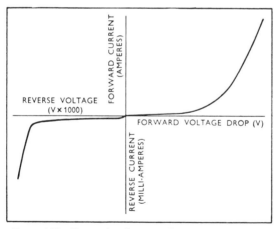

Figure 4.23   Forward and reverse characteristics of a silicon diode. Note difference in voltage scales

blast. If the atmosphere is corrosive, dusty or damp, closed-circuit air cooling or liquid cooling may be used. Heavy current rectifiers usually employ a forced liquid cooled welded-up heat sink/busbar assembly as the best means of achieving the high rating suited to the electrochemical environment.

Basic connections. The choice of single-way (half wave) or double-way (bridge) connection depends partly upon the d.c. voltage required. With single-way the d.c. passes effectively through only one diode at a time instead of through two in series as in the bridge connection. The forward voltage drop is therefore half the value of that for the bridge connection. Hence the losses are also half. However transformer losses and cost are higher for single-way operation. The voltage above which double-way connection is used is generally determined by a combination of efficiency and cost considerations.

Figures 4.24 to 4.29 show a number of different rectifier configurations. Only one diode per arm is shown although there may, in an actual system, be a number of diodes in a series-parallel arrangement for each arm.

Three-phase bridge (6-pulse). The double-way connection, Figure 4.24 gives 6-pulse rectification with a 120 degree conduction angle. The voltage regulation is virtually a straight line over the normal working range. This is the most commonly used connection for industrial power supplies.

TRANSFORMER
SINGLE SECONDARY

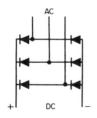

Figure 4.24   6-pulse circuit for 2-wire single bridge d.c. output

Double-star (6-pulse). With this connection the six-phase secondary windings are separated in the two opposed star groups, the neutral points of which are connected through an interphase reactor, Figure 4.25. The voltage equalising effect of the interphase reactor enables the two star groups to share the current, and

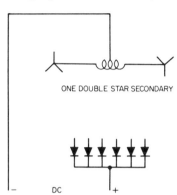

ONE DOUBLE STAR SECONDARY

Figure 4.25 Half-wave 6-pulse circuit for 2-wire d.c. output

120 degree conduction occurs at all currents above the very small value required to magnetise the interphase reactor. The voltage regulation curve is virtually a linear slope while 120 degree conduction is taking place but there is a sharp rise of about 15% as the interphase reactor becomes demagnetised. With modern core steels this point is usually well below 0.5% load and the rise is not normally objectionable or harmful. A small shunt load permanently connected or automatically switched in at low loads can be provided if necessary to eliminate this effect. This connection is primarily used for power supplies for low voltage process work of a continuous nature, where overall efficiency is of prime importance.

Multi-circuit connections. For 12-pulse working two 6-pulse rectifier equipments, two groups of equipments or two rectifier circuits need to be phase shifted 30 degrees from each other. This is usually achieved by having a star/delta relationship between transformers or transformers windings associated with the two 6-pulse rectifiers. Phase displaced rectifiers closely paralleled in this way usually require an inter-circuit reactor to equalise the

93

AND

TRANSFORMER
DOUBLE SECONDARY

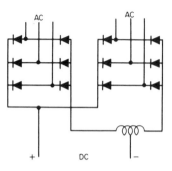

Figure 4.26 Transformer 12-pulse circuit for 2-wire d.c. output from parallel bridges wth intercircuit reactor

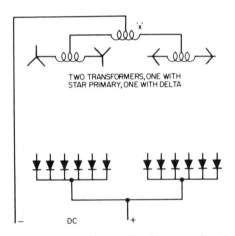

TWO TRANSFORMERS, ONE WITH
STAR PRIMARY, ONE WITH DELTA

Figure 4.27 Half-wave 12-pulse circuit for 2-wire d.c. output. The reactor X can sometimes be omitted

94

voltage between the phase-displaced groups (see Figures 4.26 and 4.27) but these are sometimes omitted if there is sufficient d.c. circuit reactance in connections or in other reactors. On very large installations, pulse numbers greater than 12 can be obtained by suitably phase shifting a number of 6- or 12-pulse rectifier equipments.

Three-wire circuits. 3-wire d.c. outputs may quite easily be obtained with several rectifier and transformer connections. Choice of a particular connection depends upon the required out-of-balance current, the pulse number of the outer to mid-wire

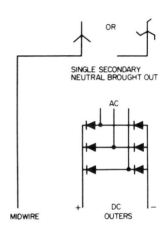

Figure 4.28 Circuit for 3-wire d.c. output with 6-pulse outers and 3-pulse mid-wire. Star secondary for up to 15% mid-wire current and interstar secondary for 100%

d.c. voltage, and the cost involved. The arrangement shown in Figure 4.28 is the most economic, comprising a normal three-phase bridge rectifier and a transformer having a star or interstar connected secondary with the neutral point brought out. This connection provides for not more than 20% out-of-balance current in the mid-wire with a simple star winding or 100% if an interstar winding is used. In such circuits the d.c. voltage waveform contains the harmonics of 6-pulse operation across the outers and 3-pulse operation from mid-wire to outers. The r.m.s. value of the harmonic voltage is about 6% of rated d.c. voltage for 6-pulse working and about 25% for 3-pulse working.

The connection shown in Figure 4.29 consists of two 3-phase bridge rectifiers connected in series and fed by a transformer having two secondaries, one star connected and the other delta connected. This connection will handle 100% out-of-balance mid-wire currents with d.c. voltage harmonics outer-to-outer of only about 3% r.m.s. (12-pulse) and the outer to mid-wire of about 6% (6-pulse).

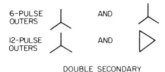

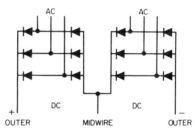

Figure 4.29   Circuit for 3-wire d.c. output with 6- or 12-pulse outers and 6-pulse mid-wire. Suitable for mid-wire current up to 100%

Supply harmonics.   It is important that the harmonics drawn from the supply by a rectifier system be kept within the limits laid down by the supply authority. Recommendations normally specify the maximum ratings of rectifiers, having a stated pulse number and assuming no phase control, which can be fed from supplies at various voltages at the point of common coupling for two or more consumers.

Table 4.1 shows the general recommendations for the UK; these requirements are usually more onerous than those in other countries where larger rectifier loadings may be permissible. It should be remembered that a three-phase bridge circuit has a pulse number of six. Pulse numbers less than six are used only on rectifiers of a few kW rating or by special arrangement with the supply authority. The table is only a guide and prospective rectifier users should always consult their local supply authorities who are aware of other consumers' rectifier plants that may already be connected to the system.

Table 4.1 Maximum permitted rectifier loadings at specified voltage in the UK

Supply system voltage (kV) at point of common coupling	Type of converter	Permissible kVA capacity and corresponding effective pulse number of three-phase installations		
		3-pulse	6-pulse	12-pulse
0.415	Uncontrolled	–	150	300
	Half controlled	–	65	–
	Controlled	–	100	150
6.6 and 11	Uncontrolled	400	1000	3000
	Half controlled	–	500	–
	Controlled	–	800	1500
33	Uncontrolled	1200	3000	7600
	Half controlled	–	1200	–
	Controlled	–	2400	3800
132	Uncontrolled	1800	5200	15000
	Half controlled	–	2200	–
	Controlled	–	4700	7500

Regulation. The inherent regulation of a rectifier equipment is defined as the rise in voltage from full load to light load, and is expressed as a percentage of the rated full load d.c. voltage. Light load is taken as 5% of rated load.

With certain rectifier circuits, the d.c. voltage can rise sharply below 5% load because of the demagnetisation of an interphase transformer. Steps can be taken to prevent this occurring.

Small rectifier equipments incorporating fuse protection will have an inherent regulation of the order of 5%. Larger equipments designed to be protected by switchgear, and having a higher rectifier transformer reactance will have an inherent regulation of approximately 8%. Lower regulation values than these can be obtained but they will usually cost more.

Rectifier protection. Rectifier equipments are protected against overcurrent and surge voltages, and are designed so that internal component failure cannot cause damage. There are three basic overcurrent schemes, circuit-breakers, fuses or a combination of both.

When fuses are used on their own they are of the fast acting type, specially designed to match the characteristics of the diodes. They are connected in series with the diodes but will also disconnect the diode should it have an internal fault, allowing the passage of current in the blocking period. When diodes are connected in parallel they normally have individual fuses. Indicator fuses, of the striker pin type, are also connected across the diode fuses to assist in identifying a blown fuse and the faulty diode. For small equipments containing six or less diode fuses it is often acceptable for these to be the sole means of protection.

Circuit-breakers are used where it is not desirable to lose the complete d.c. supply due to a feeder fault on the d.c. distribution network. This is generally applicable to large rectifier equipments. In such cases fuses are provided for protection of the diodes but d.c. feeder faults are cleared by high speed d.c. circuit-breakers. This leaves the rectifier supply available for healthy circuits.

When d.c. overcurrents are likely to occur often but are not normally of maximum severity, i.e. not short-circuits, the overcurrents may be cleared by a moulded case a.c. circuit-breaker with diode fuses as back-up protection for severe faults. Such protection applies to small equipments, up to 200/300 kW, and only where the a.c. supply voltage is under 1000 V. A slightly higher transformer reactance than normal allows the a.c. breaker to clear most faults.

Surge voltage protection is provided by capacitor/resistor networks within the rectifier equipment. These networks limit transient voltage surges and commutation voltage peaks to levels substantially below the transient rating of the diodes. In most cases the surge absorbing circuits are themselves protected by fuses. On larger equipments, local or remote indication of fuse operation can be provided.

When regeneration can occur it is necessary to protect the d.c. network by some other means than the surge absorbing circuit. Any motor attempting to feed power back into the rectifier terminals raises the d.c. voltage to a level which may cause damage. Protection for this condition can be provided by a loading resistor that can be permanently connected or switched into circuit under regenerative conditions.

Thyristors

The thyristor is a semiconductor switch either of the *pnpn* or *npnp* type, whose bi-stable action depends on regenerative internal feedback. The four-layer device (Figure 4.20) is usually silicon although germanium has been used. Devices with the two endmost layers only connected to external terminals (anode and cathode) are called four-layer diodes, those with three layers accessible (anode, cathode and *p*-gate) are called thyristors.

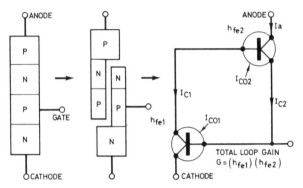

Figure 4.30   Two-transistor analogue of thyristor

The structure is best visualized as consisting of two transistors, a *pnp* and an *npn* interconnected to form a regenerative feedback pair as shown in Figure 4.30. Current gain around the internal feedback loop G is $h_{fe1} \times h_{fe2}$, where h_{fe1} and h_{fe2} are the common emitter current gains of the individual sections. If I_{co1} is the collector to base leakage current of the *npn* section and I_{co2} is the collector to base leakage of the *pnp* section, then

    for the *pnp* section: $I_{c1} = h_{fe1}(I_{c2} + I_{co1}) + I_{co1}$

    for the npn section: $I_{c2} = h_{fe2}(I_{c1} + I_{co2}) + I_{co2}$

and the total anode-to-cathode current $I_a = (I_{c1} + I_{c2})$

from which $I_a = \dfrac{(1 + h_{fe1})(1 + h_{fe2})(I_{co1} + I_{co2})}{1 - (h_{fe1})(h_{fe2})}$

With a proper bias applied, i.e. positive anode to cathode voltage, the structure is said to be in the forward blocking or high impedance 'off' state. The switch to the low impedance 'on' state is initiated simply by raising the loop gain G to unity. As this occurs the circuit starts to regenerate, each transistor driving its mate to saturation. Once in saturation all junctions assume a forward bias, and the total potential drop across the device approximates to that of a single junction. Anode current is then only limited by the external circuit.

To turn off the thyristor in a minimum time it is necessary to apply a reverse voltage and under this condition the holes and electrons in the vicinity of the two end junctions will diffuse in these junctions and result in a reverse current in the external circuit. The voltage across the thyristor will remain at about 0.7 V positive as long as an appreciable reverse current flows. After the holes and electrons in the vicinity of the two end junctions have been removed, the reverse current will cease and the junction will assume a blocking state. The turn-off time is usually of the order of $10-15\,\mu s$. The fundamental difference between the transistor and thyristor is that with the former conduction can be stopped at any point in the cycle because the current gain is less than unity. This is not so for the thyristor, conduction only stopping at a current zero.

APPLICATIONS

The chief use of diodes is as d.c. supplies for electrolytic process lines, for traction substations and for general industrial purposes. Thyristors have mainly been applied as phase-controlled converters producing variable d.c. supplies for motor drives, turbine-generator and hydro-generator excitation, vacuum-arc furnaces, electrochemical processes, battery charging etc. These are still the major applications, but increasingly thyristors and power transistors are being employed in new switching modes as d.c. choppers or as inverters producing a.c. of fixed or variable frequency. Some of the less-common applications for thyristors are given below.

Excitation of synchronous motors. Most specifications for synchronous motors state that they must be capable of withstanding a gradually applied momentary overload torque without losing synchronism. This was detailed in BS 2613, withdrawn in 1978 and superseded by BS 5000 Part 99. The latter standard does not have a similar section dealing with this matter.

In the past the overload torque requirements has been met by using a larger frame size than necessary to meet full-load torque requirements, the torque being proportional to the product of the a.c. supply voltage and the d.c. field produced by the excitation current.

With thyristor control of excitation current it is possible to use a smaller frame size for a given horsepower rating and arrange to boost the excitation by means of a controller to avoid loss of synchronism under torque overload conditions.

The excitation current of a synchronous motor may be controlled by supplying the motor field winding from a static thyristor bridge, using the motor supply current to control the firing angle, Figure 4.31. A pulse generator varies the firing angle of the thyristors in proportion to a d.c. control signal from a diode function generator. Variable elements in the function generator enable a reasonable approximation to be made to any of a wide range of compensating characteristics.

When the motor operates asynchronously, i.e. during starting, a high emf is induced in the field winding, and the resulting voltage appearing across the bridge must be limited to prevent the destruction of the bridge elements. This may be done by using a shunt resistor connected as shown.

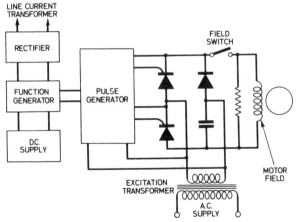

Figure 4.31   Simple compensated excitation circuit

Where more exacting requirements have to be met, current feedback can be applied to eliminate effects of non-linearity in the pulse generator and rectifier bridge and it will also improve the response of the system to sudden changes of load. Automatic synchronizing is possible without relays by incorporating a slip-frequency sensing circuit to control the gate which supplies the control signal to the pulse generator.

Variable frequency supply. It is possible to use a cycloconverter to control the speed of an induction motor. The cycloconverter is a rectifier device first developed in the 1930's but with the improved control characteristics of thyristors and better circuit techniques, a continuously variable output frequency is possible.

Figure 4.32 illustrates the process of conversion for 15 Hz output from 50 Hz supply. During the conduction cycles starting at point *a* the output voltage reaches a maximum since there is no firing delay. At point *b* commutation from phase two to phase three is slightly delayed and commutation is further delayed at point *c*.

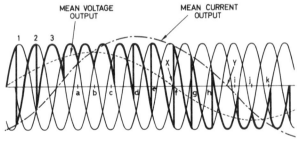

Figure 4.32   Synthesis of 15 Hz from 50 Hz (load power factor 0.6 lagging)

At time *e* the firing delay is such that the mean output voltage is only just possible. In the diagram the low-frequency load power factor is 0.6 lagging so that although the mean load voltage crosses the axis at *X* the current remains positive until *Y*. Consequently the rectifiers which conduct at instants *f*, *g* and *h* are giving positive load current and negative voltage, that is inverting. From *i* the system behaves as a controlled rectifier using the negative

102

group of thyristors, d, e and f until the mean output voltage becomes positive when an inversion period starts again.

D.C. chopper controlling a d.c. motor. The basic circuit is shown in Figure 4.33, the supply being applied to the motor by firing $T1$. At standstill the current rises rapidly to a value controlled by the circuit resistance and at a rate controlled by the inductance of the motor.

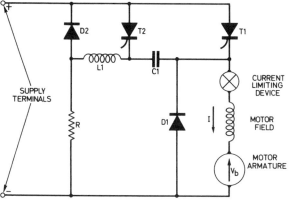

Figure 4.33   Basic circuit

After a certain interval $T1$ is switched off, the flow of current round the motor and through the diodes $D1$ being maintained by the inductive energy of the system. When this current has decayed to a certain level, $T1$ is again fired. In order to maintain the constant motor current the current limiting device provides a control signal which varies the length of the times $T1$ is on or off.

In Figure 4.33 the circuit components $T2$, $C1$, L, R and $D2$ are included for the purpose of switching off $T1$.

Control of fluorescent lighting banks. Dimming of fluorescent lighting by means of thyristors is possible, the circuit being shown in Figure 4.34. A separate transformer is necessary to maintain

103

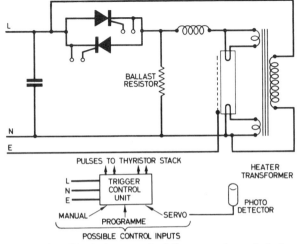

Figure 4.34 Circuit for control of fluorescent lamp, including dimming facilities

Figure 4.35 Stack assembly of thyristors showing the heat sink arrangements. It is rated to control up to 120 A on a 415 V a.c. system. (International Rectifiers)

constant preheating currents to the electrodes irrespective of the setting of the dimmer. This entails use of a 4-wire distribution system. Control is effected by simply varying the firing angle of the thyristor.

Cooling. High power thyristors have to dissipate any losses in a small volume and thus artificial means are often provided to take the heat away from the devices. It is usual therefore to mount the thyristors on aluminium alloy castings which act as heat sinks, see Figure 4.35. For very high power installations this natural cooling may be supplemented by forced draught means.

Acknowledgement. – The Editor acknowledges that much of the information on thyristors has been extracted from the I.E.E. Symposium on 'Power Applications of Controllable Semiconductor Devices.' The Section on thyristor theory has been compiled from information contained in the G.E. Handbook on thyristors. Both sections have been updated.

5

Rectifiers and converters

METAL RECTIFIERS

Three basic types of metal rectifier are available in the UK namely, selenium, germanium and silicon. The last two are generally referred to as semiconductor rectifiers and their theory of operation is described in Chapter 4. One of the major manufacturers in the UK of these rectifiers is Westinghouse Brake and Signal Co., to whom acknowledgement is made for providing information upon which this chapter is based.

Selenium rectifiers. During the past 25 years there has been a continuous development of the selenium rectifier so that the stability of the rectifier is such that it can be operated at relatively high temperatures, i.e. 120°C if required. Alternatively it can be operated for a much longer time than hitherto under normal operating conditions.

Selenium rectifiers are used widely for all low power requirements where initial cost is important and the ability to withstand substantial and repeated overloads eliminates the need for special protective devices that may be required for silicon or germanium rectifiers. Although the efficiency and performance of selenium rectifiers may be slightly inferior to the other two and the size somewhat greater, these features are often of less importance for outputs below 25 kW.

Applications include electroplating where oil-immersed units provide currents up to 200 kA. At high values of current water-cooled germanium units are used due to their small weight and the limited space which they require.

Electrostatic precipitation where d.c. voltages from 30–100 kV are required is a popular application for selenium rectifiers. Oil-immersed equipment is reliable in operation, robust and provides an efficient means of rectification. Westinghouse has manufactured equipments with outputs of 60 kV and currents from 60mA to 500mA. These equipments are transductor controlled with very rapid arc extinction.

Cinema arc power supplies can be provided by transductor or choke/capacitor constant current selenium equipments for both high and low intensity arcs and can be operated from either single-phase of three-phase supplies.

Germanium rectifiers.   Germanium rectifiers are used extensively for low power and medium power industrial applications where the ambient temperature is not high and particularly at voltages below 100V. Where high efficency at such voltages is of overriding importance, the size of the installation may be in the megawatt range. The germanium is extracted from coal and zinc deposits and refined to a high degree of purity. The resultant grey metallic material is pulled into a single crystal which is specially cut to form small wafers. By heat treatment, an indium button is welded on to and diffused into the germanium wafer. The bond between the germanium and the indium forms the rectifying junction which is mounted in a hermetically sealed housing.

Both germanium and silicon rectifiers are used extensively in equipments providing power for the electrolytic production of chlorine and hydrogen. Outputs in excess of 27kA at 120V have been provided from these equipments, germanium offering the slightly better efficiency at d.c. voltages below 100V, but they may not be economical in countries where the ambient temperature is high.

Industrial d.c. power supplies are best met by germanium and silicon semiconductor equipments although as stated earlier for powers below 25kW the selenium rectifier still remains the most attractive plant. Both types are used extensively for large telephone exchanges. Chargers for battery electric vehicles are generally of the germanium rectifier type. Welding is another area where both germanium and silicon rectifiers compete with each other.

Silicon rectifiers.   Silicon is a metal obtained from sand and is refined to a high degree of purity, drawn into large monocrystals and cut into wafers. A thin plate of aluminium is bonded to and diffused into the silicon wafer and the junction between the aluminium and the silicon forms the rectifying junction.

Many of the applications for this type of rectifier have already been outlined but this type of rectifier is considered to be the best

107

for railway traction supplies. Variable speed drives is another area that this type of rectifier is widely employed. The facility of the grid control makes it ideal for speed control of d.c. machines up to the largest ratings.

Comparison of three types. The semiconductor germanium and silicon rectifiers are, for a given output, more compact than selenium. This is due in part to the low forward resistance per unit area of rectifiers, and in part to the high voltages which they withstand in the reverse direction. Silicon will withstand a higher reverse voltage than germanium and will also operate at a higher temperature. The forward resistance of germanium is lower than that of silicon.

The d.c. voltage/current characteristics of the three types of rectifier are shown in Figure 5.1. It will be seen that with a rectifier operating at normal current density, one may expect a low voltage drop with germanium, a higher voltage drop with selenium and a still higher one with silicon. In practice it is not usual for the reverse current to exceed about one half per cent of

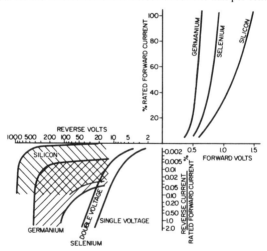

Figure 5.1   Comparison of forward and reverse characteristics of the three types of rectifier

the forward current and, therefore the selenium rectifier is capable of being operated at about 32 V, germanium 70–90 V and silicon 100–300 V. The voltage at which the rectifiers are operated is determined by many factors such as duty cycle, circuit connection, temperature, etc.

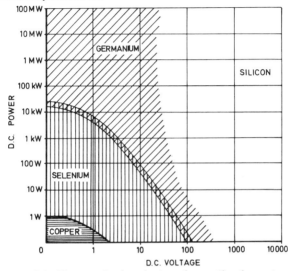

Figure 5.2   First step in the selection of a rectifier for a given application. Other factors than output voltage and rating may dictate the final choice for a specific application

In the case of the Westcode Semiconductors Westalite, the wide spread of permissible reverse voltage allows the selection of two grades of useful selenium rectifier, single- or double-voltage. The still wider spread of germanium, coupled with the higher maximum voltage operation, permits a still larger range of useful grades of rectifier. Silicon has an even wider range and consequently more electric grades of mechanically identical rectifiers are available.

The first choice of rectifier can be made from Figure 5.2 although it should be realised that the choice may well be modified by consideration of economics, duty cycle, etc.

Temperature characteristics. There is a maximum critical temperature above which each of these three rectifiers cannot safely operate. It is important therefore to ensure that the temperature rise of the rectifier and the ambient temperature together does not exceed the critical value. Forward ageing is the limit for a selenium rectifier and such units are suitable for temperatures up to 70°C with specials being available for use at total temperatures up to 130°C.

Although the maximum temperature at which germanium rectifiers can safely operate from forward and reverse ageing is 90°C, another factor limits the safe working temperature to 70°C. Due to the low thermal mass of this type of rectifier, relatively small overloads can cause rapid temperature increase. It is considered advisable therefore to limit the total operating temperature to 50°C. For similar reasons the total operating temperature of the silicon rectifier is reduced from 200°C to about 160–170°C to give an added factor of safety.

Overloads. Selenium rectifiers are formed on a substantial metal base and operate at a low current density. They can therefore withstand severe current overloads of short duration without damage. As mentioned above the low thermal mass of the other two types does not permit them to withstand current overloads.

Voltage overloads give rise to increase leakage currents and self-heating. Again the silicon and germanium devices are more sensitive to them and care must be taken to protect them from such overloads under normal operating conditions.

Parallel and series connections. With germanium and silicon rectifiers care has to be taken when connecting them up in series or parallel. Performance ratings should be closely matched or ratings should be substantially reduced. It is desirable to have individual protection of diodes by fuses or an equivalent.

Size and efficiency. A germanium rectifier is roughly one third the volume of a selenium rectifier for the same power output. A silicon rectifier is about one third the volume of a germanium device or one tenth of the volume of a selenium rectifier.

The overall efficiency of an installation is determined by a consideration of all the equipment and to some extent by the circuit connection, load etc. Tables 5.1 lists the efficiencies of the three types of rectifiers.

Table 5.1 Comparison of efficiency of rectifiers (in %)

D.C. voltage	Selenium	Germanium	Silicon
6	85	91	83
12	91	95	90
25	91	97	94
50	91	97	97
100	92	97	97
500	92	97	98
750	92	98	98

CONVERTING MACHINES

The term converting machines is used to cover those arrangements whereby a.c. is converted to d.c. by machines having rotating parts. Although their use is gradually dying out because of the advent of static converters nevertheless there are still many in operation today. There are three main types as follows.

Rotary converters. These consist of a wound rotor revolving in the field of a d.c. generator, the rotor being fitted with slip-rings at one end and with a commutator at the other. If while rotating at synchronous speed an a.c. supply is connected to the slip-rings, d.c. can be taken from the commutator. There is only one winding and the power to keep the machine running and to supply the electrical and friction losses is taken from the à.c. side.

A rotary converter will run from the d.c. side when a.c. can be taken from the slip-rings – this arrangement being called an *inverted rotary converter*. Practically all rotary converters of any size are polyphase; three-phase for small and medium outputs and six-phase for larger outputs.

Ratio of transformation. The d.c. voltage will be $\sqrt{2}$ or 1.41 times the a.c. voltage for a single-phase, and the various ratios for polyphase machines are given in Table 5.2.

Table 5.2 Transforming ratios

	Single phase	Three phase	Six phase	Twelve phase
Volts between slip-rings as a percentage of d.c. volts	70.7	61.2	35.4	18.3

The relationship is given by a.c. volts between slip-rings.

$$= \frac{\text{d.c. volts}}{\sqrt{2}} \sin \frac{\pi}{m}, \text{ where } m = \text{no. of slip-rings.}$$

As d.c. voltages are usually in the region of 220 to 240 V it will be seen that with normal a.c. supply a transformer is needed to give the required voltage for the supply to the a.c. side.

Six-phase machines are the most usual since it is fairly simple to obtain a six-phase supply from the secondary of the transformer, using a delta connection on the h.v. side.

Voltage regulation is obtained either by varying the power factor (by excitation control), thus using the reactance of the transformer, or by using a booster. Voltage control can also be obtained by using an induction-regulator.

The efficiency of a rotary converter varies from 90–94%; it has a high overload capacity, and as the power factor is under control it can be kept approximately at unity.

Normally, rotary converters are not self-starting from the a.c. side, but a starting winding can be wound on the *stator* to act as an induction motor. Other methods include starting from the d.c. side and the use of an auxiliary starting motor. With these methods careful synchronising is necessary before switching on to the a.c. supply otherwise serious damage may result.

Motor generators. These consist of two entirely separate machines (from the electrical point of view), and any two machines (e.g. a motor driving a generator) form a motor converter.

Normally they are coupled machines for converting a.c. to d.c. and consist of either an induction motor or a synchronous motor driving a d.c. generator. They also convert in the reverse direction from d.c. to a.c. when the latter is required where there is no supply. Also motor converters are used as frequency changers (i.e. an a.c. motor driving an alternator).

As there are losses in both machines the efficiency is not high and not usually above 90%. One of the advantages is that high-voltage can be taken to the a.c. side and a wider control is obtainable as to the voltage on the output side than with rotary converters. Motor generators are sometimes used as standby supplies for computer installations.

Motor converters. These consist of two machines mechanically coupled together but also connected electrically. The motor portion consists of an induction motor with a wound rotor, the rotor being connected to the winding of the rotor of the generator portion. The efficiency at full load varies from 86–92% with a power factor of 0.95 and over. They are rather more stable than rotary converters and are self-starting from the a.c. side. They are not quite so efficient at light loads. Both motor generators and motor converters are discussed in some detail under 'High integrity power supplies' on page 146.

6

Computers and programmable controllers

Today's computer systems process information and data at high speed and provide output in text, data, tabular, graphical and diagrammatical format.

They range from small personal computers used by individuals in the home or at work, to giant mainframe systems which have terminals connected to them to provide shared processing facilities for several thousand people simultaneously. And networks of computer systems can now be interconnected so that it is as easy to transfer information to a computer on the other side of the world as it is to one at the other side of the same office.

As well as handling the original computing applications such as invoicing, payroll, accounting and engineering design, the modern computer systems and associated terminals are now used extensively in schools, supermarkets, hospitals, travel agencies and tax offices, and process many activities which were previously carried out manually.

A computer may be considered to consist of two parts, the hardware and the software. The term hardware describes the electronic and electromechanical components which carry out the processing, store information, and control input, output and communications functions.

Software defines the set of instructions which control the operation of the computer, the routines developed to enable files of data to be established and interrogated at any instant, and the specific application programs which manipulate information and data to produce the desired results.

Input techniques have progressed considerably since the early days of computing. Today, the keyboard of a visual display unit or terminal is the most common means of feeding new information into a computer system. Optical character and bar code reading tech-

Figure 6.1 *The ICL Quattro personal computer. Up to four terminals can be linked to the processor simultaneously*

niques are used in sales order processing, retailing and other areas where operations are repetitive and speed is important.

Floppy disks are used extensively for storing data and programs for use on personal computers, word processors and other small systems and their contents can be transferred to and from the computer store at high speed. Fixed magnetic discs of high capacity and extremely high reading speed are used in conjunction with medium and larger systems.

Output from a computer system can be provided on a range of printers from simple dot matrix types to correspondence-quality and high speed laser systems. The visual display unit is also used extensively to obtain information from a computer, particularly where it is sufficient to view the result and thus a printed report is unnecessary.

Since 1948, when the computer first arrived on the market, there have been three generations. The first computers were built using thermionic valves. They were superseded by systems based on transistors. The third generation and latest computers are based on micro-interrogated circuits formed on silicon chips. They are compact, reliable and operate at high speeds.

115

The largest indigenous UK computer supplier is ICL, which was formed in 1968 as a result of the merger of the leading computer companies in the country. In 1984, ICL became a member of the £2 billion STC communications and information systems group.

ICL employs approximately 20,000 people and sells its products in over 70 countries. Its principal development centres and manufacturing operations are located in the UK. Overseas, ICL has a 40% stake in International Computers Indian Manufacture Limited.

A description of ICL's products and strategy is given as an indication of the type of activities of the computer industry as a whole.

For the past few years, ICL has specialised in providing integrated solutions to meet the information technology needs of users in key markets – particularly retailing, manufacturing, financial services, public administration and defence. This means that the company's activities cover the development, manufacture, and marketing of information systems and software, and the provision of consultancy, training and support services.

ICL's products range from personal computers, word processors, terminals and graphics workstations through to office and departmental systems and up to the most powerful mainframes. All are designed so that they can be linked together in networks and meet internationally-defined open systems standards to give users maximum freedom of choice.

Collaboration with other suppliers also forms part of the ICL strategy. This is arranged in order to acquire and intercept advanced technologies, and to give ICL customers access to advanced systems which complement the company's own products. For example, ICL collaborates with Fujitsu of Japan on advanced chip technology for mainframe systems, is conducting pre-competitive research in conjunction with Siemens of Germany and Bull of France, and has an agreement with Sun Microsystems of USA to market ICL developed computer-aided engineering systems based on SUN graphics workstations.

ICL has won the Queen's Award for Technological Achievement for the innovative design of its top of the range Series 39 mainframe systems. The special features include a nodal architecture which enables several processors to be linked together, fibre optic communications, and high speed integrated circuit 'chips'.

Series 39 mainframes are used extensively by commerce, industry and government in many countries around the world. Users in the UK include British Telecom, the Inland Revenue, the Automobile

116

Association and a large number of local government authorities.

All processors in the Series 39 range, which has a power span of 50:1, utilise the same ICL designed VME operating system having advanced security, transaction processing and networking capabilities.

While mainframe systems will remain the backbone of an organisation's infrastructure, the fastest growth is likely to be seen in office and departmental systems. With the advances made in the past few years it is now possible to establish an office system which mirrors the way an office works, enables information to be shared, integrates word processing with spreadsheets and illustrations, maintains personal records, diaries and address lists, and can interchange information in the form of electronic mail with other systems around the world.

ICL is one of the principal UK suppliers of computing facilities for office use and its OFFICEPOWER software provides a range of fully integrated functions covering all the essential office services. The OFFICEPOWER software is run on ICL DRS 300 distributed resource systems or CLAN department systems and is equally suitable for the small office or department of 100 or more people.

As OFFICEPOWER is based on the internationally-accepted UNIX operating system, a wide range of application packages are available for immediate use and many more are being developed by independent software suppliers.

Other advanced hardware systems supplied by ICL include the System 25 networking minicomputer, the Quattro multi-user personal computer and the DRS professional workstation.

PROGRAMMABLE CONTROLLERS

Programmable controllers (pcs), sometimes referred to as programmable logic controllers (plcs), were introduced in the late 1960s to provide an alternative to electromechanical relays, but since then have expanded their application so that now they are being widely employed by many manufacturing industries. A pc is a solid state logic control system that can be programmed with a user-orientated language so that it initiates or controls a wide range of functions in terminal equipment like motors, valves, dampers and lights.

One of the pc's widest applications in the electrical field is still as a replacement for electromechanical relays and also timers and counters. It detects the change in the state of input signals from pushbuttons, limit switches and sensors and acts on this information to produce desired output signals to operate loads such as motor controllers, contactors, solenoids and pilot lights. A pc differs from a mini-computer in a number of ways, particularly in that programming is simpler.

Figure 6.2   The MPCI expandable controller and input/output expander from Cutler-Hammer Industrial Controls

Present day pc capabilities include addition, data transfer, bit calculations, printer control instructions, data logging, information display and proportional integral derivative (pid) three-term control algorithm. The ability to function as a pid control system opens up a tremendous field for the continuous process industries where there is a mixture of sequence, analogue and digital control. An expandable pc is shown in Figure 6.2.

Components.   Components of a pc are: central processor unit (CPU), memory, input/output system, communications module, power supply facilities, programming device, equipment controls and indication. However, as far as the user is concerned a pc

118

consists mainly of a number of black boxes linked together which initiate signals to his plant to effect some pre-arranged kind of operation. It is not necessary for an engineer to understand the inner workings of a pc or the terminology associated with it in order to employ it intelligently. He can obtain guidance as to the best system to use for his own installation from manufacturers. However a description of the component parts is useful.

Central processor unit.   The processor unit manipulates data in accordance with instructions stored in the memory. Factors relevant to the user are: the speed with which the operation is performed; the format of the data and instructions, i.e. the method of programming; and the type of instruction possible.

Memory.   The memory or program store can be volatile or non-volatile with the option of the user being able to transfer from the former to the latter. Memories available include the fast but volatile random access memories (RAM), which lose their contents when the power is switched off; their contents can be read or written in a random order. A read-only memory (ROM) is used to permanently store fixed programs and data. The erasable programmable read-only memory (EPROM) stores information to control the behaviour of the pc, and is a fail-safe type, retaining its contents in the event of a mains failure. Its contents can only be erased by using ultra-violet light and can then be rewritten, although necessitating removal of the chip from the equipment for modification. The latest memory device (at the time of writing) is the EEPROM, an electrically erasable programmable read-only memory, which allows the designer to modify its operation by means of software. At the present time capacity is somewhat limited, up to 64 K compared with EPROM's 256 K.

Input/output system.   An electronic 'black box' which is used by the operator to send information to, or receive information from the pc. The signals may be d.c. or a.c. and in analogue, digital or special purpose form. Typically a simple pc input/output system changes high a.c. or d.c. voltage signals from input sensors to low d.c. logic signals used in the programmable device (or sequencer), and low d.c. logic signals to higher voltage signals to control output loads. Such a system consists of individual plug-in input and output modules on a multi-module mounting base. The logic module interfaces the sequencer to its input/output system.

Communications module. This provides information from the pc to equipment like a VDU (for the operator) and also to other processors or computers. It can also be used to operate a printer to give a permanent record of status at any required time.

Power supplies. The equipment requires a power supply which may be either a.c. or d.c. to suit the user. The preferred values are 24 V or 48 V d.c. nominal or 93–132 V and 198–264 V a.c., either 50 Hz or 60 Hz. There are limits associated with these values, and as regards the d.c. the ripple must not exceed 5% peak to peak. An essential requirement of a pc is that it must be able to accept a defined supply interruption of not more than 10 ms. If it exceeds this value then the pc must stop accepting input signals and turn-off all output signals; in other words provide a fail-safe situation. As mentioned earlier the content of a memory may be lost by a supply failure. The power supply for the pc may be contained within one of the pc components or inside a cubicle or switch panel housing the whole equipment.

Programming device. This provides the logical instructions (program) which determines the behaviour of the pc. The program may be 'executive', the name given to a fixed program entered into the pc during manufacture, or 'user' ('control') which is created by the user to suit his particular application. This can then be altered during the life time of an installation if the right type of pc is initially used.

Equipment controls and indication. These may be, for example, relay or solenoid coils on motor starters, lamps in display panels, or position or status indicators. It is the equipment or component that is operated by the output signal from the pc.

Programming language. Most pc programming language is based on relay symbology. The standard relay ladder type of instruction looks very similar to the normal control and relay circuits drawn up for switchgear and motor control gear but uses logic circuits usually incorporating 'NAND' and 'NOR' gates. A NAND gate is an 'AND' gate feeding an inverter (a 'NOT' gate) and a NOR gate is equivalent to an 'OR' gate and an inverter. All other types of gates can be derived from NAND and NOR gates. One can communicate with the pc in plain English instruction and can receive information either as printed text or in graphical form on a VDU. Symbolic language allows program alterations to be made easily.

Developments. Timers, counters and shift registers are now designed into all but the most simple pcs. Their inclusion was a simple step, exploiting inherent commands within the micro-processor.

Analogue inputs and outputs have brought the real world into the heart of pcs. It is difficult to imagine how many of today's demands, particularly in the process industries, could have been met without the continual improvement in accuracy and speed.

Digital inputs and outputs have been improved and extended. Most systems now cater for all of the internationally used voltage ranges with improved tolerance and flexibility. Noise immunity and reliability problems have been significantly reduced as designers gain more experience with improved components, which allow greater packing densities resulting in the emergence of compact input/output (I/O) boards.

Probably the single most important improvement to pcs has been brought about by the inclusion of serial communications. It has made possible the advantages of remote I/O, instrumental in significant cost reductions. These cost reductions are realised even with small pcs performing the simplest of tasks. Programmers utilise serial communications not only for connection to the pc but also to peripheral equipment such as tape units, printers, colour graphic monitors, etc. The application of interprocessor communications effecting hierarchical control schemes has changed the trend from single large processor pcs to smaller units controlling individual tasks. Computer communications have brought status reports of plant operation directly into the office, eliminating time-consuming and costly manual methods.

Languages have improved with the more popular format relay ladder being extended to include more advanced functions. Some advanced systems now incorporate higher level languages such as BASIC, reducing still further the gap between what has traditionally been called a pc and a computer.

So the pc continues to evolve into the single most important element within the control gear jigsaw. Its functions continue to expand driven by the demands of automation and increasing production speeds. The future promises to be just as eventful as the past, with pcs continually being developed and refined.

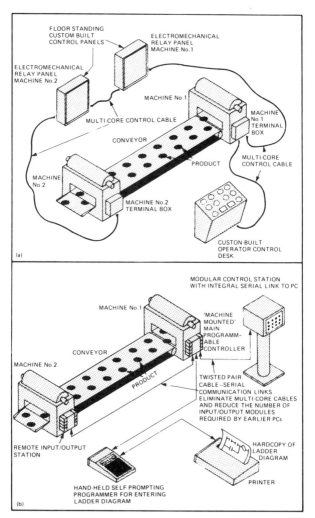

Figure 6.3a   Before pcs, controllers required the time-consuming drawing of ladder diagrams and hardwiring of relay/timer circuits
Figure 6.3b   Modern programmable controllers have their control sequences programmed using a hand-held keyboard

Arguably, however, pcs have reached a certain maturity and it is difficult to see any major fundamental changes. Figure 6.3 shows contrasting control techniques for both relay and pc schemes and indicates the advantages of the latter.

If we treat the relay control panel and pc control panel as a 'black box' very little has changed in terms of conception. However, a number of subtle changes have taken place. Typically the pc control panel is smaller, particularly if the equivalent relay panel involves large numbers of 'internal relays'. So space has been optimised resulting in a corresponding cost reduction. The number of operator inputs reduces slightly as the process machine is made more automatic. It should be noted that this does not affect the total number of inputs to the control panel as a corresponding increase in the number of machine sensors is required.

The real difference obviously exists within the respective control panels. The advantages of pcs really come into their own when modifications are likely to be needed. Fault-finding is far easier with pcs, their initial cost very often being offset on the machine's first breakdown.

Installation costs, however, have been little affected by the use of pcs. The total number of interconnecting control wires has been unaffected. But without doubt a new revolution is taking place. Not as obvious as the first pc but its implications are far more far reaching. It is epitomised by the Cutler-Hammer, S16 pushbutton expander unit, used in conjunction with the MPC1 controller, Figure 6.4.

The S16 unit looks just like any other operator station comprising a combination of pushbuttons, lights and selector switches. The S16 family of human interface panels are microprocessor based, through-panel-mounted enclosures which can be arranged to match custom-designed machine control panels utilising lighted pushbuttons, indicating lights, selector and keylock switches. The human interface panels are economical alternatives to custom-built panels with as few as eight pilot devices installed. However, cost savings are fully realised when 12 or more pilot devices are installed.

The actual configuration of the panel fronts is almost limitless since control of the I/O is handled by the programme contained within the pc. The pilot devices can be used to control multiple inputs, outputs or coils depending on the control logic.

This unit can be installed adjacent to or remotely from the pc,

Figure 6.4   Cutler-Hammer S16 pushbutton expander unit used in conjunction with the MCPI controller

allowing flexible use of space in already crowded control panels, enclosures, pedestals, etc. It plugs into the expansion chassis port of the MPC1C20 pc and acts as a remote I/O to the pc. The S16 can be connected to an adjacent MPC1C20 or located over 200 m away! Multiplexing allows two twisted pair connections instead of a large bundle of wires.

The human interface panels are applicable to many types of small to medium sized machines which can be controlled by an MPC1C20 pc and require manual setup or human control during operation. Packaging machines, machine tools, discrete products machinery, assembly machines, automation equipment, food processing equipment, automated testing equipment, etc., are application examples.

The S16 unit reduces both the installation and the pc control panel costs. Figure 6.3 shows the alternative human interface techniques for both conventional and the alternative S16 approach. In order to fully appreciate the concept behind the S16 unit, it is necessary to review the current process involved in constructing a typical operator station.

- Design control scheme specifying number of pilot devices and required pc size. Provide all necessary drawings to enable construction.
- Drill required number of holes in operator panel door ready to accept pilot devices.
- Fit and assemble pushbuttons, selector/key lock switches, indicator lights, etc.
- Assemble and fit terminal board to operator panel base.
- Wire pilot devices to terminal board including common links. Install trunking if required.
- Install multicore cable from operator panel to pc control panel. Construct conduit/trunking if required.
- Wire from pc control panel terminal board to pc intput/output modules.

Because we have become so familiar with the above process its complexity is taken for granted. Compare this with the simpler approach taken by the S16 unit.
- Cut panel door and fit S16 unit (4-hole fixing).
- Install serial communications cable directly from S16 unit to serial port on the pc.

It is not necessary to design the operator station provided a rough estimate of pilot device numbers are known, the actual function of each location in the S16 unit being controlled by software. The drilling and fitting of the various pilot devices is replaced by a single panel cut-out. No terminal boards are required as the serial communications cable, replacing the multicore cable, 'plugs' into the S16 unit and pc processor. Cost savings are realised by reducing the number of pc input boards, eliminating or reducing the size of terminal boards, reducing the pc control panel enclosure size, eliminating long runs of multicore cables, simplifying panel construction (wiring, assembly) and reducing design and draughting time. An added bonus exploits the serial communications feature. If additional pilot devices or complete S16 units are added at a later date no additional cables are required, the added information being sent down the same communication cable. The S16 unit, therefore, provides a better, more flexible and cost effective way to design the operator stations of today and tomorrow.

Applications. Some of the many applications where pcs are being used are indicated below. The list is by no means exhaustive but is typical of the wide range of operations now covered by these electronic devices.

High speed positioning.   Automation of production machines, formerly the preserve of numerical control systems is being increasingly taken over by pcs. Machine tools and flexible transfer lines are particular examples where there is a need for accurate control of traversing speed and positioning which can be handled by a pc. In a particular development described in an article in Electric Drives & Controls, linear motion control resolution of 0.001 mm and repositioning at 15 m/min is possible.

Motor control centres.   Considerable savings can be effected by using pcs in motor control centres in place of electromechanical relays. Cost savings can also be made by reducing the number of control interconnections. The distribution of control wiring centralising the control function and the positioning of outgoing terminations are major factors in design. Using pcs and multiplexing within the mcc can effect considerable savings. For example several input/output modules, each capable of handling 5–10 digital or analogue signals can be employed in an mcc, with one module for each starter compartment. The module can be connected to the external sensors through a system driver or computer to reduce the wiring requirements.

Communications between an mcc, control desks and other items of equipment using local area networks (LAN) can represent a substantial proportion of the total cost. Multiplexing systems are available to minimise these costs, an example being Cutler-Hammer's Directrol method, Figure 6.3.

Factory automation.   Factory automation systems are an area where the pc is expected to predominate. New fields such as machine tool control (as described above) open up a vast scope for certain manufacturing lines. Other industries expected to benefit by the wholesale use of pcs are petrochemicals, water and sewage treatment plants, food and beverage, materials handling, electronics and robotics.

Microprocessors.   Microprocessors are among the latest electronic devices to be introduced to the market, the technology of production and utilisation being established as recently as about 1976. A microprocessor is a programmable intelligent building brick for electronic circuits, and with a memory and input and output facilities forms a microcomputer. The rapid development of microprocessors over the past few years has brought their cost down to acceptable levels, but programming costs can be very large and must therefore be taken into account when considering any application.

Fundamentally a microprocessor is a silicon chip containing a large number of electronic components like transistors. In 1979 designers were able to mount 100 000 transistors on a single chip, and this figure is expected to rise to one million during the 1980s.

There are five types of memory associated with microprocessors as follows: (1) random-access memory (RAM); (2) read-only memory (ROM); (3) programmable read-only memory (PROM); (4) electrically alterable read-only memory (EAROM); and (5) a ferrite core. A ferrite core is not a large scale integration (l.s.i.) device but is sometimes employed because it is not corrupted by mains failure.

Applications. During the current decade many new applications are expected to be developed for microprocessors, but as a controlling device it can be used for machine tools; a wide range of industrial controls, e.g. in process plants; for cars, heating systems and domestic appliances; computer games; and in television-linked information systems.

Microprocessor system. Typical of microprocessor-based industrial control systems is the GEM80 from GEC Industrial Controls. The system is claimed to simplify the problem of managing time, energy and materials for maximum cost effectiveness. The product range is capable of performing functions such as sequence control, process control, data logging and display – all in one technology. It thus provides advanced capabilities for meeting changing plant requirements.

GEM80 is widely used in a whole range of process and manufacturing industries and to control capital plant. Based on the complete control system philosophy of distributing functional controllers around a manufacturing complex, GEM80 provides all the facilities for interlocking, sequence control, process control, computer interface, data logging, data display, distributed control systems and plant supervisory control. Among the advantages claimed by GEC for this system are flexibility, easy programming, versatility (using GEM80 control language), distributed control (throughout a plant), comprehensive plant interfacing and fault diagnosis facilities.

Figure 6.5 GEM80 control system installed in the automotive industry (GEC Industrial Controls Ltd)

The control system is already in use world-wide covering a great variety of applications. For example, it controls and monitors compressors in power stations; it is used for assembly and spraying applications in the automotive industry; it controls furnace charging and batch weighing in foundries; and it serves in food processing, steel making, coal preparation, boiler plants and systems for both nuclear power stations and offshore oil platforms. Figure 6.5 shows the use of the system in the automotive industry.

7

Power supply

POWER GENERATION

In the previous edition of this Pocket Book it was comparatively simple to describe the different methods by which power is generated in the U.K. because it was a nationalised body with basically three organisations, the Central Electricity Generating Board (CEGB), South of Scotland Electricity Board (SSEB) and the North of Scotland Hydro-Electric Board (NSHEB). Now that the industry has been split up for privatisation there are six different power generating companies (described later), with the transmission being controlled by the National Grid Company plc, which is owned by the twelve regional electricity companies (RECs) (formerly the Area Boards). A number of other independent companies are also building power stations.

Under the old system of electricity supply the CEGB and the SSEB were mainly thermal generating bodies (including nuclear power stations) with a small amount of hydro power, the NSHEB being basically hydro and pumped storage. This easy division is not now possible but nevertheless one can still refer to the different methods of power generation but not relate them to the individual generating companies.

Generating bodies. Power generation is now the responsibility of a number of separate companies formed from the CEGB, SSEB and NSHEB together with some independently owned companies which are building power stations. The separate companies are: National Power plc (30 GW); PowerGen plc (19 GW); Scottish Power plc; Scottish Hydro-Electric plc; Nuclear Electric plc (8 GW); and Scottish Nuclear plc. The two nuclear companies will continue to be owned by the Government and therefore remain in public owner-ship. Much of the following information and that on renewable sources is published by permission of the Electrical Review and Inside Energy.

129

A number of power stations are already under construction by the following companies, most of which are independent organisations: Corby Power (350 MW CCGT (combined cycle gas-turbine)); Lakeland Power (220 MW CCGT); National Power (650 MW CCGT); Peterborough Power (350 MW CCGT); PowerGen (900 MW CCGT); Scottish Hydro (230 MW CCGT): Teeside Power (1,725 MW CCGT and combined heat and power). Lakeland Power was commissioned towards the end of 1991; the next one to be on line is Scottish Hydro in April 1992.

In addition there are nearly 40 other independently owned generating companies which are planning to erect power stations, with some still awaiting planning permission. The Inside Energy publication suggested, in August 1991, that the following independent power stations will go ahead: AES Electric's 700 MW station at Medway; Keadby Power's 720 MW project on Humberside; IVO Energy's 1,100 MW project also on Humberside; Eastern Generation's 380 MW project at Lawford; Rugby Power's 380 MW station at Rugby; at least one further plant from ICI and Envon; and International Power Generation's proposed power station, details of which have not yet been published; National Power's 1,500 MW Didcot Power Station and PowerGen's 1,350 MW Connah's Quay project also appear likely to go ahead. By the second half of the 1990s there is likely to be around 6.5 GW of independent generating capacity in England and Wales.

Marketing power. Power in England and Wales is sold through an energy trading pool administered by the National Grid Company which has the responsibility for the bulk transmission of electricity throughout the area. Each day the generating organisations submit a bid for the sale of electricity from each of its available generating units for the 24 h period from 5.00 am the following day. The National Grid Company ranks the generating units in order of the bids submitted and then calls on them as the demand rises and falls. For all power sold through the pool for a particular half-hour period, the companies are paid at the highest priced generating unit scheduled to run at that time. They also receive capacity payments which vary according to the balance of plant supply and electricity demand. Additional payments are made for other services.

Coal- and oil-fired stations. The balance of the UK power generation system for many years has been based on coal- and oil-fired boilers producing superheated steam to drive steam turbine generator units.

The coal-fired plant is still the major producer of electrical energy and although oil-fired generation continues to make an important and strategic contribution to the electrical energy requirements this has now been displaced from second place by nuclear generation. Standard ratings of the largest turbine generator sets are 500 MW and 660 MW and units of this size now account for more than 50% of the system installed capacity.

Steam can be delivered to the turbine stop valve at pressures exceeding 70 bar and 566°C. Greater pressures, up to 350 bars, are being contemplated in the UK and other countries, with temperatures up to 649°C. These higher temperatures associated with the higher pressures require special austenitic steels and many problems need yet to be resolved.

Gas turbines. Gas turbines were introduced many years ago as peak lopping generators but in the form of CCGT stations now under construction will be used as base load generating plant. Gas used in a CCGT produces virtually no sulphur dioxide, a quarter of the nitrogen oxides and half the carbon dioxide compared with an equivalent capacity coal-fired station. They are also much more thermally efficient, being up to 50% compared with about 38% for the most efficient coal and oil-fired plant. A 900 MW CCGT power station is being built by PowerGen at Killingholme on Humberside and when commissioned in 1992 will be the first of this type in the UK in operation. The station will be fed from the Pickerill gas field in the North Sea.

PowerGen is also examining the feasibility of using topping cycle technology whereby a gas turbine is added to an existing plant and its hot exhaust gases are fed into the coal-fired boiler. This will not only improve overall thermal efficiency but also reduce emission of the three harmful gases mentioned above. Gas is the preferred fuel of the future for power generation, mainly because of its environmental attractions.

Nuclear power. Britain has been implementing a substantial nuclear power programme since the mid-1950s and commercial nuclear power stations are now producing electricity at costs comparable to oil- and coal-fired stations.

With the splitting of the supply industry nuclear power generation has been separated from all other forms of generation and is the responsibility of Nuclear Electric plc and Scottish Nuclear Ltd, both being answerable to the Government. Three smaller experimental/research nuclear power stations are controlled by the AEA Technology or British Nuclear Fuels Ltd and they also feed power into the National Grid.

Under the first commercial programme, nine gas cooled (Magnox) stations with a total design capacity of 4,800 MW were commissioned between 1962 and 1971. Since then two of them, Berkeley and Hunterston A, have been taken out of service. The remaining seven with their electrical outputs are: Bradwell – 245 MW, Hinkley A – 470 MW, Trawsfynydd – 390 MW, Dungeness A – 424 MW, Sizewell A – 420 MW, Oldbury – 434 MW, Wylfa A – 840 MW. Bradwell continues to generate, subject to annual review.

The present nuclear programme is based on the advanced gas cooled reactor (AGR) of which five are controlled by Nuclear Electric and two by Scottish Nuclear. Electrical outputs of these AGR stations are: Hinkley B – 1,120 MW, Dungeness B – 720 MW Hartlepool – 1,020 MW, Heysham I – 1,020 MW, Heysham II – 1,020 MW and, in Scotland, Torness – 1,250 MW and Hunterston B – 1,120 MW.

Nuclear Electric is building a 1,188 MW pressurised water reactor (PWR) at Sizewell. This is Britain's first PWR station and is estimated to cost £2.03 billion (1991 price). Fuel loading is scheduled for late 1993 and the station should become fully operational by mid-1994.

In May 1988 the then Central Electricity Generating Board (CEGB) published a case for another PWR station at Hinkley Point C for submission to a public inquiry. Planning permission has now been granted with the funding permission being delayed pending the 1994 Government Review. Permission to build two more PWR stations at Wylfa B and Sizewell C has been withdrawn by Nuclear Electric, who took over the responsibilities of these stations from the CEGB when the industry was privatised.

AEA Technology, formerly the UKAEA, operates the 250 MW fast breeder reactor (FBR) at Dounreay, and British Nuclear Fuels Ltd (BNFL) is responsible for the Magnox stations at Calder Hall (200 MW) and Chapel Cross (196 MW). In the UK there is now a total of about 13 GW of nuclear capacity in operation or under construction.

Pumped-storage schemes. Pumped-storage schemes consist of two reservoirs, one at high level and the other at low level, so that

132

during the day hydro-electric generation takes place, allowing the lower reservoir to be filled from the high-level reservoir; during the night the water from the lower level is pumped back to the higher level. Reversible pump turbines are used for this purpose.

This system of generation has been developed to operate in conjunction with the high merit plant of large coal-fired nuclear power stations, which act as base load units. When the demand is low during the night, the outputs from nuclear power stations are utilised to transfer water from the lower to the high level reservoir.

Hydro-electric and pumped-storage generation. Hydro-electric and pumped-storage generation in Scotland is divided between Scottish Hydro-Electric plc and Scottish Power plc. Hydro-electric generation is mostly confined to the more mountainous and wetter regions in the West and North. Scottish Hydro-Electric operates 54 conventional hydro stations with a number of smaller stations with a total installed capacity of 1,064 MW. The largest single station, with an installed capacity of 130 MW is at Sloy with Fasnakyle second at 66 MW. The gross heads for these two stations are 277 m and 159 m respectively.

Scottish Power has two conventional hydro stations and one pumped storage scheme, Cruachan, about which more is said later. The two conventional stations are at Galloway and Lanarkshire (Falls of Clyde), their combined electrical outputs being 125 MW. Although the nameplate ratings of the machines at Galloway total 104 MW, they have been uprated to a new figure of 109 MW.

The Galloway scheme became fully operational in 1936 and was conceived partly to provide employment at that time as well as to contribute to local and national supply resources. There are five hydro stations forming the scheme, at Tongland, Glenlee, Kendoon, Earlstown and Carsfad.

There are two power stations associated with the Lanarkshire scheme, at Bonnington and Stonebyres, of which Bonnington is much the larger, the total output of both being 16 MW. This scheme was the first of Scotland's hydro-electric projects, and utilises the River Clyde waters in the Falls of Clyde section near Lanark.

National Power operates four hydro plants with a total installed capacity of just over 30 MW. By far the largest is Dolgarrog with 27 MW capacity, the other three being Chagford (less than 1 MW), Mary Tavy (2 MW), and Morwellham (1 MW). PowerGen is responsible for Rheidol which has an installed capacity of 53 MW. Nuclear Electric owns Maentwrog (24 MW).

There are a number of pumped storage schemes in the UK and they are divided up among some of the generating companies and also the National Grid. These are used to provide a rapid source of power to cope with sudden changes in demand for electricity.

The National Grid owns and operates the pumped storage schemes at Dinorwig and Ffestiniog, both in North Wales. Dinorwig can produce around 1,320 MW in approximately 12 s. Between them the two stations have a combined output of 2,088 MW and a storage capacity of 10,600 MWh. This storage capacity ensures that the speed of response can be followed by a sustained supply of power for the transmission system.

Scottish Power has just one pumped storage scheme, the 400 MW installation at Cruachan where the operating head is 365 m. The four vertical reversible Francis pump/turbines and motor/generators have ratings of 100 MW each when generating and absorb 110 MW when pumping. The Foyers pumped/storage scheme at Great Glen, operated by Scottish Hydro-Electric has a generating capacity of 300 MW and absorbs 305 MW when in the pumping mode.

Renewable sources of energy. Because of the world's dwindling energy resources many countries are examining the possibilities of developing renewable sources of energy. Research is being carried out into a number of areas including the use of: wind, landfill, geothermal, tidal, wave, solar, waste, tyres, sewage gas and chicken litter. The Government's policy (1991) is to work towards the generation of 1,000 MW of electricity from renewable resources by the year 2000. An Order laid before Parliament on 18 September, 1990 has set a Non-Fossil Fuel Obligation (NFFO) for the initial levels required for renewable sources of electricity. The 12 Regional Electricity Companies are expected to achieve these levels between 1 October 1990 and 31 December 1998. Net capacity in the first block is to be 102 MW by April 1995. The Department of Energy is not, at the present time (July 1989) supporting research in four renewable sources: active solar heating, photovoltaics, large scale offshore wave energy and geothermal aquifers. So much is happening in the field of renewable sources of energy that one cannot hope to be up-to-date in a book, but the methods mentioned will alert the reader to the possibilities that exist.

Wind energy. As a source of energy wind power on land sites is considered promising but uncertain, and offshore it is less attractive. Two types of wind turbine are employed, with either a vertical axis

(VAWT) or an horizontal axis (HAWT). A number of commercial installations have been operating for some time, for example at Burger Hill in Orkney (3 MW), at Fair Isle (50 kW) while experimental installations include a 300 kW machine at Carmarthen Bay and two 3 MW sets at Richborough, Kent. In 1990 a 600 kW wind turbine was commissioned in Foula Island, 50 km west of Lerwick in the Shetland Isles. When demand is low, the energy is used to pump water into a small reservoir serving a hydro-electric plant.

Many private companies are involved in developing and installing wind turbines, at least ten being underway at the time of writing. National Power is proposing to build 23 turbines, each rated at 350 kW at Cold Northcott, Cornwall and four other power projects with a combined capacity of 13 MW. UK Wind Farms has more than 10 projects in hand totalling up to 60 MW for Kirby Moor, Cumbria and Commaes Valley, Powys.

Landfill schemes. Landfill sites generate predominantly a mixture of methane and carbon-dioxide and they have been exploited as long ago as the 1970s but only recently have they become key issues as sources of energy for generation purposes. The UK is the second largest exploiting nation (after the USA), some 50% of the gases being used for electricity generation and the rest being employed by industry to fire kilns and boilers, etc. There are about 5,000 landfill sites in the UK, of which only about 450 are producing enough methane for power generation, according to a survey carried out by the Energy Technology Support Unit (ETSU) of Harwell, a body set up by the Department of Energy.

Ten companies are involved in schemes covering 22 landfill sites, the largest having a capacity of 2.25 MW at Chorley, Lancashire which has just been commissioned (mid-1991) by Biffa Waste Services. The Company applied for another scheme, also in 1991, which will build up to about 4 MW over several years. ARC has ordered the generating sets for a 1 MW site at Huntingdon, Cambridgeshire, with at least two more projects also totalling about 1 MW being planned. Norweb and Land Fill Gas has six approved projects in the north of England totalling 7.29 MW. Another seven (about 7 MW) have been submitted for approval.

Geothermal. Geothermal energy derives from the heat flowing outwards from the earth's interior, and in some places from the decay of long-lived radioactive isotopes of uranium, thorium and potassium. It can be tapped from hot dry rocks (HDR) and aquifers,

135

with most effort being concentrated on HDR. The Department of Energy has funded hot water aquifer research for 10 years to 1986, but results overall have shown that there is little prospect of them being exploited economically as a heat source on their own. That programme has been curtailed although some financing of the Southampton City Centre borehole is being provided. This is a joint private sector/local authority district heating scheme.

HDR research has been under way since 1976 by a team from the Cambourne School of Mines at Rosemanowes in Cornwall in three phases. Phases I and II are completed (depths of 300 m and 2,500 m) and the main feature of Phase III is the development, with industrial companies, of a conceptual design for a 6,000 m deep commercial-size prototype HDR system. Work in connection with Phase III has been redirected to the European HDR programme which involves the construction of a pilot plant at Rosemanoires, Alsace, France or Bad Urach, Germany.

Tidal energy. Tidal energy in Britain does not rank alongside wind energy in its potential as a renewable energy source for electricity generation. The current programme aims at reaching the point where decisions can be made whether to go ahead with the construction of a scheme. The largest part of the programme is devoted to the Severn Barrage, jointly funded by the Department of Energy and the Severn Tidal Power Group. The latter company is about half-way through a 21 month study (September 1991) which primarily concentrates on the estuary's sediments, particularly in the Bridgwater Bay area. It will also refine the barrage's potential output. Present power estimates for a 16 km barrage between Cardiff and Western-Super-Mare are 8,640 MW from 216 bulb-turbines.

Other areas being studied include the Mersey Estuary (700 MW), the Conway Estuary, North Wales (35 MW) and the Wyre Estuary, Lancashire (50 MW).

Wave Energy. The Department of Energy has scaled down its programme on shoreline wave energy but work is continuing on small shoreline devices. Support is being given to an experimental installation, based on the Queen's University, Belfast concept, utilising a Wells turbine driving a 75 kW marine wound rotor induction generator on the island of Islay in the Inner Hebrides. The equipment is located in a natural rock gully north of Portnahaven and consists essentially of a concrete box, internally 10 m long, 4 m wide and 9 m high built over the gully. The wave entering and receding in the gully forces the column of water in the chamber to oscillate, driving the air above through the Wells turbine coupled to the generator. The turbine efficiently converts the pneumatic power of the air flow into rotational motion at a speed compatible with the 4-pole generator. Average power levels of about 20 kW/m are experienced. Present energy costs (1991) are estimated to be slightly less than 8 p/kWh assuming a payback period of 30 years and a discount rate of 5%.

Solar energy. The main R & D emphasis of the Department of Energy is on passive solar building designs, utilising the form and fabric of the building to capture solar radiation and so reduce the need for artificial light, heat and cooling. Active solar heating employs collectors to capture and store the sun's heat primarily for space and water heating. Photovoltaics is a method to convert the sun's energy into electricity by means of semiconductor devices. Costs at present are high but are expected to fall with time; however to compete with other renewable sources, costs would need to fall by a factor of 10 to 20. Active solar heating is unlikely to become cost-effective in the UK except for special applications and so the Department of Energy has halted its R & D work in this field.

Other renewable energy sources. The gases produced from incinerating municipal waste can be used for electricity generation. SELCHP has already bought a site and will burn waste from three London Boroughs. Some 30 MW is expected to be generated. Hampshire County Council is interested in a 30 MW scheme. A scheme in its early stages in south east London is being proposed by Cory Environmental and has a possible capacity of 96 MW.

South West Water and North West Water have schemes which will utilise sewage gases for generation purposes. The former already has two CHP schemes under construction.

Elm Energy & Recycling UK is planning to produce 22 MW through a tyre burning scheme due to come on stream by mid-1992. It will consume 22 million tyres a year.

Chicken litter is another source of energy being utilised by Fibropower at Eye in Suffolk, and Glandford, Humberside, each project being capable of producing 14 MW. The Suffolk plant is expected to come on stream in July 1992.

SMALL GENERATING PLANTS

Small generating plants are available for base-load operation, as standby power supplies or as temporary mobile power sources.

They are generally diesel driven sets although some of the larger ratings, above about 1 MW may be gas turbine driven.

Base load diesel generating sets are used as sources of prime power for remote areas at considerable distances from the national supply. Pumping stations for water and oil pipelines, oil fields, lighthouses, sewage schemes and defence installations are examples.

Some islands such as Guernsey and overseas countries sometimes use diesel generating sets for their national supplies, but these are considerably larger, of the order of 5 MW or more, and are not considered in this chapter.

The most important application for small generating plants is undoubtedly as a source of standby power, emphasised particularly during 1971 and 1972 when the UK experienced wide spread power cuts.

Where loss of supply can endanger life e.g. in hospital operating theatres, the installation of alternative power sources is a necessity. Such power is usually provided by an automatically started diesel generating set. Industry too can suffer badly from the effect of a supply interruption. There is a consequential loss of production, and in many cases, damage to process plant. It is not surprising that where continuous processes are involved, manufacturers install standby power plant.

Temporary supplies are generally provided by the public networks for use on construction sites to enable work to be carried on during the dark periods of winter working days, or for 24 hours a day when there is a requirement so to do. Temporary power is sometimes required to provide electricity to ships in docks when their own power sources are inoperative, or for ships at sea when their main or standby power supplies have failed. When the public supply is too expensive or not available, recourse can be made to temporary mobile power sources.

Construction. The standard range of generating sets manufactured by Dale Electric of Great Britain Ltd, Filey, Yorkshire, is indicative of the type of equipment available from a number of UK companies. The basic generating set comprises of an engine coupled to an alternator, mounted upon a bed plate, with a control panel.

Diesel engines are the most commonly used because of their heavy-duty build, low running costs and fast starting qualities. They can also include turbo-charging which provides higher power ratings. Every diesel engine should include radiator or air cooling, protective guards over moving parts, a governor to regulate the engine speed and thus the frequency of the alternator, and fuel, oil and air filters which can be replaced during servicing.

Most sets have an electric starting system which usually comprises starting controls, battery and axial starter. Some are started by a key switch on the panel, others are fully automatic on mains failure. Larger diesel engines however (above 2 MW) are started by an air compressor. A typical fully automatic diesel generator is shown in Figure 7.1.

Figure 7.1   Typical installation of a fully automatic Dale generating set. This one is powered by a Perkins 6.3544, rated at 50 kVA. It shows the control panel with cable to the load, starter battery and leads, air-inlet ducting to the radiator, exhaust outlet pipe and silencer, air filters, steel guards around moving parts and anti-vibration mountings

Gas turbine engines are lighter, cleaner and virtually vibration free making them ideal for rooftop installations, but they are more expensive then diesel engines. The engine is coupled to an alternator which is usually brushless, self-exciting and self-regulated. Its output voltage is maintained within close limits of ±2%. To comply with British Standards it is screen protected, drip-proof and fan ventilated. All windings are generally insulated with varnish suitable for use in hot climates.

Control panels vary considerably in content but the basic requirements are voltmeter and selector switch, ammeter and selector switch, engine temperature and oil pressure gauges, fuses, indicator lamps and a circuit-breaker complete with overload trip. More advanced control systems have facilities for synchronisation and load sharing of two or more sets.

Each generating set should include a fuel tank and pipes, engine exhaust silencer, foundation bolts to secure the baseplate or anti-vibration mountings, engine tools and operating manuals. Fully automatic sets should also include a separate contactor panel which provides the automatic changeover of the load from the mains to the generating set and vice versa.

Where a customer relies on independent power generation, or on standby power in the event of mains power failure, the cheapest installation is not always the best investment. When the power has gone, it is too late to discover that the generating set is not up to the necessary standards.

The more reputable UK companies are members of the Association of British Generating Set Manufacturers (ABGSM). This Association promotes and encourages technical and commercial standards within its membership. Customers are well advised to only deal with ABGSM members.

Typical control system. The Series 3 control system from Newage Engineers has been developed for the automatic control of brushless a.c. generators. It consists of a shaft-mounted permanent magnet generator and a voltage control unit which comprises circuits for voltage regulation and machine protection. The components are shown in Figure 7.2.

The control unit is interlinked with the machine main stator, the permanent magnet generator and the exciter stator winding, to give closed-loop control of output voltage with load regulation within the limits of ±2% from no load to full load including cold to hot variations at any power factor between 0.8 lagging and

140

unity and inclusive of a speed variation of 4½%. Closer regulation is available.

The solid-state voltage control unit, although being powered by the permanent magnet stator windings, derives a sample voltage from the main stator output for voltage control purposes. In response to this sample voltage, the unit controls the power fed to the exciter, hence the main rotor, and finally the machine output, in such a way as to maintain the output voltage within the specified limits.

The protection circuit is incorporated as a form of safety against malfunction of the generator or voltage control unit and provides back-up protection for overload conditions external to the generator.

Voltage control unit.   The Newage voltage control units consist essentially of four sub-circuits: a reference and comparator which compares a sample voltage derived from the main stator, with a fixed reference to produce an error voltage proportional to the difference in these voltages; a firing circuit which triggers the exciter stator control thyristor, the conduction angle of which is

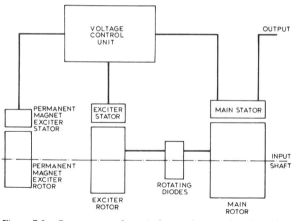

Figure 7.2   Components of a typical control system for a brushless alternator (Newage Engineers)

proportional to the comparator error voltage; a stabilisation circuit which modifies the above error voltage proportional to the rate of change of exciter stator current; and power components which form the basic power supplies for field and control circuits.

Reference and comparator circuit. The reference and comparator circuit comprises components: R1–R13, R46, R58 (R46 = voltage range control), C1, C3, C18, D1–D4, D6, D21, TR1 and T1. 'R' is the optional 'Hand Trimmer' used for external fine adjustment of machine voltage. See Figure 7.3.

A proportion of the machine output voltage is attenuated by the Network R1–R5, R41 and R58 and is applied to the primary winding of the sensing circuit isolation transformer T1. The secondary voltage of the isolating transformer is then rectified and filtered to give a d.c. voltage which is again proportional to machine output voltage. Zener diode D21 and series diode D6 are fed via R10 from a smoothed d.c. rail to provide an accurate, temperature stable reference voltage against which the above filtered voltage is to be compared. The comparison is achieved simply by electrically subtracting the feedback portion of the output voltage from the reference voltage across D21/D6 giving an error voltage of approximately 1.45 V. This voltage has a small a.c. component which is then further filtered to yield a steady d.c. error signal which controls the base of transistor TR1.

Resistors R11, R12 and R13 form a gain stabilisation and level shifting network in order that the current in R14 and R47 can be accurately controlled. The gain control resistor, R47, determines the amount by which the collector current of TR2 varies for a known change in the error voltage at the base of TR1.

Firing circuit. The firing circuit comprises components: R14–R19, R40, R47, R60 (R47 = gain control), C4, D7, TR2, TR3 and CSR2.

With the anode gate-circuit of the programmable unijunction transistor reverse biased, all the current from the collector of TR2 flows into the timing capacitor C4. The charging rate of this capacitor is thus dependent upon the error between output voltage and the appropriate reference requirements.

If the error increases (machine voltage falls), the charging is more rapid. If the error decreases (machine voltage rises), the charging rate is less rapid. The programmable unijunction transistor CSR2 is connected as a level detector and is triggered into

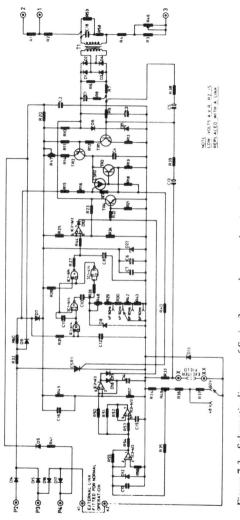

*Figure 7.3   Schematic diagram of Series 3 control system showing voltage control unit and protection circuit
(Newage Engineers)*

143

conduction when the voltage across C4 exceeds the voltage at the junction of R16 and T17 by approximately 0.6 V. The voltage appearing at the junction of R16 and R17 is fixed by the values of R15 to R19 and is synchronised to the clamped square wave at the anode of D7.

On reaching the trigger level, C4 is rapidly discharged through TR3 and R19 creating a voltage pulse across the latter which, in turn, triggers the exciter stator control thyristor CSRI through R40 (gate current limiting resistor). With CSR1 in its conducting state, further pulsing of the CSR2/TR3 combination is inhibited and C4 is kept discharged. Synchronisation of the trigger circuit is achieved when the exciter stator control thyristor stops conduction and the voltage at the anode of D7 rapidly rises to its clamping level. This allows C4 to accept another charge via the collector of TR2. It can therefore be seen that the thyristor CSR1 is caused to conduct at a specific point in time determined by the charging rate of C4 and hence the error between the actual and required machine voltages. The resultant variations in 'conduction angle' adjust the machine exciter stator current and thus closed loop control is achieved.

Stabilisation circuit. The stabilisation circuit comprises components: R38, R39, C5 and C13.

As already described, the signal resulting from the comparison of machine and reference voltages has both d.c. and a.c. components. The majority of the a.c. ripple is filtered out due to the presence of C13. At the same time, the negative terminal of C13 is connected to the negative end of a power resistor through which is flowing the machine exciter stator current. This current is essentially pure d.c. due to the high inductance of the stator windings and the inclusion of a 'free wheel' diode D13. If the stator current now changes (e.g. due to a machine load change), C13 will charge (or discharge) modifying temporarily the error voltage at the base of TR1, in such a direction as to oppose this initial change. This circuit adds damping into an otherwise underdamped system.

Components C5 and R38 form a 'lead compensation' circuit which is incorporated to improve the stability/performance characteristics of machines with larger kVA ratings.

Frequency detection circuit. The frequency detection circuit comprises components: R26–R31, R42, R43, R48, C8–C12, D8 and IC1.

The two-input nand gate IC1/a (here connected as an inverter) 'squares up' the clamped waveform at the anode of D7. The resultant waveform appearing at the output of this gate has a very sharp and clean trailing edge which is co-incident with the rising edge of the synchronising waveform across CSR1. Gates IC1/b and IC1/c, resistors R28 through R30, R42 and R43, variable resistor R48, capacitor C1 and diode D8 form a negative edge triggered monostable whose output 'off' time is adjusted by R43 and is subsequently inverted by gate IC1/d. With the 'mark' of the monostable fixed by the setting of R48 and the frequency jumper link, a change in frequency results in a change in the d.c. voltage appearing across the filter capacitor C8. The conversion is linear in that for a 5% decrease in frequency, the output voltage (of the frequency detection circuit) also falls by 5% and so on.

Exciter stator voltage monitoring circuit.   The exciter stator monitoring circuit comprises components: R34–R37, R50–R56, R49, C15, C17, D12 and IC3.

A resistive divider chain formed by R34, R36, R37 and variable resistor R49 is arranged such that a fixed percentage of the machine exciter stator voltage is tapped off and fed to the low pass filter network R35 and C17. Integrated circuit amplifier IC3/a is connected as a unity gain inverter and translates the negative going signal at the wiper of R49 into a positive going signal of the same magnitude. The circuit formed by components R51–R53, C15 and IC3/b is a 'capacitance multiplier' and as far as its input terminal (junction R50, 51, 53, 54) is concerned, acts just like a low Q capacitor of high capacitance value. This 'capacitor' is charged via R54 and gives the protection circuitry its inverse time characteristic. The voltage appearing on this 'capacitor' is compared by IC2/a with the fixed level at the junction of R45 and R47 and on reaching this level causes the output of IC2/a to go high thus activating the reference clamp circuit.

Reference clamp circuit.   The reference clamp circuit comprises components: R21–R25, R44, R45, R57, C14, C16, D10, D11, D18, TR4 and IC2.

The reference clamp (which cuts off the machine excitation supply) is activated from two separate sources.

(a) An over-excitation condition is sensed by IC2/a and the output of this device switches TR4 fully on (via D11, R22) thus reducing the reference voltage level to zero. The machine output voltage must follow this level and hence it also falls to zero.

(b) A fall in output frequency causes the voltage on C8 to fall below the threshold of IC2/b set by resistors R24 and R25. This in turn causes the output of IC2/b to rise turning on TR4 (via D10, R22) and thus reducing the reference voltage. However, unlike the over-excitation circuit, negative feedback is incorporated to prevent TR4 turning full on and the reference voltage is reduced at a rate proportional to the decrease in frequency. By carefully setting the nominal value of voltage appearing on C8 for a fixed input frequency, the 'cut in' point of the underfrequency protection is also set.

Power circuit. The power circuit comprises components: R32, R33, R41, D5, D9, D13–D17 and CSR1.

The power supply for both exciter stator and control circuits is derived from the three phases of the permanent magnet generator stator winding. Diodes D14 through D17, connected to the regulator terminals P2, P3 and P4 rectify the three-phase output in such a way as to provide a d.c. voltage (nominally 145 V) which has a 'gap' of sixty electrical degrees in its waveform. The purpose of the gap is to enable the exciter stator control thyristor to turn off at the end of its conducting period and to provide synchronisation of the trigger circuit. The low voltage supply for the control circuit is derived via dropper resistors R32 and R41, diodes D5 and D9 and is regulated by zener diode D22. A synchronisation signal is provided for the trigger circuit and the Under Frequency Protection Circuit by resistor R60 and is clamped by diode D7.

The permanent magnet generator is an eight pole device and thus the d.c. supply rail executes two complete cycles for every full output cycle of the four pole main machine.

(Note: For a six pole main machine the P.M. executes only 1.33 cycles/cycle)

HIGH INTEGRITY POWER SUPPLIES

The maintenance of a continuous supply from public authorities cannot be guaranteed, and the increasing use by industry and commerce of equipment that cannot tolerate a long interruption (or in some cases, no interruption) in the supply or requires a high quality supply has led to the development of alternative sources of power. There are basically four different power supply systems providing the following: load isolation; uninterruptible power supplies; standby power supplies; and frequency conversion facilities. They can be provided by one of two systems, dynamic or static, and their functions are as indicated below.

Load isolation. Some of the sophisticated electronic aids in use today must be protected from transients, spikes or other disturbances that occur in the public supply network and power supply systems are available that isolate the mains from the load and themselves provide a 'clean' supply.

The dynamic option can offer additional protection against blackouts (complete mains failure) of up to 500 ms (typical) by using the inertia of the rotating mass to provide the energy. This inertia is often provided by a flywheel which forms part of the system.

Outputs are typically: dynamic 5–50 kVA single-phase, 5–750 kVA three-phase; static 1–40 kVA single-phase, 20–500 kVA three-phase.

Uninterruptible power supply (UPS). Where no break in the public supply to the load is admissible it is necessary to install a UPS or no-break system. In general the mains feeds the UPS system, which incorporates a battery back-up, and the load is supplied by the UPS equipment. In the event of a mains failure the battery provides the power through the UPS system to the load. Depending on the circumstances a standby diesel-generator may form part of the installation, its purpose being to replace the battery as the source of power once it has started up and is ready to take the load. This enables the battery capacity to be kept within economic limits because it can be sized to power the load for say 15–30 minutes until the standby diesel generator has started up. Pages 134–137 discuss standby diesel generators in some details. Figures 7.4 and 7.5 are typical rotary no-break systems.

Outputs are typically: dynamic 15–500 kVA three-phase; static 1–40 kVA single-phase, 20–500 kVA three-phase.

Short-break systems. As the name implies a standby or short-break supply restores power to the load in the event of a mains

Figure 7.4   A 440 V 60 kVA 60 Hz Arcontrol no-break motor-generator set with its control board which also houses the battery charging equipment

failure, Figure 7.6. It does not provide an uninterrupted supply and the break between mains failure and restoration may be measured in seconds if an automatic start-up method of a standby diesel-generator is employed or several minutes if a manual one is adopted.

Outputs are typically: dynamic, 20–150 kVA three-phase; static 1–25 kVA single-phase, 20–250 kVA three-phase.

Frequency converters.   There is often need to provide the load with a different frequency to that of the supply and still maintain a

Figure 7.5 This rotary no-break motor-alternator set from Mawdsley's Ltd incorporates a flywheel which is enclosed under a cover for safety reasons. The drive is a 150 kVA d.c. motor. Control equipment is in the cubicle behind the set

high quality. The dynamic option offers additional protection against blackouts of up to 500 ms (typical) duration by using the energy of the rotating masses, which may include a flywheel. Frequency combinations met with in practice are (in Hz) 50/60, 60/50, 50/400, 60/400, 50/450 and 60/420. Figures 7.7 and 7.8 illustrate two different types of rotary frequency converters.

Outputs are typically at 50/60 Hz: dynamic 5–50 kVA single-phase, 20–500 kVA three-phase; static 1–25 kVA single-phase, 20–500 kVA three-phase. Similar outputs are available at 400 Hz for telecommunications and ground power supplies and 450 Hz for computers.

Rotary systems. As the name implies a rotary system incorporates rotary machines, i.e. some form of motor generator set, and by their very nature they provide total electrical isolation from the public supply network. No mains-borne interference can be transferred to the load. A recently introduced system replaces

Figure 7.6 Arcontrol short-break motor-generator set before shipment to Ecuador where it supplies an air traffic control installation

the drive motor by a quick start diesel engine and this is described in some detail later. Usually a rotary system has a flywheel which provides a ride through capability of several hundred milliseconds, the duration depending on the inertia of the rotating masses.

With the motor generator systems three types of drive are available, each having its own characteristics. The first is a

Figure 7.7   A 50/441 Hz 75 kVA silenced rotary frequency charger for computer supplies (Mawdsley's Ltd)

synchronous motor, generally a synchronous reluctance machine, which starts as a conventional cage induction motor but operates at synchronous speed. Interruptions in the mains supply of up to about 100 ms can occur without loss of synchronism and departure from the agreed tolerances.

The second form of drive is an asynchronous motor, usually a low slip cage induction machine which accepts interruptions in mains supplies of up to three seconds while maintaining the output frequency to within 10%. Finally, the third type of drive machine can be a d.c. motor. This is necessary if the output

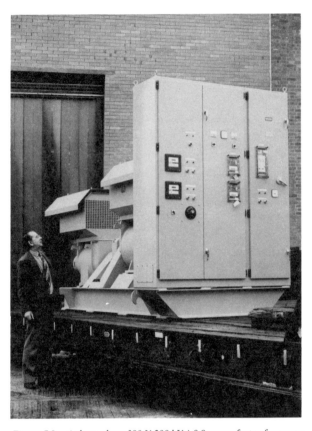

Figure 7.8   A three-phase 380 V 200 kVA 0.8 power factor frequency charger being shipped to Saudi Arabia (Mawdsley's Ltd)

frequency tolerances of the alternator are to be less than the input tolerances.

UPS arrangements (rotary).   Principal components are shown in Figures 7.9, 10 and 11. The simplest no-break (UPS) system is shown in Figure 7.9 where the battery provides the standby power and therefore it would not be used where a continuous standby supply was required. The addition of a static or electromechanical

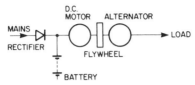

Figure 7.9   A simple UPS standby system with a d.c. motor drive

bypass switch Figure 7.10 allows maintenance to be carried out on the rotating machines without loss of supply to the load. By duplicating the motor-alternator sets as shown in Figure 7.11 redundancy is provided without loss of reliability. A bypass switch can be incorporated in each motor-alternator circuit if required and a separate or common battery system utilised.

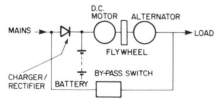

Figure 7.10   A UPS system with a bypass switch. This switch may be electromechanical or static

When parallel redundancy is employed it is usual to size each motor-alternator so that it can meet the total load requirement. Sometimes it may be possible to reduce the power needed by the connected loads when there is a mains failure enabling the two sets to be of smaller output and so keeping the cost of the overall installation down. Under these circumstances each set may be capable of providing half of the total load and so have to be operated in parallel in normal conditions. If one of the UPS sets fails the other will still be able to provide all critical loads:

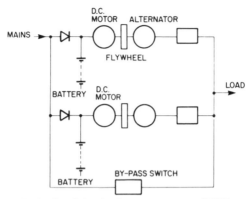

Figure 7.11   Parallel redundant arrangement of UPS system

furthermore if the mains supply fails both sets can still operated in the parallel mode to provide power as required with a back-up facility for the critical loads. If any of these systems were required to provide a continuous standby supply then a diesel generator would be included in the system.

Similar supply systems can be provided using a diesel engine to drive the motor generator set rather than a battery. This is an economic proposition if the user wants to have a continuous standby supply. Two arrangements are shown in Figures 7.12 and 7.13.

Diesel no-break set.   Arcontrol has recently introduced a range of microprocessor-controlled Hitzinger diesel no-break systems with ratings from 40–400 kVA, Figure 7.14. Like the systems

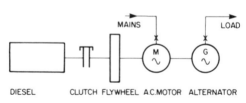

Figure 7.12   No-break diesel system

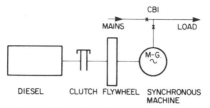

Figure 7.13   Short-break diesel system

described earlier it is based on the use of a flywheel coupled with a rapid starting diesel engine, and has been developed particularly for computer supplies. There are several versions of the system. One maintains the transient frequency disturbance following a mains failure to ±1% and the voltage within ±5% recovering to normal in 300 ms. Another design for general industrial applications keeps the transient frequency disturbance within ±5%. A further version enables 50% of the load to be provided in UPS mode with additional loads such as lighting and air conditioning

Figure 7.14   The Arcontrol diesel no-break system is particularly suitable for computer installations. The separate floor-standing control cubicle is not shown

being connected within 5–10 sec after mains failure. All components are mounted on a common bedplate with a separated floorstanding control cubicle.

Static systems. Static systems are available to provide load isolation, UPS, standby power or frequency conversion facilities. Here we are concentrating on their function as UPS or standby systems and the different arrangements that are available. However, it should not be forgotten that they can also be supplied to meet the other requirements like load isolation and frequency conversion.

At the time of writing, static inverters are available with 50 Hz power output ratings ranging from 100 VA (or even less) up to 40 kVA single-phase and around 750 kVA three-phase as a single module. Larger three-phase outputs may be obtained by paralleling units. Most UK industrial requirements are for 50 Hz outputs (or 60 Hz for American equipment) but some computers may require 400, 415 or even 441 Hz frequencies.

Operating modes. Basically there are four different operating modes, although minor variations and modifications to them increase the number of possible arrangements. These are: (1) single inverter system with an electromechanical means of changeover (a standby supply), Figure 7.15; (2) single inverter

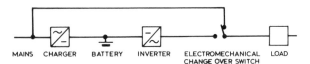

Figure 7.15   A standby supply

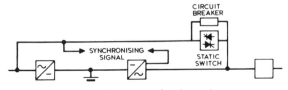

Figure 7.16   A no-break supply

system with either a static or electromechanical changeover switch, Figure 7.16 (a no-break supply); (3) parallel non-redundant system Figure 7.17; (4) parallel redundant system Figure 7.18.

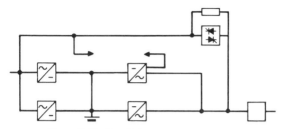

Figure 7.17   Parallel non-redundant system

From Figure 7.15 it will be seen that the a.c. load is fed permanently from the mains through a changeover contactor and the static inverter is not energised. If the mains fails the inverter is automatically switched on and the load transferred to it through the changeover contactor. When the mains supply is restored the system reverts back to its original condition. There is of course an interruption of up to 500 ms in the supply depending on the speed of switching operations.

In Figure 7.16 the changeover switch is of static form the load being fed from the inverter so that mains failure does not affect the supply to the load. If the inverter develops a fault the static switch connects the load to the mains within a few milliseconds. With the static switch in the bypass line (rather than in the inverter circuit) and a circuit-breaker shunting it as shown all units in the system can be maintained with the load still energised. A synchronising circuit ensures that the inverter and bypass supplies are kept in synchronism.

Parallel non-redundant operation is shown in Figure 7.17 and can take several forms. In the system shown parallel operation provides greater output but the regulation and voltage control must be such as to ensure equal sharing of the load. Each inverter supplies its own share of the load and may have its own battery system but the problems associated with the master/slave technique have to be considered. Failure of an inverter means a reduction in power output; alternatively arrangements can be made to transfer the whole load to the mains bypass circuit shown in the diagram. The faulty inverter can then be repaired or replaced and the whole system switched back to the inverter.

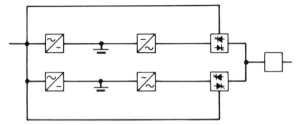

Figure 7.18   Parallel redundant system

The fourth operating mode, that of parallel redundancy is shown in Figure 7.18. Two inverters are employed, one of which is used to supply the load while the other acts as a standby. Should the active inverter fail the other is automatically switched into circuit. The switch may be either a contactor or a static device depending on whether a small break in supply can be tolerated.

With this arrangement work can be carried out on an inverter without affecting the supply to the load. The mains acts as a third source and can be switched on to the load by either an electromechanical or static switch.

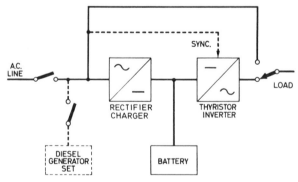

Figure 7.19   Static standby power system

Most computers and their peripheral equipment can accept a break in supply of up to 10 ms and static switches operate within this limit. Battery capacity determines the length of time that an inverter can supply its load following a mains failure, this time generally ranging from 10 to 30 minutes except in the case of some emergency lighting units which may have to function for one or more hours. For very large UPS systems, which are often backed up by diesel generators, the battery may be called upon to supply the load for a short time only, based on the time it takes to start the diesel generator and for it to accept the load. Figure 7.19 shows how a diesel generator is connected to a static power supply system.

SOLAR ENERGY

Although it has been appreciated by the scientists for a very long time that solar energy represented a vast source of supply, very little has been done in the past to tap it. Within the last few years, however, interest has quickened and today there are many practical devices in operation.

The main component is the solar cell. This is a thin disc of pure silicon containing a minute quantity of boron (or similar substance) to give the silicon a negative potential. A layer of p-type material a few microns thick is diffused on to the upper surface of the disc and the ends lapped over the perimeter.

This portion is then enclosed in a containing case with a glass face, the top surface of the disc being filled with silicon grease to prevent loss by reflection.

As will be appreciated from the description, the arrangement broadly speaking is similar to that found in a transistor where electrons flowing from the n-plate to the 'holes' of the p-plate constitute a current. In the solar cell, power is produced by this process at the barrier junction.

Solar energy has been the primary source of power for space craft since 1958. The basic components of a solar cell are shown in Figure 7.20. A slice of single crystal silicon, typically 20 mm $\times$ 200 mm and 300 μm thick, forms the energy conversion component. In the n on p configuration a shallow junction is formed by diffusing phosphorous into the boron-doped crystal. Metal contacts are plated or evaporated on the front and back of the cell and the active surface is coated with silicon oxide or titanium oxide anti-reflective layer.

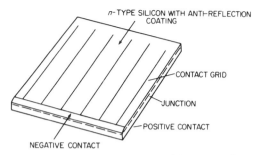

n-TYPE SILICON WITH ANTI-REFLECTION COATING

CONTACT GRID

JUNCTION

POSITIVE CONTACT

NEGATIVE CONTACT

Figure 7.20   Solar cell as used for spacecraft

A solar cell of this nature operating at 250°C in normal sunlight above atmosphere has a short-circuit current of 140–150 mA and an open-circuit voltage between 530 mV and 580 mV; the latter is independent of area. Maximum power is between 55 mW and 65 mW and is obtained between 400 mV and 500 mV. Output falls as the cell is turned away from the sun approximately as the cosine of the angle of incidence. A rise in temperature causes a sharp fall in conversion efficiency which is about 11% maximum. An 80°C rise will halve the output.

As bare silicon is a poor emitter, a cell is covered with glass or fused silica. This cover, together with a highly emissive back surface limit the steady-state temperature of a sun-orientated array to about 60°C. In space applications the cover also provides some protection against radiation and micrometeorites.

Many other solar cell materials have been studied for space applications but the only serious contenders are gallium arsenide and polycrystalline cadmium sulphide. Both have failings compared with silicon. However cadmium sulphide cells are used in photo-electric devices, see Section 4.

Although work has been carried out on attempting to harness solar energy for large scale power generation no real success can be reported to provide this power at an economic price. Focussing of the sun's rays to produce concentrated heat energy is the main method adopted (as discussed on page 137), one large installation being in Switzerland. There has been talk of building up an installation of this nature in space and beaming down the energy, but this is not practical at present.

160

8

Transmission and distribution

Two-wire D.C. Referring to Figure 8.1, the volt drop in each conductor $= IR$, therefore the total volt drop $= 2IR$. The voltage drop will therefore be given by $E - V = 2IR$. The power loss in each conductor $= I^2R$. Therefore total power loss $= 2I^2R$.

$$\text{Efficiency} = \frac{\text{output}}{\text{input}} = \frac{VI}{EI} = \frac{EI - 2I^2R}{EI}$$

$$= \frac{VI}{VI + 2I^2R}$$

$$\text{Voltage regulation} = \frac{E - V}{V} = \frac{2IR}{V}$$

Single-phase A.C. Referring to diagram in Figure 8.2, the constants are shown as X and R, where X is the reactance of the conductor and R is its resistance (capacitance is neglected here).

Taking the power factor of the load as $\cos \phi$, the relation between the volts at the receiving end V, and the sending end E, will be given by

$$E = \sqrt{[(V \cos \phi + 2IR)^2 + (V \sin \phi + 2IX)^2]}$$

The volt drop and regulation can be found from the values of E and V.

An approximate value for the volt drop per conductor is given by $IR \cos \phi + IX \sin \phi$. So the total volt drop will be $2(IR \cos \phi + IX \sin \phi)$.

The power loss per line is I^2R giving a total power loss of $2I^2R$. The power factor ($\cos \phi_s$) at the supply end is found from

$$\tan \phi_s = \frac{(V \sin \phi + 2IX)}{(V \cos \phi + 2IR)}$$

and the efficiency will be found by

$$\frac{VI \cos \phi}{VI \cos \phi + 2I^2R}$$

Three-phase A.C. Neglecting capacitance, the line constants will be as shown in Figure 8.3 and the following details refer to a balanced delta connected load.

Reactance and resistance drops per conductor will be IX and IR. But for three-phase reactance and resistance drops per phase will be $\sqrt{3}IX$ and $\sqrt{3}IR$. The relation between V and E will then be given by

$$E = \sqrt{[(V \cos \phi_1 + \sqrt{3}IR)^2 + (V \sin \phi_1 + \sqrt{3}XI)^2]}$$

The power factor at the supply end can then be obtained from
$$\tan \phi = \frac{(V \sin \phi_1 + \sqrt{3}IX)}{(V \cos \phi_1 + \sqrt{3}IR)}$$

The loss in each line will be I^2R, the total loss in this case being $3I^2R$. The efficiency can be found from

$$\frac{\sqrt{3}VI \sin \phi_1}{\sqrt{VI} \cos \phi_1 + 3I^2R}$$

The voltage regulation of the line will be found from $\dfrac{E - V}{V}$.

The vector diagram for a three-phase circuit is shown in Figure 8.4, and this can be used for single phase by omitting the $\sqrt{3}$ before the IR and IX.

Kelvin's Law. In any transmission line it can be shown that the maximum economy is obtained when the annual capital cost of the line equals the cost of the energy loss in transmission during the year. This is known as Kelvin's Law and is used as a guide for determining the size which should be used for a transmission line. The result obtained by applying Kelvin's Law must be considered also from the point of view of volt drop, current-carrying capacity and mechanical construction.

The capital cost of a line is the cost (usually taken over a year) for the interest on the capital expended, plus depreciation and maintenance. Usually a figure of between 10 and 20% of the capital cost is taken to cover these items. The energy loss in the

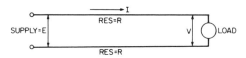

Figure 8.1   D.C. 2-wire supply

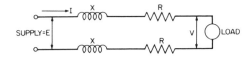

Figure 8.2   A.C. single-phase supply

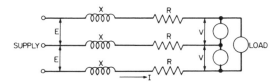

Figure 8.3   A.C. three-phase supply

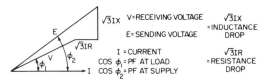

Figure 8.4   Vector diagram for three-phase line

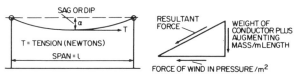

Figure 8.5   Overhead lines

line during the year can only be estimated and the following equation can be used

$$\frac{eBs}{100} = \frac{mI^2Rp \times 8760}{1000 \times 240}$$

where e = interest and depreciation in percentage per annum

B = cost per km of line per square millimetre of cross-section s in £

m = number of conductors

I = rms value of the current taken over a year

R = the resistance of one conductor per km

p = the cost in pence for energy per unit

s = section of line in square millimetres

From the above equation the ideal section for any transmission line can be obtained and the nearest standard size larger should be first considered.

BRITISH REGULATIONS FOR OVERHEAD LINES

The Electricity (Overhead Lines) Regulations, 1970 permit the exclusion of ice as a loading on the conductor, but increases a wind pressure to 760 N/m on the projected area of bare conductor and the factor of safety to 2.5. This applies to h.v. light construction lines with a conductor size less than 35 mm² copper equivalent and voltage exceeding 650 V but not greater than 33 kV. The minimum temperature assumed is −5.6°C.

For heavy construction lines, i.e. for conductors larger than 35 mm² copper equivalent at voltages exceeding 650 V and for all lines not exceeding 650 V, the factor of safety is 2.0 assuming the same minimum temperature as for light lines but with a wind force of 380 N/m on the projected area of augmented mass of conductor; the augmenting diameter for lines exceeding 650 V is 19 mm and for lines not exceeding 650 V is 9.5 mm.

The factor of safety for supports for h.v. light construction lines is 2.5 with no wind pressure acting on the supports. For h.v. heavy construction lines and all lines not exceeding 650 V the factor of safety for supports is 2.5 with a wind pressure of 380 N/m² acting on the supports the steelworks and insulators, etc.

The wind pressure on the lee side members of lattice steel or other compound structures including A and H shall be taken as one half the wind pressure on the windward side members. The factor of safety is calculated on the crippling load of struts and upon the elastic limit of tension members.

Mechanical strength of overhead lines. Referring to Figure 8.5, the tension in an overhead line conductor will be found from the formula:

$$\text{Tension in conductor} = T = \frac{wL^2}{8s}$$

where W = total equivalent weight of conductor in newtons per metre

L = span length in metres

s = sag in metres

T = tension in newtons

The conductor must then be designed to withstand this stress or tension allowing for the necessary factor of safety (this varies from 2.0 to 2.5). The cross-sectional area of the conductor can be derived from the formula, solving for a:

$$T = a \times f_t \times \frac{1}{\text{factor of safety}} \text{ newton}$$

and when temperatures are defined, T_1 can be calculated from the following formula:

$$Ea\alpha(t_2 - t_1) + \frac{w_1^2 L^2 Ea}{24T_1^2} - T_1 = \frac{w_2^2 L^2 Ea}{24T_2^2} - T_2$$

where E = modulus of elasticity of conductor in MN/m^2

a = total cross-sectional area of conductor in mm^2

α = linear coefficient of expansion, per °C

t_1 = initial temperature in °C

t_2 = final temperature in °C

w = weight in N/m

w_1 = initial weight in N/m

w_2 = final weight in N/m

L = span length in m

T = tension in N

T_1 = initial tension in N

T_2 = final tension in N

s = sag of conductor in m

d = conductor diameter in mm

r = radial thickness of ice in mm

p = wind pressure in N/m^2

f_t = breaking tension in conductor in N/mm^2

Allowance for augmented mass. The Electricity (Overhead Lines) Regulations 1970 make allowance for ice on the line which

165

is known as the augmented mass. It is derived from:

Mass of ice $= w_I = w_i \times r \times (d + r)$ kilograms/metre

where w_i = weight of ice in kg
$\quad\quad r$ = radius thickness of ice in mm
$\quad\quad d$ = diameter of conductor in mm

Allowance for wind. Wind pressure is expressed in N/m^2. For overhead lines covered by the Electricity (Overhead Lines) Regulations 1970, Schedule 2 Part 1 for line conductors not exceeding $35\,mm^2$ of copper equivalent cross-sectional area, where the voltage of the system exceeds 650 V but not 33 000V the wind force is taken to be $760\,N/m^2$. Schedule 2 Part 1 of the 1970 Regulations applies to line conductors other than those covered in Part 1, and the wind force in these cases is taken as $380\,N/m^2$.

Allowance for effective weight. In many calculations account has to be made for the allowances for wind and augmented mass acting on the line conductor. The effective weight, w, is the resultant of consideration being taken of the weight of the conductor and wind loading acting on the conductor.

$$w = \sqrt{[(\text{weight of conductor} + \text{ice})^2 + (\text{wind load})^2]}$$
$$= \sqrt{\{(w_c + [w_i \times r \times (d + r)]^2 + [p \times (d + 2r)]^2\})}$$

In many cases the cross-sectional area (a) of the conductor is fixed, and it is necessary to find the amount of sag (s) for a stated span length (L) for a specific value of T.

$$T = \frac{wL^2}{8s} \quad\quad \therefore s = \frac{wL^2}{8T}$$

For light construction h.v. lines the wind pressure is taken as $760\,N/m^2$ acting on the bare conductor.

Resultant $w = \sqrt{[(w_e + w_i)^2 + (w_w)^2]}$

where w_e = weight of conductor per metre in grams
$\quad\quad w_i$ = augmenting mass in grams
$\quad\quad w_w$ = pressure of wind in gram/metre^2 on augmented diameter.

EFFICIENCY OF TRANSMISSION AND DISTRIBUTION SYSTEMS

The normal method of comparing the efficiency of any transmission or distribution system is to compare the weight of copper required to transmit a certain load at a given voltage with the

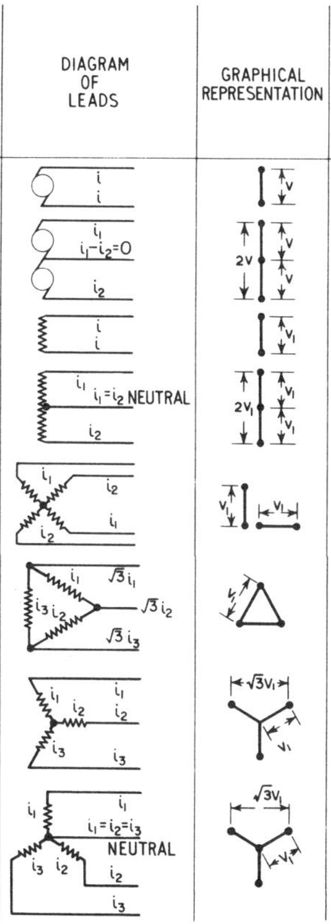

NAME	DIAGRAM OF LEADS	GRAPHICAL REPRESENTATION	EQUIVALENT WEIGHTS OF COMBINED CONDUCTORS FOR THE SAME PERCENTAGE LOSS AND THE SAME MAXIMUM VOLTAGE TO EARTH PF = 1·0
d.c. 2—WIRE	i i	v	100
d.c. 3—WIRE	i_1 $i_1 - i_2 = 0$ i_2	$2v$ v v	31·25
SINGLE PHASE 2—WIRE	i i	v_1	200
SINGLE PHASE 3—WIRE	i_1 $i_1 = i_2$ NEUTRAL i_2	$2v_1$ v_1 v_1	62·5
TWO PHASE 4—WIRE	i_1 i_2 i_1 i_2	v_1 v_1	200
THREE PHASE "MESH" (OR Δ)	i_1 $\sqrt{3} i_1$ i_3 i_2 $\sqrt{3} i_2$ $\sqrt{3} i_3$		50
THREE PHASE "STAR" (OR Y)	i_1 i_1 i_2 i_2 i_3 i_3	$\sqrt{3} v_1$ v_1	50
THREE PHASE 4—WIRE	i_1 $i_1 = i_2 = i_3$ NEUTRAL i_3 i_2 i_2 i_3	$\sqrt{3} v_1$	58·3

Figure 8.6 Comparison of systems. Neutral is taken as half size.
In a.c. circuits V is r.m.s. value. Efficiencies based on same power
transmitted.

167

same loss in transmission. For this purpose d.c. 2-wire is often taken as a standard and the other systems compared with it as regards the total weight of copper necessary.

Referring to Figure 8.6, the d.c. 2-wire system is taken as 100% and the weight of copper required is indicated for each different system. It is important to note that the calculations are made on the basis of the same maximum voltage to earth. In the case of the a.c. systems this means that $V/\sqrt{2}$ (i.e. the rms value) has to be used in the calculations. For the three-phase a.c. system V_1, is equal to $V/\sqrt{2}$ where V is numerically equal to the d.c. voltage.

In these comparisons the power factor of an a.c. load is taken as unity and it is assumed that in a 2-, 3- and 4-wire system the loads are balanced. It will be seen that except for the d.c. 3-wire system the three-phase 3-wire system scores in that a less total weight of copper is required than for any of the other systems illustrated.

Size of neutral. In the 3- and 4-wire systems employing a neutral, the size of the neutral conductor can be either equal to the 'outers' or half the size of the 'outers'. For the calculations in Figure 8.6 the neutral has been taken as half-size, and for the case where the neutral is full size allowance must be made for the increase in weight of copper. If a full-size neutral is used for a three-phase 4-wire system the total weight of copper is increased by one-seventh, making the comparative figure 67% compared with d.c. 2-wire.

Minimum ground clearances. There is no stipulated maximum working temperature for conductors, it being left to the supply authority to determine its own. The regulations call for a likely minimum ground clearance of a conductor 'at its likely maximum temperature (whether or not in use).' See Table 8.1.

Table 8.1 Minimum ground clearances

Voltage (kV)	At positions access-ible to vehicular traffic (m)	At positions not accessible to vehicular traffic (m)
Not exceeding 33	5.8	5.2
Exceeding 33 but not 66	6.0	6.0
Exceeding 66 but not 132	6.7	6.7
Exceeding 132 but not 275	7.0	7.0
Exceeding 275	7.3	7.3

168

9
Cables

UNDERGROUND CABLES

Until the 1970s the type of cable used predominantly for underground power distribution for public supply in the UK at voltages up to 33 kV was the impregnated paper insulated lead-sheathed cable. The standard constructions and requirements for this type of cable are specified in BS 6480.

Later developments, such as use of aluminium sheathing, increased adoption of PME systems by the electricity supply industry and, especially, the world-wide trend towards the greater use of cables with extruded insulations of synthetic materials, have resulted in a substantial reduction in the use of this type of cable for new installations, but it still constitutes a major proportion of the cable already installed.

Solid-type cables. The type of cable described in BS 6480 has traditionally been known as 'solid type', to distinguish it from the gas-pressure and oil-filled types of paper-insulated cables used for voltages above 33 kV and as alternatives to solid cables at 33 kV. The term is perhaps becoming out-dated in view of the growing use of the cables with extruded insulations, which might be regarded as more solid than impregnated paper.

The cable conductors are generally of stranded copper or stranded aluminium, although there has been some use of solid aluminium conductors for the 600/1000 V rated cables. A stranded conductor consists of a number of wires assembled together in helical layers around a central wire or group of wires, providing flexibility for drumming, undrumming and handling generally.

The insulation is applied as paper tapes in layers up to the required thickness, determined by the voltage rating. The paper is impregnated with an insulating compound, usually by the process known as 'mass impregnation', which is carried out after the paper tapes have been applied. The cable is dried and evacuated in a sealed vessel to which the hot impregnant is then admitted. An alternative, less used, method involves pre-impregnation of the papers before they are applied to the conductors.

The type of impregnant once used, consisting usually of mineral oil thickened with resin, has now been largely replaced in the UK by non-draining compounds, which contain a proportion of high melting point micro-crystalline wax. Under some conditions of use the fluid oil-resin compounds gave rise to problems due to migration. The non-draining compounds, while fluid at the temperatures employed for impregnation, are, over the normal operating temperature range of the cables, solids of a sufficiently plastic consistency to provide satisfactory bending performance. After its introduction for the 600/1000 V rated cables towards the end of the 1940s the mass-impregnated non-draining cable, abbreviated to MIND cable, was developed gradually for increasing voltages until in the UK it has become virtually the standard paper-insulated solid-type cable over the voltage range.

The impregnated paper insulated lead-sheathed solid-type cable in its simplest form as a single-core cable, is illustrated in Figure 9.1a. A lead alloy sheath is extruded over the insulated core and this is protected by an extruded PVC oversheath. The metal sheath is an essential component to exclude water, which, in quantity, would destroy the insulating properties of the impregnated paper. The PVC oversheath is the standard form of protection for single-core cables because they are frequently used for interconnectors installed at least in part inside buildings, such as at sub-stations, where a clean finish which will not readily propagate fire is preferable.

Single-core cables have limited use, and (except for very high currents, demanding large conductors, and for fairly short interconnectors) multicore constructions are normally used. The cable of 600/1000 V rating which embraces the 240/415 V standard voltage for domestic supplies has four conductors, three for the phase currents and a neutral. The typical design is illustrated in Figure 9.1b. The conductors, except for the smallest sizes, have a shaped cross-section, so that when the cores (the insulated conductors) are laid up together they form a compact circular cable with minimum spaces in the centre and at the sides at the rounded corners of the cores, to be filled with paper strings or jute yarns. It also reduces the cable diameter and therefore the amount of lead sheath and armouring.

170

Further insulating papers are applied over the laid-up cores, these constituting the 'belt' insulation. The lead sheath, applied after the cable has been impregnated, is protected by paper tapes overlaid with fibrous materials, generally hessian and/or cotton, all impregnated with bitumen. This provides a bedding for the steel tape armour, which in turn is covered with bitumen and further layers of bituminised fibrous material.

Steel tape armour was conventionally used for 600/1000 V cables for the electricity supply industry and galvanised steel wire armour for higher voltage cables, partly to provide identification. For 600/1000 V cables, however, wire armour is used where longitudinal strength is required (e.g. for cables pulled into long lengths of ducting), as well as protection against impact and abrasion.

Bituminised fibrous materials have proved generally adequate to protect lead-sheathed cables from corrosion, the metal itself not being susceptible to corrosion in most underground conditions, but in aggressive environments extruded coverings, usually PVC, may be used.

For higher voltages the cables are generally three-core. The construction for voltages up to 11 kV is similar in principle to that of the 600/1000 V cable. The insulation thickness is greater, of course, and the manufacturing processes differ in detail to provide for the higher operating electrical stresses.

For voltages above 11 kV, the belted construction gives place to the 'screened' cable as the standard. At 11 kV both types are provided for in the standards, but the belted type has the greater usage.

In the belted cable half of the thickness of insulation required between conductors is applied to each core and, with the cores laid-up, the balance of the required thickness to earth is applied as the belt. In the screened cable the whole of the required insulation to earth is applied to each core, on to which is then lapped a thin metal tape or a laminate of paper and aluminium foil, known as 'metallised paper'. The laid-up screened cores are bound by a tape which includes a few copper wires in the weft, so that when the lead sheath has been applied the screens will be in electrical contact with it. In the screened cable the fillers, which are electrically weak compared with the lapped insulation, are excluded from the electric field and the direction of the field in the cores is radial across the paper thicknesses. Screened cables are sometimes described as 'radial-field cables'.

A typical three-core screened 11kV cable is illustrated in Figure 9.1c. The carbon paper screen applied to each conductor is a standard feature for cables rated at 11kV and above in BS6480. It is to reduce the electrical stress at the conductor surface by smoothing out the profile and to exclude from the field the small spaces between the wires of the outer layer, which can otherwise be sites for discharge.

A carbon paper screen is also applied over the insulation of single-core 11kV cables and over the belt of 11kV belted cables to eliminate discharge in spaces between the outside of the insulation and the inside of the lead sheath in places where the latter does not make close contact.

Another form of three-core screened cable for 22 and 33kV is the 'S.L.' type. This has circular cores each separately lead sheathed. The sheathed cores are bound together and the assembly armoured and served. This design virtually eliminates the possibility of breakdown between phases and the cores can be terminated individually. It tends to be more costly than three-core cable under a common lead sheath and has not had much use in the UK.

Consac cable. In the UK the four-core paper insulated lead-sheathed cable has been replaced to a large extent by CNE cables on public supply networks. The adoption of protective multiple earthing (PME) by the electricity boards has been a major factor in this change. In the main distribution cables it is no longer necessary to keep the neutral conductor separated from earth and in CNE cables (Combined Neutral and Earth) there is effectively a saving of one conductor by combining one of the functions performed by the sheath of the lead-sheathed cable, provision of the earth return path, with the function of the neutral conductor.

The Consac cable, now the subject of BS 5593, has three shaped solid aluminium phase conductors, impregnated paper insulation and an extruded aluminium sheath of dimensions adequate to give a conductance at least equal to that required for the phase conductors. The aluminium sheath is the combined neutral and earth conductor as well as the barrier to water. This type of cable is illustrated in Figure 9.1d. The bituminised fibrous materials generally used to protect lead sheaths are not adequate for aluminium, which is particularly susceptible to corrosion in underground conditions. The Consac cable has an extruded PVC oversheath applied over a layer of bitumen which seals the interface between the aluminium and the PVC.

172

Figure 9.1a   Single core 600/1000 V 300 mm² lead alloy sheathed, PVC oversheath cable.

1. Circular stranded conductor   3. Sheath, lead alloy
2. Impregnated paper insulation   4. PVC oversheath

(Figures 9.1a to g are reproduced by courtesy of B.I.C.C. Ltd.)

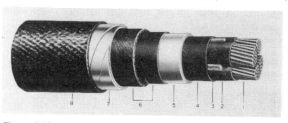

Figure 9.1b   Four-core lead sheathed cable, steel tape armour and served. Suitable for 600/1000 V 3-phase 4-wire systems.

1. Shaped stranded conductor   5. Sheath, lead or lead alloy
2. Impregnated paper insulation   6. Bedding
3. Filler   7. Steel tape armour
4. Impregnated paper belt   8. Serving

This is a very economic form of l.v. distribution cable, but less convenient for the tee-jointing of service cables than the type with extruded insulation and waveform concentric wire neutral described later.

11 kV aluminium-sheathed cables   The availability of presses for extrusion of aluminium sheaths has led also to the adoption of 11 kV aluminium sheathed cables by the electricity boards in the interests of economy. Although the 11 kV lead-sheathed cable has not been completely replaced, the aluminium-sheathed type

Figure 9.1c Three-core 11kV 150mm² lead sheathed screened cable; single wire armoured and served.

1. Shaped stranded conductor
2. Carbon paper screen
3. Impregnated paper insulation
4. Screen of metal tape intercalated with paper tape
5. Filler
6. Copper woven fabric tape
7. Sheath, lead or lead alloy
8. Bedding
9. Galvanised steel wire armour
10. Serving

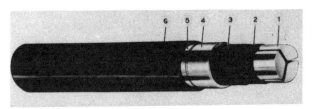

Figure 9.1d Typical Consac cable for use in low voltage PME systems.

1. Solid aluminium conductors
2. Paper core insulation
3. Paper belt insulation
4. Extruded smooth aluminium sheath
5. Thin layer of bitumen containing a corrosion inhibitor
6. Extruded PVC or polythene oversheath.

constitutes the bulk now purchased. Smooth and corrugated sheaths have been used, protected by PVC oversheaths applied on a layer of bitumen. There are benefits and disadvantages in each type of sheath, but the greater flexibility of the corrugated sheath has been the major factor in causing it to be generally preferred by the users.

174

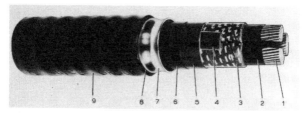

Figure 9.1e 11 kV belted type cable with corrugated aluminium sheath.

1. Shaped stranded conductor
2. Carbon paper screen
3. Impregnated paper insulation
4. Filler
5. Impregnated paper belt
6. Carbon paper screen
7. Corrugated   aluminium sheath
8. Bitumen containing corrosion inhibitor
9. Extruded PVC oversheath

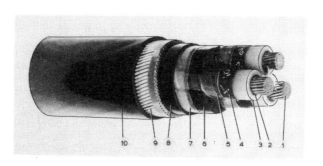

Figure 9.1f XLPE cable.

1. Circular stranded conductor
2. Semiconductor XLPE screen
3. XLPE insulation
4. Semiconducting tape screen
5. Copper tape screen
6. PVC filler
7. Binder
8. Extruded PVC sheath
9. Galvanised   steel   wire armour
10. Extruded PVC oversheath

175

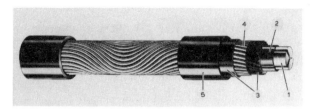

Figure 9.1g Typical Waveconal cable as used in low voltage p.m.e. systems.

1. *Solid aluminium conductors*
2. *XLPE core insulation*
3. *Rubber anti-corrosion bedding*
4. *Aluminium wires*
5. *Extruded PVC oversheath*

The corrugated aluminium-sheathed 11 kV cable is illustrated in Figure 9.1e. Up to the sheath it is similar to a cable for lead sheathing, but partial filling is needed of the spaces under the sheath resulting from the corrugation. Both belted and screened designs are supplied, the belted having the greater usage.

PVC insulated cables. Well before 1970 PVC insulated wire armoured cables to BS 6346 had become the established type for industrial installations and power stations for voltages up to 3.3 kV. PVC compounds can be formulated to give a range of flexibility and hardness, but, being thermoplastic materials, they soften at elevated temperatures. In applications where the maximum loads to be carried are precisely known and fairly close protection against overload can be provided by fuses or other devices, to ensure that damaging temperatures are not reached, this is not a significant disadvantage. For the l.v. public supply system, however, there is preference for cable which will tolerate overloads of greater magnitude and duration and PVC insulated cables have only limited uses in that area.

A particular advantage of the extruded types of insulation, such as PVC, is that they are much less affected by moisture than is paper, and terminating is simplified on this account. There are no metal sheaths to be plumbed and there is no need to enclose the ends in water-resisting compound.

The conductors for this type of cable are stranded copper or solid aluminium. Stranded aluminium conductors are quite feasible and are sometimes supplied when requested by overseas users, but BS 6346 does not include them on the basis that, if the more economic aluminium conductors are to be used, then the fullest economy is achieved by the use of the solid form, which is particularly suitable for PVC insulation.

As with paper-insulated cables, the conductors for multicore cables, except for the small sizes, are of shaped cross-section. The PVC compound is extruded on to the conductor by a technique allowing a uniform thickness to be applied to the shaped profile. The insulant is extruded in a hot plastic state and cooled by passage through a water trough. The requisite number of cores are assembled together with a spiral lay and with non-hygroscopic fillers when required to give a reasonably circular laid-up cross section. Either PVC tapes or an extruded layer of PVC compound is applied over the assembled cores to serve as a bedding for the armour. PVC tapes are the cheaper alternative for the cables with shaped conductors and are more often used for cables to be installed in air, unless it is required that the terminating gland should seal onto the bedding, when the extruded form is more suitable. Extruded beddings are also preferable for cables to be buried direct in the ground. The armour may be galvanised steel wires or aluminium strips and, in addition to giving mechanical protection provides earth continuity in the way that the metal sheath does on the paper insulated cable.

The National Coal Board use wire-armoured PVC cables with certain special design features for 3.3 kV supplies in the mine tunnels and shafts. An important reason for this choice is the better resistance to rock falls of the resilient PVC insulation compared with paper.

XLPE insulated cables. XLPE is the recognised abbreviation for cross-linked polyethylene. This and other cross-linked synthetic materials, of which EPR (ethylene propylene rubber) is a notable example, are being increasingly used as cable insulants for a wide range of voltages.

Polyethylene has good electrical properties and in particular a low dielectric loss factor, which gives it potential for use at much higher voltages than PVC. Polyethylene has been and still is used as a cable insulant, but, as a thermoplastic material its applications are limited by thermal constraints. Cross-linking is the effect produced in the vulcanisation of rubber and for materials like XLPE the cross-linking process is often described as 'vulcanisation' or 'curing'. Small amounts of chemical additives to the polymer enable the molecular chains to be cross-linked into a lattice formation by appropriate treatment after extrusion.

The effect of the cross-linking is to inhibit the movement of molecules with respect to each other under the stimulation of heat and this gives the improved stability at elevated temperatures compared with the thermoplastic materials. This permits higher operating temperatures, both for normal loading and under short-circuit conditions, so that an XLPE cable has a higher current rating than its equivalent PVC counterpart. The effects of ageing, accelerated by increased temperature, also have to be taken into account, but in this respect also XLPE has favourable characteristics.

BS 5467 specifies construction and requirements for XLPE and EPR insulated wire-armoured cables for voltages up to 3.3 kV. The construction is basically similar to that of PVC cables to BS 6346, except for the difference in insulant. Because of the increased toughness of XLPE the thicknesses of insulation are slightly reduced compared with PVC. The standard also covers cables with HEPR (hard ethylene propylene rubber) insulation, but XLPE is the material most commonly used.

From 3.8 kV up to 33 kV, XLPE and EPR insulated cables are covered by BS 6622 which specifies construction, dimensions and requirements. The polymeric forms of cable insulation are more susceptible to electrical discharge than impregnated paper and at the higher voltages, where the electrical stresses are high enough to promote discharge, it is important to minimise gaseous spaces within the insulation or at its inner and outer surfaces. To this end XLPE cables for 6.6 kV and above have semi-conducting screens over the conductor and over each insulated core. The conductor screen is a thin layer extruded in the same operation as the insulation and cross-linked with it so that the two components are closely bonded. The screen over the core may be a similar extruded layer or a layer of semiconducting paint with a semiconducting tape applied over it. Single-core and three-core designs are employed, and there is scope for constructional variation depending on the conditions of use, subject to the cores being surrounded individually or as a three-core assembly by a metallic layer, which may be an armour, sheath or copper wires or tapes. A typical armoured construction which has been supplied in substantial quantities is shown in Figure 9.1f.

In the UK this type of cable, mainly in single-core form, is favoured for power station cabling, where lightness and convenience of terminating are major considerations. Three-core designs are also used for site supplies.

For underground distribution at 11 kV, the XLPE cable does not compete economically with the paper-insulated aluminium-sheathed cable, but work is in progress on standardising and assessing XLPE cable design, including trial installations, in preparation for any change in the situation.

Overseas, where circumstances are different, XLPE cable is the type in major demand. With manufacturing facilities increasingly orientated to this market, XLPE insulated cables constitute a large proportion of UK production.

Aluminium waveform cable. The aluminium waveform cable, often described as 'Waveconal', is a type of CNE 600/1000 V distribution cable for public supply utilizing XLPE as the insulation. This and the Consac cable now constitute the largest part of the cables purchased by the electricity supply industry for this purpose. Like the Consac cable it has three solid shaped aluminium phase conductors, but the insulation is of XLPE, with HEPR as an alternative. The laid-up cores are bound with an open-lay tape and covered with an unvulcanised rubber compound into which are partly embedded aluminium wires applied in a waveform constituting the combined neutral and earth conductor. A further layer of the rubber compound is applied over the aluminium wires so that it is pressed between the gaps to amalgamate with the underlying layer. The concentric wires are thus effectively sandwiched in the rubber compound to protect them from corrosion and prevent water spreading between them in the event of the PVC oversheath extruded overall becoming damaged. The cable is illustrated in Figure 9.1g.

The waveform application of the concentric conductor allows the wires to be lifted from the underlying cable without cutting to give access to the phase conductors to make service joints. This is particularly convenient for maintaining the neutral/earth continuity during live jointing, which can be carried out at the 240 V phase-to-earth voltage when adding services.

Pressurised cables. The electrical strength of solid-type cable is limited by the possibility of drainage of fluid impregnants and by the effects of thermal expansion and contraction which result in the formation of small voids within the insulation. Insulation thicknesses have to be great enough to ensure that electrical stress will not cause severe and destructive ionization within the voids. Consequently, although solid type cables are occasionally used for voltages as high as 66kV, they are generally uneconomic or impracticable above 33kV.

For 33kV and higher voltages, pressurised cables have been developed wherein ionization does not occur even when the electrical stress on the insulation is three or four times as great as the maximum permissible in solid-type insulation.

Oil-filled cables. The earliest form of pressurised cable was the so-called low pressure oil-filled cable, which remains the predominant type for super-tension service, being employed at voltages up to 525kV.

Void formation is prevented by the use of a very low viscosity impregnating oil and the provision of external oil-feed tanks whereby the insulation is always maintained in a fully-impregnated state. Channels are incorporated in the cables to permit oil flow resulting from changes in cable temperature.

Oil-filled cables are available in 3-core form at voltages up to 150kV. Single-core cables are employed for higher voltages than this and are also used for terminating 3-core cables; single-core cables are also used in the 33kV to 132kV range when conductor sizes greater than about 630mm^2 are required.

The oil-filled cable is designed to operate with an oil pressure within the range 30 kN/m^2 to 525 kN/m^2 for normal installations. To withstand the internal oil pressure, the lead alloy sheathed cable requires to be reinforced with metal tapes; no reinforcement is required for the corrugated aluminium sheath design of cable. Oil feed tanks are provided to sustain the oil pressure within the design range. Several types of oil-filled high voltage cables are shown in Figures 9.2a to d.

Internal gas pressure cables. Whereas in the types of press-urised cable described above the occurrence of voids is prevented, in the internal gas pressure cable ionisation within voids is suppressed by the introduction of nitrogen which permeates the insulation at 1,400 kN/m^2 and increases their breakdown strength to such an extent that their presence no longer imposes any severe restrictions on operating stress.

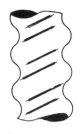

Corrugation of
aluminium sheath

*Figure 9.2a   Construction of oil-filled cables. Ductless, fillerless
oval-conductor construction used for 3-core 33 kV cables.*

1. Conductor
2. Insulation paper
3. Core screen metallised paper
   and cotton woven fabric tape
4. Laid up cores
5. Aluminium sheath
6. Extruded outer corrosion
   protection

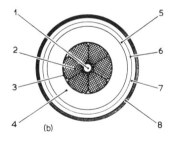

Figure 9.2b   Make up of 400 kV cable.

1. Oil duct
2. Six-segment copper conductor
3. Screening tapes
4. Oil-impregnated paper insulation
5. Screening tapes
6. Lead alloy sheath
7. Tin-bronze reinforcing tapes
8. Cotton binding tapes and
   PVC or PE sheath

181

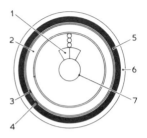

Figure 9.2c Hollow self-supporting conductor for single-core cables.

1. *Conductor, including screen*
2. *Insulation paper*
3. *Core screen*
4. *Copper woven fabric tape*
5. *Aluminium sheath*
6. *Polythene or PVC outer sheath*
7. *Oil duct*

Figure 9.2d 3-core ducted type served cable.

All diagrams by courtesy of BICC Ltd.

The paper insulation may be mass-impregnated (as in a solid-type cable) or alternatively, the insulation may be built up by use of pre-impregnated paper tapes.

Operating electrical stresses. The insulation of solid-type cables is of such thickness that the maximum electrical stress seldom exceeds about 4.5 MV/m. Pressurised cables, depending on the cable type and system voltage, etc., may operate with electrical stresses of the order of 15 MV/m. For 33-kV service the requirements of impulse strength limit the design stress to about 8.5 MV/m, but even with this modest stress the savings in insulation thickness and cable diameter are sufficient to make pressurised cables economic.

182

Current-carrying capacities. The current-carrying capacity of a cable is controlled by the necessity for dissipation of the heat generated by the power losses in conductors, insulation and sheath and the maximum temperature at which its insulation can safely be operated (65–90°C depending on voltage, type, etc.) and by the manner in which it is installed. Generally, the current rating of a cable buried in the ground is rather less than when the cable is installed in air, although in some cases the reverse is true. When cables are in buried, unfilled ducts instead of being buried direct in the ground, heat dissipation is hindered and current ratings are decreased.

Aluminium sheaths. Smooth profile aluminium sheaths are widely used nowadays on pressurised cables, particularly gas pressure cables, because they are sufficiently strong to withstand high internal gas-pressure without reinforcement. Corrugated seamless aluminium sheath is also available and is particularly appropriate for use on low-pressure oil-filled cables. The corrugation design enables sheath thickness to be reduced with resulting cost savings and improved bending performance and in conjunction with modern protections has led to an increasing use of aluminium sheathed oil-filled cables.

Acknowledgement for the information on cables and for the illustrations used in this section is made to BICC Ltd.

UNDERGROUND CABLE CONSTANTS

Insulation resistance. The insulation resistance is not directly proportional to the radial thickness of insulation and can be found from

$$R = \frac{\rho}{2\pi} \log_e \frac{R}{r}$$

where ρ is the specific resistance of the insulating material. This is more conveniently expressed in ohms or megohms per km. This is given by

$$R = 1.43\rho \log_{10}\frac{R}{r} \times 10^{-12} \text{ megohms per km}$$

183

Capacitance. The capacitance of a single-core cable is given by

$$C = \frac{0.024k}{\log_{10}\dfrac{R}{r}} \text{microfarads per km}$$

where k is the permittivity

Permittivity of impregnated paper insulation is usually about 3.5; that of polythene 2.3; rubber and PVC 5 to 8.

Voltage gradient. The question of voltage or potential gradient in insulated cables is important, especially in the design of oil-filled and gas-pressure cables. Most supertension cable designs are based on selecting an appropriate maximum voltage gradient having regard to the characteristics of the insulation and to the voltages, particularly impulse voltages, to which the cable will be

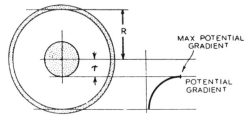

Figure 9.3   Potential gradient in single-core cable

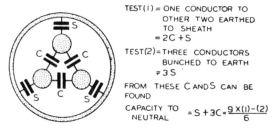

TEST(1) = ONE CONDUCTOR TO OTHER TWO EARTHED TO SHEATH
= 2C + S

TEST(2) = THREE CONDUCTORS BUNCHED TO EARTH
≑ 3S

FROM THESE C AND S CAN BE FOUND

CAPACITY TO NEUTRAL $= S + 3C = \dfrac{9 \times (1) - (2)}{6}$

Figure 9.4   Capacitance in 3-core cables

subjected. The general shape of the voltage gradient curve is shown in Figure 9.3, the maximum gradient occurs at the surface of the conductor and its value is given by:

$$g_{max} = \frac{E}{r \log_e \dfrac{R}{r}} \text{ V/cm}$$

Economical cable design involves use of the maximum voltage gradient or design stress that can be used without risk of electrical failure. When the values of g_{max} and E are fixed it will be found that a minimum value of R is obtained when $\log_e \dfrac{R}{r} = 1$, i.e. when $R = 2.718r$. If the smallest possible cable diameter is desired the diameter over the insulation should be 2.718 times the conductor diameter and the radius of the conductor should be such as to satisfy the equations

$$g_{max} = \frac{E}{r} \text{ V/cm}$$

For system voltages up to about 132 kV, this approach to cable design is seldom practicable because the value of r so obtained would be too small to allow the use of a reasonable conductor cross-section. At system voltages such as 275 or 400 kV the $R = 2.718r$ relationship is frequently employed in cable design; in the case of oil-filled cables with small copper sections, the required value of r may be achieved by making the internal oil duct of larger diameter than would otherwise be necessary.

WIRING CABLES

There are a number of l.v. cables used in the wiring of domestic, commercial and industrial dwellings. In addition the supply to house services and street lighting schemes may be by cables other than PVC or paper insulated solid type cables. These are described in the following paragraphs.

Service cables. The service cables which carry supplies to houses and other premises with small loads are usually single-phase, connected by service joints to one phase and the neutral of the main l.v. distribution cables laid in the streets or pavements. For larger loads three-phase service cables may be used.

One type of service cable is a small version of the Waveconal cable. For single-phase supply it has only one circular phase conductor insulated with XLPE with the waveform neutral wires embedded in their rubber protection laid around it, the whole being protected by a PVC sheath. For three-phase and neutral supplies the cable is very similar to Figure 9.1g except that the three phase conductors are circular.

Alternative types are insulated with PVC, one of the few uses of PVC insulation for distribution cables for public supply. The concentric conductors of these are copper wires applied helically without the rubber protection against corrosion required for aluminium wires. The phase conductors are usually copper also, but may be of solid aluminium.

One type is a CNE cable; in this the wires of the concentric conductor, which serves as neutral and earth, are not individually covered. In the other type, known as 'split concentric', the wires to be used as the neutral conductor are individually covered with a thin layer of PVC while the wires of the earth continuity conductor are bare. The two types are illustrated in Figure 9.5. The single-phase split concentric type is covered by BS 4553.

Figure 9.5   Two types of concentric service cable. Left: split concentric. Right: combined neutral and earth (CNE) concentric

Combined neutral and earth service cables are similar to split cables in design except that all the outer conductors are bare copper wires as shown in Figure 9.5 (right).

Both designs have a black extruded PVC oversheath and are available in sizes ranging from 4 mm² to 50 mm².

Split concentric cables have a stranded copper or solid aluminium central conductor, insulated with red PVC and a concentric layer of bare and insulated conductors as indicated above.

Copper conductor cables. Small wiring cables, suitable for lighting and power services in buildings are generally copper conductor with PVC insulation for conduit and trunking use. For two core and cpc cables a PVC sheath is provided over the insulated cores. The cpc is usually bare copper. Sizes range from $1 mm^2$ up to about $16 mm^2$ for domestic and commercial use.

When metric cables were first introduced into the UK, $1 mm^2$, $1.5 mm^2$ and $2.5 mm^2$ sizes had solid copper conductors for both single and multicore cables. The stiffness of the two-core and cpc cable at $2.5 mm^2$ has resulted in these sizes being made also available in stranded form.

Due to the fluctuating and sometimes high price of copper experiments have been carried out over a period of years to see whether it was possible to use aluminium as a conductor material. To date termination problems, resulting in overheating of accessories, has precluded this material as a satisfactory conductor for house wiring cables.

Mineral insulated cable. This type of cable is used extensively for general lighting and power circuits, fire alarms and emergency supplies in most types of buildings and industrial installations. A range of cables is available which is BASEC approved to BS 6207 Part I. Mineral insulated cables are recommended for use in hazardous areas and a full range of BASEEFA approved terminations is available. The cable will neither burn nor support combustion and will not emit smoke or toxic gas. It will continue to operate even when fire occurs in its vicinity thus maintaining essential services such as fire alarms and emergency lighting.

Mineral insulated cable consists of copper conductors embedded in densely compacted magnesium oxide insulation and contained within a copper sheath which acts also as an excellent circuit protective conductor.

These cables have higher current ratings, size-for-size, than organic insulated cables and because they are constructed from inorganic materials, they do not deteriorate with age. They are available in a range of single-core conductor sizes from $6 mm^2$ to $240 mm^2$ and in 2, 3, 4 core sizes from $1.5 mm^2$ to $25 mm^2$. There are also some conductor sizes available as 7, 12 and 19 core cables. There are two voltage ratings available: 600 V and 1000 V.

As an option, the sheath may have an overall covering to provide protection in environments corrosive to copper. This outer covering may be of a halogen free material with extremely low smoke emission and flame propagation characteristics.

When the cable is terminated, it is necessary to fit a seal to prevent the magnesium oxide insulant absorbing moisture. A full range of seals, glands and tools is available from the cable manufacturer. The cable is ideally suitable for use on TN-C systems since the copper sheath provides an excellent combined neutral and earth (PEN) conductor.

Elastomer-insulated cables. Where high ambient temperatures are encountered elastomer insulated cable and flexible cable is available for power and lighting circuits up to 1000 V.

10

Transformers and tap-changers

TRANSFORMERS

The static transformer is based on the mutual induction between two coils wound on a closed iron circuit. The one, called the primary, is connected to an alternating supply and the other, the secondary, has an emf induced due to the changing flux established by the primary winding.

The simple single-phase transformer is illustrated diagrammatically in Figure 10.1, the primary and secondary windings being marked.

The voltage ratio depends on the number of turns on the primary and secondary, and will be given by $\dfrac{V_1}{V_2} = \dfrac{N_1}{N_2}$.

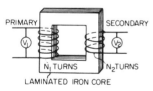

PRIMARY   SECONDARY

N_1 TURNS N_2 TURNS

LAMINATED IRON CORE

Figure 10.1 Single-phase transformer

Thus the ratio of the number of turns will give the voltage ratio of the transformer.

The main type of transformer employed for power and distribution networks is shown in Figure 10.3.

Voltage and flux considerations. In the ideal transformer it is assumed that the whole of the flux generated by the primary emf

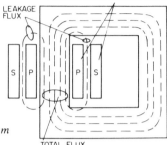

Figure 10.2 Diagram showing leakage flux

links all the secondary turns. The relationship between voltage and flux is given by the fundamental expression:

$$V = 44.4\, f\, N\, \phi_m \times 10^{-4}$$

where f = frequency in Hz; N = number of turns, primary or secondary and ϕ_m = maximum value of the flux in tesla.

In practice when the transformer is under load, all the flux does not link the secondary. This leakage flux, as it is called, is shown on Figure 10.2, the amount increasing with increasing load. This flux produces a reactance between primary and secondary which is known as the leakage reactance. It can be expressed either in ohms or as a percentage of the transformer rating.

Voltage regulation of the transformer. The voltage ratio already given represents the ratio on no load, and when the

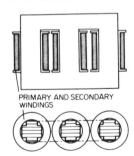

Figure 10.3 Core-type 3-phase transformer

transformer is on load there will be a fall in output terminal voltage due to the resistance of the primary and secondary windings and also due to leakage reactance. The fall in output terminal voltage is known as regulation and increases as the load increases.

Equivalent circuits. The actual circuit of a transformer can be assumed to be as shown in Figure 10.4, X_1 and R_1 being the primary values and X_2 and R_2 the secondary values. These can be formed into one circuit and the values referred to either primary or secondary. Referring the values to the primary we get

$$R = R_1 + R_2 \left(\frac{N_1}{N_2} \right)^2$$

$$X = X_1 + X_2 \left(\frac{N_1}{N_2} \right)^2$$

Referring to values to the secondary we get

$$R = R_1 \left(\frac{N_2}{N_1} \right)^2 + R_2$$

$$X = X_1 \left(\frac{N_2}{N_1} \right)^2 + X_2$$

To get the equivalent impedance Z we have

$$Z = \sqrt{(R^2 + X^2)}$$

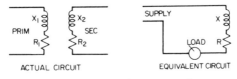

Figure 10.4   Equivalent circuit values

The values of R and X in the equivalent circuit can be obtained from the short-circuit test, which consists of passing a full load current through the transformer with either the primary or the secondary short circuited. A reduced voltage will, of course, be required owing to the short circuiting of one of the windings. Taking a 400 V/100 V transformer, the short-circuit test values

191

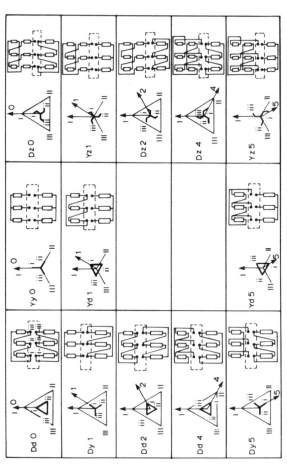

Figure 10.5 Designation of connections of separate winding three-phase transformers by connection symbols (from BS171: Part 4)

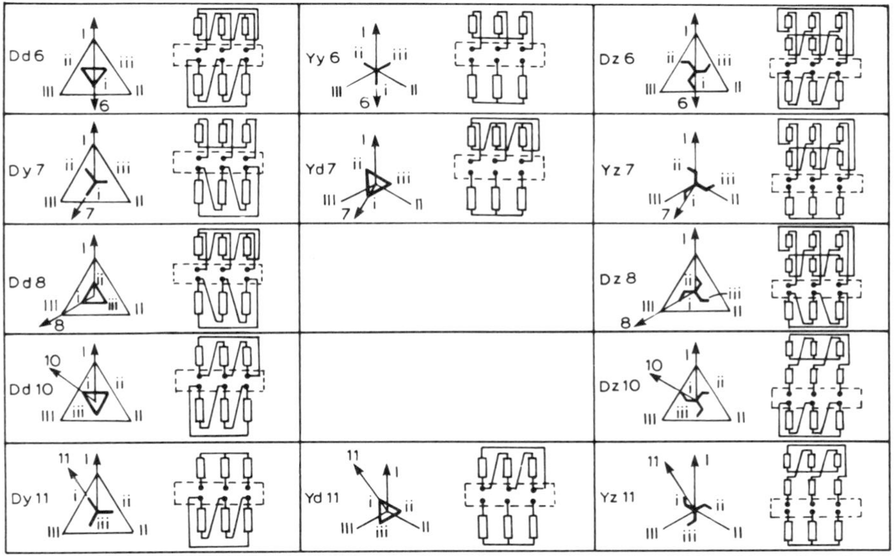

Figure 10.6 Designation of connections of separate winding three-phase transformers by connection symbols (from BS171:Part 4)

193

measured on the secondary side might be as follows: 50 A (full load current) at 5 V with a power input of 100 W. From these figures, power factor = $\dfrac{100}{50 \times 5}$ = 0.4 = cos ϕ'. (ϕ' = internal phase angle.)

$$Z = \frac{V}{I} = \frac{5}{50} = 0.1\ \Omega$$

$$R = Z \cos \phi' = 0.1 \times 0.4 = 0.04$$
$$X = Z \sin \phi' = 0.1 \times 0.916 = 0.0916$$

Approx. volt drop = $IR \cos \phi + IX \sin \phi$.

Values for unity power factor (cos ϕ = 1).

$$\text{Voltage drop} = 50 \times 0.04 \times 1.0 + 0$$
$$= 2 \text{ volts}$$

$$\text{Regulation} = \frac{2}{100} \text{ or } 2\%$$

Values for 0.8 power factor (cos ϕ = 0.8)

Voltage drop = $50 \times 0.04 \times 0.8 + 50 \times 0.0916 \times 0.6$
$$= 4.36 \text{ volts.}$$

$$\text{Regulation} = \frac{4.36}{100} = 4.36 \text{ per cent}$$

Efficiency.

$$\text{Efficiency} = \frac{\text{output}}{\text{input}}$$

and this can also be written

$$\frac{\text{input} - \text{losses}}{\text{input}}$$

or

$$\frac{\text{output}}{\text{output} + \text{losses}}$$

Input, output and losses are in kVA.

The losses in a transformer are essentially very small and the efficiency of large transformers is over 99%. Even with small transformers efficiencies of 97% or 98% are usual. There are two main losses in a transformer, namely, *the iron losses* and *the copper losses*. The iron losses are normally taken as constant at all loads whereas the copper loss is proportional to the square of the load.

Taking the case of a 100kVA transformer the full load losses might be taken as 800W iron losses and 1200W copper losses. The efficiency at any other load can be obtained by making the copper losses proportional to the square of the actual load. Thus, on half-load the copper losses would be $1200 \times (\frac{1}{2})^2 = 300W$.

The total losses = 800 + 300 = 1100; therefore efficiency

$$= \frac{50 \times 1000 \times 100}{50 \times 1000 + 1100} = 97.84\% \text{ for half load}$$

This compares with $\dfrac{100 \times 1000 \times 100}{100 \times 1000 + 2000} = 98.04\%$ for full load.

Connections of phase windings. The star, delta or zigzag connection of a set of phase windings of a three-phase transformer or of windings of the same voltage of single-phase transformers associated in a three-phase bank are indicated by the letters Y, D or Z for the high-voltage winding and y, d or z for the intermediate and low-voltage windings. If the neutral point of a star-connected or of a zigzag-connected winding is brought out, the indication is YN or ZN and yn or zn respectively. (From BS 171: Part 4.)

Parallel operation. For satisfactory operation of transformers the following points must be watched:
A. The same voltage ratio.
B. The same phase displacement.
C. The same impedance drop (or voltage regulation).
D. As nearly as possible the same internal phase relationship between resistance and reactance.

It will, of course, be evident that transformers in parallel must have the same secondary voltage for a common primary input.

The question of phase displacement will be seen from Figures 10.5 and 10.6. A pair of three-phase transformers of similar characteristics and having the same connection symbols can be operated in parallel by connecting together the terminals which correspond both physically and alphabetically. For example DdO, YyO and DzO.

The question of voltage regulation is, of course, important, as this decides the proportion of the total load which is taken by each transformer.

195

The proportion of the load carried by each transformer also depends on the ratio betwen R and X in the two transformers.

When joining up transformers in parallel or paralleling supplies from two separate transformers it is, of course, important to check the phase rotation. This can be done by a phase-rotation-indicating instrument.

AUTOMATIC TAP-CHANGERS

Electricity supply practice necessitates that the voltage appearing at a consumer's terminals must be maintained within declared limits. This is an important issue today as most households have a built-in monitor in the form of a television screen and consequently the general public is more sensitive to voltage fluctuations and load shedding.

This means that a method must be employed to obtain control of the voltage on transmission and distribution networks. Due to its comparatively low cost, reliability and ease of operation, on-load tap changing on the transformer has become the accepted means of doing this.

By this means the turns ratio of the transformer winding is altered. Tappings are brought out from one of the windings and by appropriate connections the number of turns on that winding is altered. In the United Kingdom the tappings are nearly always on the high voltage winding to take advantage of the lower current conditions.

There are various types of tap changing mechanisms in use, but all have two fundamental parameters to meet, namely:

1. The load current must not be interrupted during a tap change.
2. The tap change must be carried out without short-circuiting a tapped section of the winding.

To meet both criteria means that some form of bridging or transfer impedance is required during the transitional stage. Many tap-changers made prior to and immediately after the second world war, used reactors for the bridging impedance. In recent years, the reactor type of on-load tap-changer has been largely superseded by the high speed resistor transition type and all major developments have been concentrated on resistor transition.

Traditionally, the tap-changer has consisted of two main units. The tap selector switch is the unit responsible for selecting the tap on the transformer windings, but does not make or break current. The diverter switch is where the actual switching of the load takes place.

The operation of the diverter switch is extremely fast, generally in the order of 3–3½ Hz, so that the time during which the resistors are in circuit carrying current is very small, of the order of one to one and half cycles at transition. This means that the resistors need only be short-time rated, and are thus usually small enough to be accommodated on the diverter switch structure.

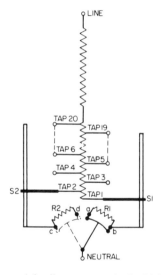

Figure 10.7 Electrical switching sequence resistor tap changing transformer used where load must not be interrupted

A leading company in the United Kingdom in the maufacture of tap-changers, achieves the short time of operation by using a stored energy mechanism, the construction and operation of which is such that it is virtually impossible for the gear to stop between taps.

A typical diagram showing the electrical switching sequence of a linear tap-changer is shown in Figure 10.7. For simplicity only a single phase is shown.

In the initial position selector switch S1 is on tap 1 and S2 on tap 2. The diverter switch connects tap 1 to the neutral point of the

197

transformer winding. The sequence of operations in changing to tap 2 is as follows:

1. As the stored energy mechanism operates the moving contact starts to travel from one side of the diverter to the other; contact b is opened and the load current flows through resistor $R1$ to contact a.

2. The moving contact d then closes. Both resistors $R1$ and $R2$ are now in series across taps 1 and 2 and the load current flows through the mid-point of these resistors.

3. Further travel of the moving contact opens contact a and the load current then passes from tap 2 through resistor $R2$ and contact d.

4. Finally, when the moving contact reaches the other side of the diverter switch, contact c is closed and resistor $R2$ is shorted out. Load current from tap 2 now flows through contact c, the normal running position for tap 2.

The change from position 1 to 2 as described involves no movement of the selector switch. If any further change in the same direction is required i.e. from 2 to 3 the selector switch $S1$ travels to tap 3 before the diverter switch moves and the diverter switch then repeats the above sequence but in the reverse order. If a change in the reverse direction, the selector switches remain stationary and the tap change is carried out by the movement of the diverter switch only.

Moving coil voltage regulator. The basic winding arrangement of a typical moving coil voltage regulator consists of two fixed coils wound on the upper and lower halves of a magnetic core and connected in series opposition. A third coil of the same length is short-circuited upon itself and is free to move over the other two coils. The moving coil is entirely isolated electrically so that no flexible connections, slip rings or sliding contacts are required.

The division of the voltage between the two fixed coils is determined by their relative impedances, and these are governed entirely by the position of the moving coil. With the moving coil in the position shown in Figure 10.8 the impedance of coil a will be small and that of b large. If a voltage is then applied across the two coils connected in series the greater part of the voltage will appear across coil b and a small part across coil a.

When the moving coils is at the bottom of the leg as in Figure 10.9 the relative impedances of coils a and b will be reversed, and the greater part of the voltage will now appear across coil a. With an arrangement similar to that shown in Figures 10.8 and 10.9, a range of voltage variation of practically 0–100% can be obtained with smooth infinitely variable control.

For applications on transmission and distribution systems voltage variations from 10–25% are normally sufficient, and these can be obtained from a Ferranti moving coil voltage regulator by the use of additional windings.

In these arrangements, the voltage variations obtained as shown in Figures 10.8 and 10.9 are changed to the desired values by additional coils connected either to buck or boost a voltage in series with the line.

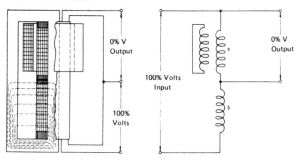

Figure 10.8 Typical moving-coil voltage regulator. Coil (a) impedance small; (b) impedance large

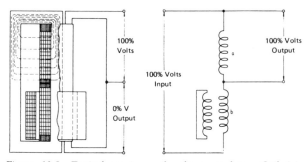

Figure 10.9 Typical moving coil voltage regulator. Coil (a) impedance large; coil (b) impedance small

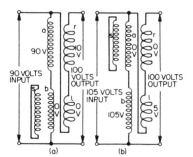

Figure 10.10 Typical moving coil voltage regulator with additional coils to buck or boost the line voltage

Figure 10.10 shows an arrangement incorporating two additional coils, *r* and *l*, which would be suitable for providing a constant output of 100% voltage whilst the input varied between 90% and 105% of the nominal value. Any desired value of buck or boost can be provided by choosing a suitable number of turns on the two coils. The position of the moving coil is altered by a small motor operated by a sensing device connected across the output.

AUTOMATIC VOLTAGE REGULATORS

Modern automatic voltage regulators used on large generating plant employ semiconductors to provide maximum reliability, minimum maintenance requirements, very fast response, close voltage regulation and no 'dead band'.

GEC Industrial Controls has two different systems, the first for generators up to about 300 MVA and the second up to 100 MVA. A simplified circuit for the smaller machines is shown in Figure 10.11. It incorporates a Gecostat C10/120 automatic voltage regulator and is designed primarily for use with brushless a.c. generators.

Silicon semiconductor devices are used throughout enabling the equipment to operate in high ambient temperatures. Main power controlling devices are thyristors. A single-phase a.c. supply is obtained from a permanent magnet generator pilot exciter.

Generator output voltage is sensed via a single-phase voltage transformer 11 and fed through a sensing circuit. This circuit rectifies and smooths the voltage which is then compared with a reference voltage developed across a zener diode. Any error

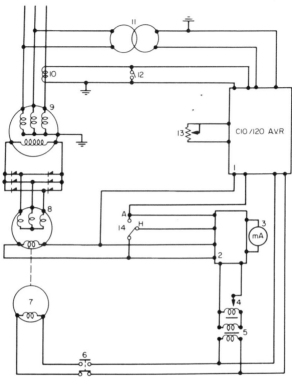

Figure 10.11 Simplified circuit diagram incorporating Gecostat C10/120 avr.

1. Automatic voltage regulator
2. Hand control unit.
3. Null balance meter.
4. Hand regulator.
5. Supply transformer.
6. Latched excitation contactor.
7. Pilot exciter (PMG)
8. Main exciter (AC)
9. AC generator.
10. Compounding current transformer.
11. Sensing voltage transformer.
12. Compounding switch.
13. External auto volts trimmer.
14. Hand-auto changeover switch

voltage produced is amplified by an integrated circuit, the output of which controls the rate at which a capacitor is charged.

When the voltage reaches a fixed level the unijunction fires and pulses are applied to the gate circuits of the thyristors via an isolating transformer. These pulses control the firing point of the thyristors and thus the flow of excitation power. The field supply is thus controlled and the generator output voltage is rapidly adjusted to $\pm1\%$ of the nominal, over the full load range of the generator.

A stabilising circuit is included around the transistor amplifier to prevent hunting and to produce an optimum response characteristic. The nominal regulated voltage may be adjusted using the 'auto volts' potentiometer on the avr.

Parallel operation.   A circuit is included in the avr which can provide a drooping characteristic to ensure that generators in parallel will share reactive load. A compounding current transformer 10 is needed to provide the compounding signal if this feature is required. The compounding switch 12 is required when a generator is run either singly or in parallel with other generators and it is desired to switch out the compounding to give better voltage regulation when operating singly.

External auto-volts trimmer.   An external auto-volts trimmer provides the facility of trimming the generator voltage on auto control at a position remote from the avr. This trimmer may be motorised if control of it is required from a position remote to its mounting place.

Hand control.   A manually controlled source of excitation is provided as an alternative to the automatically controlled source provided by the avr. The supply transformer 5 converts the output from the permanent magnet generator pilot exciter 7 to a suitable voltage. The hand regulator 4 fed from the supply transformer, provides the facility for manually varying the input voltage to the hand control unit 2 from zero to the voltage fed by the supply transformer. The hand control unit converts the variable a.c. supply from the hand regulator to a d.c. supply for excitation, which is fed to the field winding of the main a.c. exciter 8 via the hand/auto changeover switch 14.

To ensure a smooth changeover from hand to auto or vice versa a null balance meter is provided to work in conjunction with the

202

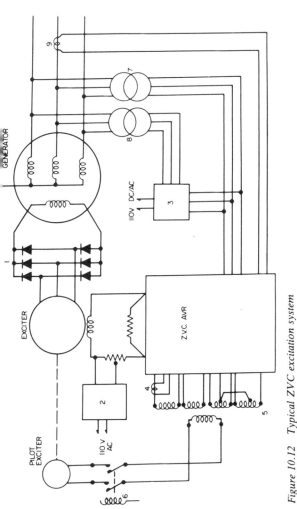

Figure 10.12 Typical ZVC excitation system

1. Rotating rectifiers
4. Rotor current CT
7. Sensing VT

2. Rotating rectifier failure unit
5. Supply transformer
8. Comparison VT

3. Fuse failure unit
6. MCB with trip coil
9. Stator CT

203

hand control. The hand regulator may be motorised to allow operation from a remote position.

Latched excitation contactor. The latched excitation contactor 6 can be operated from the generator protection in order to interrupt the main exciter field and thus the generator excitation. The use of the permanent magnet pilot exciter in the system ensures the maintenance of output even when the generator output is short-circuited and fault clearing can therefore be facilitated.

ZVC regulator system. For larger sets GEC has developed the type ZVC regulator system. It works in conjunction with the Emoreg thyristor converter that is widely used in d.c. motor drive applications. A typical excitation circuit is shown in Figure 10.12 and the ZVC comprises a thyristor converter power output stage, a thyristor firing pulse unit and the sensing and control circuits, a $\pm 15\,\text{V}$ power supply for the electronics circuits contained in the basic unit and Bin A, automatic setting control and manual control and a balance indicator. Space does not allow a full description of the operation of the system but it should be pointed out that by utilising a fully controlled bridge a rapid and effective reduction in output current is possible by operating the thyristors in the inversion region. This allows the current in the inductive exciter field winding to regenerate back into the a.c. power source.

In both avr systems described above it should be noted that the avr takes its supply from a pilot exciter. GEC Industrial Controls also manufacturers avrs which draw the excitation supply from the machine terminals through current and voltage transformers.

Thyristor divert automatic voltage regulator. This is a static unit designed to control the field current of an exciter for an alternator and is marketed by Brush Electrical Machines. The exciter may be either a brushless a.c. generator or a commutator type d.c. generator.

The field current of the exciter is supplied from a high impedance a.c. source through a single-phase bridge rectifier. The avr controls the output of the rectifier using a thyristor which is switched off every half cycle by the inductance of the exciter field. The method of operation of the thyristor divert is illustrated in Figure 10.13, a typical half-cycle being divided up into three operational phases as shown.

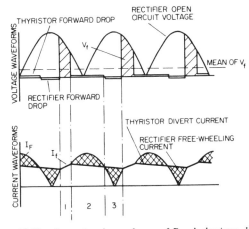

Figure 10.13   Operational waveforms of Brush thyristor divert

Phase 1 rectifier operates normally so that rectifier input current is the same as the rectifier output current and exciter field current. It is approximately constant due to the inductance of the field. Current in the choke L (Figure 10.14) is therefore substantially constant and the rectifier input voltage V_F is therefore equal to the supply voltage.

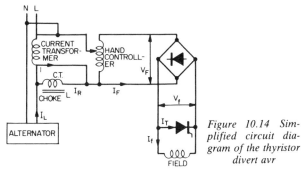

Figure 10.14   Simplified circuit diagram of the thyristor divert avr

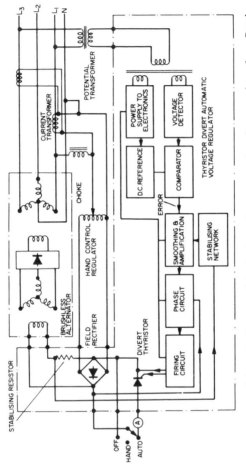

Figure 10.15 Schematic diagram of the thyristor divert automatic voltage regulator from Brush Electrical Machines

206

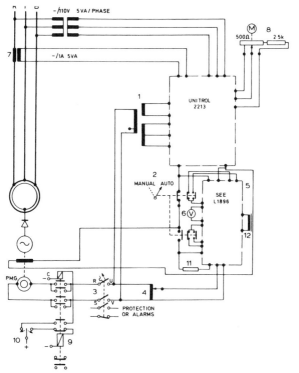

Figure 10.16 British Brown-Boveri Ltd static automatic voltage regulation of brushless alternator with p.m. generator and hand control.

1. Supply transformer
2. Changeover switch (m-b-b)
3. Protective m.c.b.
4. Variable transformer
5. Rectifier and balancing panel
6. Balance voltmeter (15–0–15 V)
7. C.T. for parallel operation only
8. Separate voltage setting potentiometer (if required)
9. Latched excitation contactor
10. Excite/de-excite switch
11. Limiting resistance for manual control
12. Choke for rectifier and balancing panel

207

Phase 2 thyristor conducts and hence diverts current from the exciter field by short-circuiting the rectifier.

Phase 3 rectifier input current drops below the exciter field current and the thyristor switches off and the rectifier 'free-wheels'.

The high impedance source may comprise simply a choke in series with the alternator output voltage but generally two current transformers are added to sustain the excitation when the alternator is subjected to short-circuit.

A schematic diagram of a thyristor divert avr is shown in Figure 10.15 having hand or auto control, applied to a brushless alternator.

Brown Boveri Unitrol System. The Brown Boveri Unitrol system of automatic voltage regulator is shown in Figure 10.16. The avr senses the regulated voltage through a three-phase star-star or single-phase voltage transformer. The current transformer 7 provides compensation to give stable parallel running or compounding for line or transformer drop. The deviation or error signal is fed to the PID unit. If an external setting potentiometer is required say to facilitate automatic synchronising, then the internal one must be disconnected. In the PID unit the error signal is amplified, thus ensuring a high accuracy independent of disturbing influences such as frequency, drift, temperature and load. Accuracy of regulation is ±0.5% when running singly under steady state conditions.

The grid pulse control unit in the avr shifts the pulses, depending on the output of the regulator, to switch the thyristors on in a fully controlled mid-point connection in the power unit. The shifting of the firing point in relation to each voltage half wave, influences the excitation current and with it the generator voltage.

The avr is fed from the permanent magnet pilot exciter 9 via a protective mcb 3 and a three-winding voltage transformer 1. Thus the system is completely independent of any station auxiliary supplies.

Manual control of the generator voltage is achieved by the variable transformer 4 and the unit 3 which rectifies the a.c. output from the pilot exciter. When manual control is required the variable transformer setting is raised until the centre zero voltmeter 2 falls to zero. At this point the automatic and manual excitation settings are the same and a smooth changeover is achieved.

11

Tariffs and power factor

TARIFFS

As electricity is not a commodity which can be stored and used as required, the flat-rate basis was found unsatisfactory and two-part or 2 and 3 block tariffs have replaced fixed-rate tariffs for domestic and commercial establishments.

The two-part tariff is based on two costs of which the first part is covered by an annual amount and the second part by a charge per unit used. Block tariffs are related to maximum load, with different unit rates related to specified unit usage per quarter.

Industry generally operates on a maximum-demand system, while there are other forms of tariff for farms. Specialised tariffs include off-peak, Economy 7 and other time of day.

Load factor. This can be defined as the average load compared with the maximum load for any given period. It can be calculated as follows:

$$\frac{\text{Actual energy consumed}}{\text{Maximum demand} \times \text{Time in hours of period}}$$

The load factor of a consumer may vary from as low as 5% to as high as 80%, but usually it ranges from 10% (for lighting only) to 40% (for industrial or heating loads). Some industries are able to offer a 24-hour load and it is in these cases that very high load factor figures are obtained.

Owing to the two-part nature of the cost of supplying electrical energy, the actual load factor has a direct effect on the cost per unit since the fixed or standing charge to cover the first cost is divided into all the units used during that period. The more units used (and the higher the load factor), the less will be the fixed cost per unit. On this account it is the aim of every supply engineer to make his load factor as high as possible. As will be explained, special inducements are generally offered to consumers who will enable him to do this.

Diversity. The diversity of the supply load is given by the *diversity factor,* which is found from

$$\frac{\text{Sum of consumers' maximum demands}}{\text{Maximum demand on system}}$$

and it will be seen that

$$\frac{\text{System load factor}}{\text{Average consumer's load factor}} = \text{Diversity factor.}$$

Note. The average consumer's load factor must be calculated with reference to actual consumption and not merely as a numerical average.

Tariffs. (*Based on the Electrical Times Handbook, 1988*) These are usually of three different types, *industrial, commercial* and *domestic.* Although the Area Boards have now been privatised to form distribution companies the tariffs have not been altered much and the following examples are still generally valid. Most of the distribution companies have introduced two changes. They offer a seasonable time of day tariff which does not have a maximum demand component but has a higher unit cost in winter than in summer. Also all customers with a demand in excess of 1 MW can negotiate a price with any electricity supplier, not only the one serving their area.

An industrial two-part tariff is always based on the maximum demand – either in kW or in kVA – and in many cases time of the year. A typical l.v. industrial tariff (for London Electricity) is therefore:

A fixed charge of £0.90 per kVA service capacity per month, £15 per month, £15.50 with night units, plus a maximum-demand charge per kW in each month as follows: April to October inclusive £0.10, November and February £4.10, December and January £7.00 and March £1.70. There are then unit charges which vary according to the time of day.

Most companies offer a wide range of options based on maximum demand and the supply can be taken at high voltage or low voltage. For example, Merseyside & N. Wales offers an l.v.

supply based on a monthly maximum demand if the load is over 10 kW as follows: fixed charge £11.60 per month, November–February 1st 10 kW £8.60 per month. March–October 1st 10 kW £0.30 per month with a day unit charge of 4.19p. Most tariffs include a fuel price adjustment clause.

It is more usual today to base the fixed charge on a stated sum per month or per annum with a penalty charge for low power factor; it therefore pays the consumer to install power factor correction capacitors to lift a low power factor to a value in excess of 0.9.

The maximum demand figure is obtained by means of a maximum demand indicator which gives the highest load (either in kW or kVA according to the tariff) which occurs for a given period – such as 15 or 30 minutes. Special tariffs are in many cases offered to consumers with favourable loads.

Domestic tariffs for the London area are based on a fixed charge of £8 per quarter with a unit charge of 5.45p per unit.

There are two forms of commercial tariffs, block or two-part, and both are available from the South Eastern Electricity. The following offer only a block tariff, Southern, Eastern, East Midlands, Midlands, South Wales, Yorkshire, North Western and South of Scotland, London, South Western, Merseyside & North Wales, North Eastern and North of Scotland offer only a two-part tariff.

It will be appreciated that the 1988 copy of the *Electricity Supply Handbook*, from which all the tariff figures are taken gives, in fact, the situation existing in 1987. For up-to-date figures the reader must approach the relevant electricity authority. Eastern Electricity has issued a completely new arrangement of tariffs (from 1 April 1988) and a summary is given here. There are three domestic tariffs D1 – domestic general; D2 – domestic Economy 7; and D3– domestic night and day. Five different small supplies tariffs (SS1-SS5) replace the rates of the combined premises, farm and block tariffs. Maximum power required must not exceed 40 kVA with certain exclusions. Some of these tariffs incorporate Economy 7, evening weekend, and night and day facilities. Three restricted hours (RH1-RH3) and three MD tariffs (MD1, MD2 and ST1) complete the range of tariffs available.

The commercial block tariff of Eastern Electricity has been changed in April 1988 to a simple 1-block system with three separate tariffs. It is typical of what has been or will be, introduced by the other distribution companies.

SS1 tariff. Standing charge of 53p per week. First 77 units per

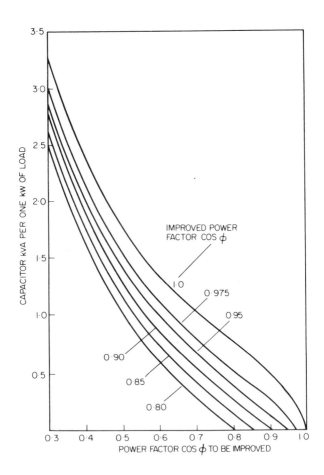

Figure 11.1   Graphical means of determining kVAr required when raising power factor from one value to a higher value.

week unit charge is 7.51p; remainder at 5.61p per unit. Covers both day and night.

SS2 tariff. Standing charge of 84p per week. First 77 units per week unit charge 7.51p; remainder at 5.61p per week. There is also a 7h night unit charge of 2.70p.

SS3 tariff. Standing charge of 72p per week. First 77 units per week unit charge is 8.66p (from 7 a.m. to 7 p.m.); remainder at 7.61p. Other times the unit charge is 3.64p and there is a night price (7h) of 2.07p per unit.

If the property is part house and part commercial the first block of 77 units is reduced to 38 units.

The commercial block tariff of the Eastern Electricity is of 3-block form for loads up to 40 kVA, the assessment being on the maximum power required. There is a fixed charge of 49p per week with the first 20 units at 9.12p for the first 5 kVA and the next 13 units at the same rate for the next 5 kVA. Secondary units rate is 5.76p for the first 58 units of the first 5 kVA with 41 units for the extra 5 kVA. Final unit charge is 4.23p.

Farm tariffs are generally of the 2-block or 2-part form. South Eastern Electricity does not offer consumers a special farm tariff, the basis of charges being the same as for the commercial block system.

Fixed simple flat-rate charges are still in use for small consumers in some areas of the country.

POWER FACTOR CORRECTION

Many tariff charges encourage the user to maintain a high power factor (nearly unity) in his electrical network by penalising a low power factor. The power factor can be improved by installing power factor correction equipment, the capital cost of which is often recovered in a few years by the savings made in reduced electricity bills.

Low power factors are caused mainly by induction motors and fluorescent lights and compensation may be applied to individual pieces of equipment, in stages by automatic switching, or in bulk at the supply intake position. Advice on the most economic system for a given installation is available from specialist firms.

There are a number of way in which power factor correction can be provided and these are described below.

By capacitor. The kVA required for power factor correction will be found by reference to Figure 11.2. The load current is represented by OI_L lagging by angle ϕ_1, such that $\cos\phi_1$ is the power factor of the load.

Assuming that it is desired to improve the power factor to $\cos\phi_2$ by means of capacitors, the resultant current must be represented by OI_R in Figure 11.2. The method employed is the constant kW one.

To obtain this amount of correction the capacitor current of OI_C must equal $I_L - I_R$, and this value will be given by $OI_C = OI_L \sin\phi_1 - OI_R \sin\phi_2$.

The vector diagram is drawn for current, but is also applicable to kVA since the current is directly proportional to the kVA. Thus OI_L, OI_C and OI_R can be taken to represent the kVA of the load, the capacitor and the resultant kVA respectively.

In this case the initial conditions would be:

$$\cos\phi_1 = \frac{kW}{kVA_L}$$

$$\tan\phi_1 = \frac{kVAr_L}{kW}$$

$$kVAr_L = kW\tan\phi_1$$

The improved conditions would be:

$$\cos\phi_2 = \frac{kW}{kVAr_R}$$

$$\tan\phi_2 = \frac{kVAr_R}{kW}$$

$$\therefore kVAr_R = kW\tan\phi_2$$

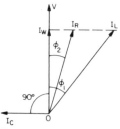

OI_L load current
OI_C capacitor current
OI_R resultant current
$\cos\phi_1$ load power factor
$\cos\phi_2$ final power factor
OI_W energy component
$OI_C = OI_L \sin\phi_1 - OI_R \sin\phi_2$

Figure 11.2   Diagram for capacitors

214

Capacitor kVAr required to improve factor from $\cos \phi_1$ to $\cos \phi_2$

$$= (kVAr_L - kVAr_R)$$
$$= kW(\tan \phi_1 - \tan \phi_2).$$

Actual capacity required. It may be necessary to transform capacitor kVA to microfarad capacity and the following relationship shows how this should be done.

Single phase. Current in capacitor is given by

$I_C = 2\pi f CV$
I_C = current in amperes
f = frequency
C = rating of capacitor in farads
V = voltage
(*Note.* 1 farad = $10^6 \mu f.$)

Three phase. The total line current taken by three capacitors in delta as shown in Figure 11.4 is given by

line current = $\sqrt{3}$ phase current in each capacitor.
Total line current = $\sqrt{3}(2\pi f CV)$.

The kVA is $\sqrt{3} \ VI \times 10^{-3}$ so that the kVA is given by

$$kVA = \frac{3(2\pi f CV^2)}{1000}$$

The C used in the above formula is the rating of one of the three capacitors forming the delta and so the total rating is $3C$. This gives us the formula:

Rating of each capacitor = $C = \dfrac{kVA \times 1000}{3(2\pi f. V^2)}$

$\therefore$ Total rating = $3C = \dfrac{kVA \times 1000}{2\pi f. V^2}$ F

Synchronous motor correction. A synchronous motor can be made to take a leading current (a current at leading power factor) by over-exciting it, and in so doing can be used to provide power factor correction.

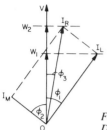

OI_L Main load current
OI_M Synch. motor current
OI_R Resultant current
$\cos \phi_1$ Load power factor
$\cos \phi_2$ Synch. motor power factor
$\cos \phi_3$ Final power factor
$OW_1 \propto$ original load, kW
$OW \propto$ final load, kW

Figure 11.3
Diagram for synchronous motor

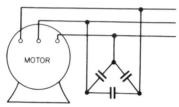

Figure 11.4   Capacitors connected in delta for three phase

Referring to Figure 11.3 the current required for the synchrnous motor cannot always be fixed by the desired amount of power factor correction, as in this case it is driving a load and the actual current will be fixed by the load on the synchronous motor and the power factor at which it is working.

It is impracticable to give formulae for working out these values as it is much better to start with the possible main load and the variable load which can be used for the synchronous motor.

Referring to the vector diagram, if values are taken either for currents as shown in the vector diagram or their proportionate kVA, their resultant current or kVA can be obtained as follows:

$$OI_R = \sqrt{[(OI_L \cos \phi_1 + OI_M \cos \phi_2)^2 + (OI_L \sin \phi_1 - OI_M \sin \phi_2)^2]}$$

Resultant power factor can be obtained from

$$\tan \phi_3 = \frac{OI_L \sin \phi_1 - OI_M \sin \phi_2}{OI_L \cos \phi_1 + OI_M \cos \phi_2}$$

216

Table 11.1 Wattless and power components for
various power factors

Power factor $\cos \phi$	Angle (degrees)	Per kVA		Per kW	
		Power	Wattless	kVA	Wattless component
1.0	0	1.0	0	1.0	0
0.98	11.48	0.98	0.20	1.02	0.20
0.96	16.26	0.96	0.28	1.04	0.29
0.94	19.95	0.94	0.34	1.06	0.36
0.92	23.07	0.92	0.39	1.09	0.43
0.90	25.83	0.90	0.44	1.11	0.48
0.88	28.37	0.88	0.48	1.14	0.54
0.86	30.68	0.86	0.51	1.16	0.59
0.84	32.87	0.84	0.54	1.19	0.65
0.82	34.92	0.82	0.57	1.22	0.70
0.80	36.87	0.80	0.60	1.25	0.75
0.78	38.73	0.78	0.63	1.28	0.80
0.76	40.53	0.76	0.65	1.32	0.86
0.74	42.27	0.74	0.67	1.35	0.91
0.72	43.95	0.72	0.69	1.39	0.96
0.70	45.57	0.70	0.71	1.43	1.02
0.68	47.15	0.68	0.73	1.47	1.08
0.66	48.70	0.66	0.75	1.52	1.14
0.64	50.20	0.64	0.77	1.56	1.20
0.62	51.68	0.62	0.78	1.61	1.27
0.60	53.13	0.60	0.80	1.67	1.33
0.58	54.55	0.58	0.82	1.72	1.40
0.56	55.93	0.56	0.83	1.79	1.48
0.54	57.32	0.54	0.84	1.85	1.56
0.52	58.66	0.52	0.85	1.92	1.64
0.50	60	0.50	0.87	2.00	1.73

If in any given case there is a fixed main load at a stated power factor plus a given kW load for the synchronous motor it is advisable to calculate the resultant power factor by working this out for various leading power factors for the synchronous motor.

It should be borne in mind that synchronous and synchronous induction motors will not work satisfactorily at a very low power factor. Values between 0.6 and 0.9 leading are usually taken for satisfactory results.

12

Requirements for electrical installations (BS 7671)

**IEE WIRING REGULATIONS
(SIXTEENTH EDITION)**

This Pocket Book is based on BS 7671: 1992, the amended form of
the sixteenth edition of the IEE Wiring Regulations. It takes into
account all the changes contained in Amendment No. 1 (issued in
December 1994). BS 7671 should be referred to in matters related
to electrical installations but it is in order to refer to the IEE Wiring
Regulations, or the Regulations, as these terms for a sub-title to the
British Standard. Throughout this chapter therefore references to
the IEE Wiring Regulations are taken to mean BS 7671:1992.
Installations which are designed after 1 July 1995 should take
account of Amendment No. 1 of the Standard. We also extract
information from the Guidance Notes listed on page 278. The
issuing of the first amendment to BS 7671:1992 coincides with
the change in the nominal value of the supply voltages as
follows: single-phase will reduce from 240 V+6% to 230 V
+ 10%–6%; three-phase supplies will change from 415 V+6% to
400 V+10%–6%. Domestic supplies in the UK will stay at 240 V
since this value falls within the new limits.

In the new edition account has been taken of the technical
substance of the parts of IEE Publication 364 so far published and of
the corresponding agreements reached in CENELEC. In addition it
also takes into account the following CENELEC Harmonization
Documents:

Cenelec Harmonization Document Reference		Part of the Regulation
HD 193	Voltage bands	Part 1 and Definitions
HD 308	Identification and use of cores of flexible cables	Part 5

Where the CENELEC work is still in the course of preparation the corresponding parts of this edition are based on the IEC documents. The Regulations will be amended from time to time to take account of further progress of the international work and other developments, the arrangement of parts, chapters, and sections being intended to facilitate this. The opportunity has also been taken to revise certain regulations for greater clarity or to take account of technical developments.

Considerable reference is made throughout the Regulations to publications of the British Standards Institution, both specifications and codes of practice. Appendix 1 in the Regulations lists these publications and gives their full titles whereas throughout the Regulations they are referred to only by their numbers. Nearly $8\frac{1}{2}$ pages are included involving 66 different British Standards. Where they appear in the Regulations is also noted. Where a reference is made to a British Standard in the Regulations, and the British Standard concerned takes account of a CENELEC Harmonization Document, it is understood that the reference is to be read as relating also to any foreign standard similarly based on that Harmonization Document, provided it is verified that any differences between the two standards would not result in a lesser degree of safety than that achieved by compliance with the British Standard (see Section 511 of the Regulations).

A similar verification should be made in the case of a foreign standard based on an IEC standard but as national differences are not required to be listed in such standards, special care should be exercised.

Notes on layout of 16th edition.　In the numbering system used in the 16th edition (which is quite different to that of the previous edition), the first digit signifies a Part, the second a Chapter, the third a Section, and the subsequent digits the regulation number. For example, the Section number 413 is made up as follows:

Part 4 – Protection for safety
Chapter 41 (first chapter of Part 4) – Protection against electric shock
Section 413 (third section of Chapter 41) – Protection against indirect contact

Part 1 sets out fundamental requirements for safety that are applicable to all installations.

Part 2 defines the sense in which certain terms are used throughout the Regulations.

The subjects of the subsequent parts are as indicated below:

Part No.	Subject
3	Identification of the characteristics of the installation that will need to be taken into account in choosing and applying the requirements of the subsequent Parts. These characteristics may vary from one part of an installation to another, and should be assessed for each location to be served by the installation.
4	Description of the basic measures that are available, for the protection of persons, property and livestock and against the hazards that may arise from the use of electricity.
	Chapters 41 to 46 each deal with a particular hazard. Chapter 47 deals in more detail with, and qualifies, the practical application of the basic protective measures, and is divided into Sections whose numbering corresponds to the numbering of the preceding chapters; thus Section 471 needs to be read in conjunction with Chapter 41, Section 473 with Chapter 43, and Section 476 with Chapter 46.
5	Precautions to be taken in the selection and erection of the equipment of the installation.
	Chapter 51 relates to equipment generally and Chapters 52 and 56 to particular types of equipment.
6	Special installations or locations – particular requirements.
7	Inspection and testing.

The sequence of the plan should be followed in considering the application of any particular requirement of the Regulations. The general index provides a ready reference to particular Regulations by subject, but in applying any one Regulation the requirements of related Regulations should be borne in mind. Cross-references are provided and the index is arranged to facilitate this.

In many cases a group of associated Regulations is covered by a side heading which is identified by a two-part number, e.g. 547–03. Throughout the Regulations where reference is made to such a two-part number, that reference is to be taken to include all the individual Regulation numbers which are covered by that side heading and include that two-part number.

Scope of Regulations. The Regulations apply to the design, selection, erection, inspection and testing of electrical installations other than those excluded by Regulation 110–02. This Regulation includes 'supplier's works', railway traction equipment, equipment of motor vehicles except caravans, ships, aircraft, mobile and fixed offshore, mines and quarries, radio interference suppression equipment except where it affects the safety of the installation, lightning protection of buildings covered by BS 6651 and those aspects of lift installations covered by BS 5655.

In some cases the Regulations may need to be supplemented by requirements or recommendations of British Standards or of the person ordering the work. Installations falling into this category include emergency lighting to BS 5266, installations in explosive atmospheres to BS 5345 and fire detection and alarm systems in buildings to BS 5839. Other cases include installations subject to the Telecommunications Act 1984, BS 6701 Part 1 and electric surface heating systems to BS 6351. The Regulations do not apply to ten different types of installations and these are listed in BS 7671. They include railway traction equipment, installations on ship, and on mobile and fixed offshore installations.

Voltage ranges. Installations operating at the following levels are covered.

(1) Extra-low voltage – normally not exceeding 50 V a.c. or 120 V ripple free d.c. whether between conductors or to earth.

(2) Low voltage – normally exceeding extra-low voltage but not exceeding 1,000 V a.c. or 1,500 V d.c. between conductors, or 600 V a.c. or 900 V d.c. between conductors and earth.

Equipment. The Regulations apply to items of electrical equipment only so far as selection and application of the equipment in the installation are concerned. They do not deal with requirements for the construction of prefabricated assemblies of electrical equipment, which are required to comply with appropriate specifications.

Contents of Regulations

Statutory regulations and associated memoranda.   In Great Britain the following classes of electrical installations are required to comply with the statutory regulations indicated below. The regulations listed represent the principal legal requirements. Information concerning these regulations may be obtained from the appropriate authority also indicated below.

Provisions relating to electrical installations are also to be found in other legislation relating to particular activities.

(i)	Electricity suppliers, installations generally, subject to certain exemptions.	Electricity Supply Regulations 1988 (as amended).	President of the Board of Trade, Secretary of State for Trade and Industry and Secretary of State for Scotland.
(ii)	Building generally (for Scotland only), subject to certain exemptions.	Buildings Standards (Scotland) Regulations 1990.	Secretary of State for Scotland.
(iii)	Work activity. Places of work. Non-domestic installations.	Electricity at Work Regulations 1989.	Health and Safety Commission.
(iv)	Cinematograph installations.	Cinematograph Regulations 1955, made under the Cinematograph Act, 1952.	The Secretary of State for the Home Department, and Secretary of State for Scotland.

| (v) | Agricultural and horticultural installations. | Agricultural (Stationary Machinery) Regulations 1959 as amended. | Health and Safety Commission. |

Two new additions are (vi) Theatres and other places of entertainment and (vii) High voltage luminous tube signs.

Failure to comply, in a consumer's installation in Great Britain, with the requirements of Chapter 13 of BS 7671 Requirements for Electrical Installations (the IEE Wiring Regulations) places the supplier in the position of not being compelled to commence or, in certain circumstances, to continue to give, a supply of energy to that installation.

Under Regulation 29 of the Electricity Supply Regulations 1988 (as amended), any difference which may arise between a consumer and the supplier having reference to the consumer's installation shall be determined by a person nominated by the Secretary of State in the application of the consumer or consumer's authorised agent or the supplier.

Where it is intended to use Protective Multiple Earthing the supplier and the consumer must comply with the Electricity Supply Regulations 1988 as amended.

For further guidance on the application of some other of The Electricity at Work Regulations reference may be made to the following publication:

(i) Memorandum of Guidance on the Electricity at Work Regulations 1989. (HS(R)25) ISBN 011 8839632

For installations in potentially explosive atmospheres reference should be made to:

(i) The Electricity at Work Regulations 1989 and HSE guidance booklet HS(G)22 'Electrical apparatus for use in potentially explosive atmospheres' ISBN 011 883746X

(ii) The Highly Flammable Liquids and Liquified Petroleum-Gases Regulations 1972

(iii) The Petroleum (Consolidation) Act 1928

(iv) relevant British Standards.

Under the Petroleum (Consolidation) Act local authorities are empowered to grant licences in respect of premises where petroleum spirit is stored and as the authorities may attach such conditions as they think fit, the requirements may vary from one local authority to another. Guidance may be obtained from the Health and Safety Executive (Guidance Note HS(G)41. Petrol filling stations: Instructions and Operation).

For installations in theatres and other places of public entertainment, and on caravan sites, the requirements of the licensing authority should be ascertained. Model Standards were issued by the Department of Environment in 1977 under the Caravan Sites and Control of Development Act 1960 as guidance for local authorities.

The Electrical Equipment (Safety) Regulations, administered by the Department of Trade and Industry, Consumer Safety Unit, contain requirements for safety of equipment designed or suitable for general use. Information on the application of the Regulations is given in guidance issued by the Department of Trade and Industry.

The Plugs and Sockets, etc. (Safety) Regulations 1994 (SI 1994/1768 ISBN 0 11 044768 9) of the Consumer Safety Act 1978, administered by the Department of Trade and Industry, containing requirements for the safety of plugs, sockets, adaptors and fuse links, etc. designed for use at a voltage of not less than 50 V.

Where a pictographic safety sign is used for a caution of risk of electric shock, the Safety Signs Regulations (SI 1980 No. 1471), administered by the Health and Safety Executive, are applicable.

The Electrical Appliance (Colour Code) Regulations 1969 (SI 310) makes requirements for the colour coding of flexible cables and flexible cords to electrical appliances.

The Management of Health and Safety at Work Regulations 1992 implements European Directives 89/391/EEC and 91/383/EEC and require employers and self-employed persons to assess risks to workers and others who may be affected by their undertaking. (An Approved Code of Practice made under Section 16(1) of the HSW Act 1974).

Provision and Use of Work Equipment Regulations 1992 (SI 2932) implements European Directive 89/655/EEC and requires employers to ensure that all work equipment is suitable for the purpose for which it is used, is properly maintained, and appropriate training is given (see HSE Guidance Booklet L22). The Electromagnetic Compatibility Regulations 1992 (SI 1992 No 2372) provide requirements for electrical and electronic products for electromagnetic compatibility. Other Regulations relevant to electrical installation include:

The Personal Protective Equipment at Work Regulations 1992 (European Directive 89/656/EEC. HSE Booklet L25); The Workplace (Health, Safety and Welfare) Regulations 1992 (European Directive 89/654/EEC. HSE Booklet L24); The Manual Handling Operations Regulations 1992 (European Directive 90/269/EEC, HSE Booklet L23).

DEFINITIONS

These Regulations include a number of definitions and some of these are included here. The well known definitions, familiar to electrical contractors, are not reproduced. These definitions indicate the sense in which the terms defined are used in the Regulations. Some of the definitions are in line with those given in BS 4727 'Glossary of electrotechnical, power, telecommunications, electronics, lighting and colour terms'. Other terms that are not defined in the Regulations are used in the sense defined in that British Standard.

Arm's reach. Zone of accessibility to touch, extending from any point on a surface where persons usually stand or move about, to the limits a person can reach with a hand in any direction without assistance. Three diagrams in the Regulations illustrate the 'zone of accessibility.'

Barrier. A part providing a defined degree of protection against contact with live parts, from any usual direction of access.

Basic insulation. Insulation applied to live parts to provide basic protection against electric shock. Basic insulation does not necessarily include insulation used exclusively for functional purposes.

Bonding conductor. A protective conductor providing equipotential bonding.

Cable ducting. A manufactured enclosure of metal or insulating material other than conduit or cable trunking, intended for the protection of cables which are drawn-in after erection of the ducting.

Circuit. An assembly of electrical equipment supplied from the same origin and protected against overcurrent by the same protective device(s). For the purposes of Chapter 52 of these Regulations certain types of circuit are categorised as follows:

Category 1 circuit. A circuit (other than a fire alarm or emergency lighting) operating at low voltage and supplied directly from a mains supply system.

Category 2 circuit. With the exception of fire alarm and emergency lighting circuits, any circuit for telecommunication (e.g. radio, telephone, sound distribution, intruder alarm, bell and call, and data transmission circuits) which is supplied at extra-low voltage.

Category 3 circuit. A fire alarm circuit or an emergency lighting circuit.

Circuit protective conductor. (CPC) A protective conductor connecting exposed conductive parts of equipment to the main earthing terminal.

Class I equipment. Equipment in which protection against electric shock does not rely on basic insulation only, but which includes means for the connection of exposed conductive parts to a

227

protective conductor in the fixed wiring of the insulation. For information on classification of equipment with regard to the means provided for protection against electric shock see BS 2754.

Class II equipment.   Equipment in which protection against electric shock does not rely on basic insulation only but in which additional safety precautions such as supplementary insulation are provided, there being no provision for the connection of exposed metal work of the equipment to a protective conductor, and no reliance upon precautions to be taken in the fixed wiring of the installations. (See BS 2754.)

Class III equipment.   Equipment in which protection against electric shock relies on supply at SELV and in which voltages higher than those of SELV are not generated (see BS 2754).

Double insulation.   Insulation comprising both basic insulation and supplementary insulation.

Earthed concentric wiring.   A wiring system in which one or more insulated conductors are completely surrounded throughout their length by a conductor, for example a metallic sheath, which acts as a PEN conductor.

Electrical installation (abbreviated *installation*).   An assembly of associated electrical equipment to fulfil a specific purpose and having certain co-ordinated characteristics.

Extraneous-conductive-part.   A conductive part liable to introduce a potential, generally earth potential and not forming part of the electrical installation.

Final circuit.   A circuit connected directly to current-using equipment or to a socket-outlet or socket-outlets or other outlet points for the connection of such equipment.

Isolation.   A function intended to cut off for reasons of safety the supply from all, or a discrete section, of the installation by separating the installation or section from every source of electrical energy.

Neutral conductor.   A conductor connected to the neutral point of a system and contributing to the transmission of electrical energy. The term also means the equivalent conductor of an IT or d.c. system unless otherwise specified in the Regulations.

PEN conductor.   A conductor combining the functions of both neutral conductor and protective conductor.

Protective conductor.   A conductor used for some measures of protection against electric shock and intended for connecting together any of the following parts: exposed conductive parts; extraneous conductive parts; the main earthing terminal; earth electrode(s); the earthed point of the source, or an artificial neutral.

A diagram in the Regulations (page 13) shows an example of earthing arrangements and protective conductors.

Reinforced insulation. Single insulation applied to live parts, which provides a degree of protection against electric shock equivalent to double insulation under the conditions specified in the relevant standard. The term 'single insulation' does not imply that the insulation must be one homogeneous piece. It may comprise several layers which cannot be tested singly as supplementary or basic insulation.

Residual current device. A mechanical switching device or association of devices intended to cause the opening of the contacts when the residual current attains a given value under specified conditions.

Residual operating current. Residual current which causes the residual current device to operate under specified conditions.

Ring final circuit. A final circuit arranged in the form of a ring and connected to a single point of supply.

Simultaneously accessible parts. Conductors or conductive parts that can be touched simnultaneously by a person or, where applicable, by livestock. Simultaneously accessible parts may be live parts, exposed conductive parts, extraneous conductive parts, protective conductors or earth electrodes.

Skilled person. A person with technical knowledge or sufficient experience to enable him to avoid dangers which electricity may create.

Stationary equipment. Equipment which is either fixed, or equipment having a mass exceeding 18 kg and not provided with a carrying handle.

Supplementary insulation. Independent insulation applied in addition to basic insulation in order to provide protection against electric shock in the event of a failure of basic insulation.

Switch. A mechanical device capable of making, carrying and breaking current under normal circuit conditions, which may include specified operating overload conditions, and also of carrying for a specified time currents under specified abnormal circuit conditions such as those of short-circuit. A switch may be also capable of making, but not breaking, short-circuit currents.

Switchgear. An assembly of main and auxiliary switching apparatus for operation, regulation, protection or other control of electrical installations.

System. An electrical system consisting of a single source of electrical energy and an installation. For certain purposes of the Regulations types of system are identified as follows, depending upon the relationship of the source, and of exposed conductive parts of the installation, to earth:

TN system. A system having one or more points of the source of energy directly earthed, the exposed conductive parts of the installation being connected to that point by protected conductors. Three types of TN system are recognised, as follows:

TN–S a system having separate neutral and protective conductors throughout the system (Figure 12.1).

TN–C a system in which neutral and protective functions are combined in a single conductor throughout the system (Figure 12.2).

TN-C–S a system in which neutral and protective functions are combined in a single conductor in part of the system (Figure 12.3).

TT system. A system having one point of the source of energy directly earthed, the exposed conductive parts of the installation being connected to earth electrodes electrically independent of the earth electrodes of the source (Figure 12.4).

IT system. A system having no direct connection between live parts or earth, the exposed conductive parts of the electrical installation being earthed (Figure 12.5).

Trunking (for cables). A manufactured enclosure for the protection of cables, normally of rectangular cross-section of which one side is removable or hinged.

PART 3. ASSESSMENT OF GENERAL CHARACTERISTICS

Every installation shall be divided into circuits to avoid danger and minimize inconvenience in the event of a fault, and facilitate safe operation, inspection, testing and maintenance (Reg. 314–01–01).

Where an installation comprises more than one final circuit, each final circuit shall be connected to a separate way in a distribution board. The wiring of each final circuit must be electrically separate from that of every other final circuit to prevent indirect energisation of a final circuit intended to be isolated (Reg. 314–01–04).

A requirement of the 16th edition is that of making an assessment of the frequency and quality of maintenance an installation will be expected to receive during its intended life. This implies that any periodic inspection, testing, maintenance and repairs likely to be necessary during the installation's intended life can be readily and safely carried out. Protective measures for safety must also be

230

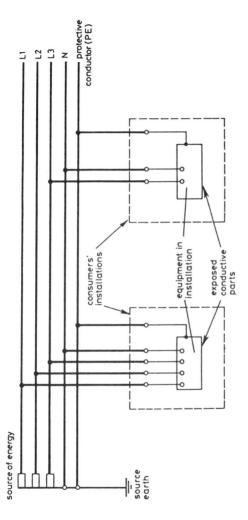

Figure 12.1 TN–S system. Separate neutral and protective conductors throughout the system. The protective conductor (PE) is the metallic covering of the cable supplying the installations or a separate conductor. All exposed conductive parts of an installation are connected to this protective conductor via the main earthing terminal of the installation. (Reproduced, by permission, from the 16th edition of the IEE Wiring Regulations).

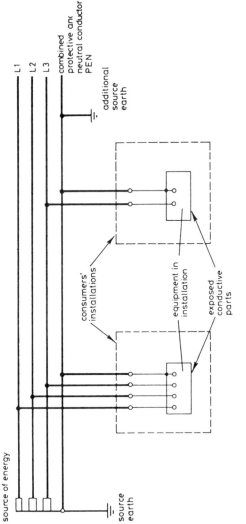

232

Figure 12.2 TN-C system. Neutral and protective functions combined in a single conductor throughout system. All exposed conductive parts of an installation are connected to the PEN conductor. An example of the TN-C arrangement is earthed concentric wiring but where it is intended to use this, special authorisation must be obtained from the appropriate authority. (Reproduced, by permission, from the 16th edition of the IEE Wiring Regulations.)

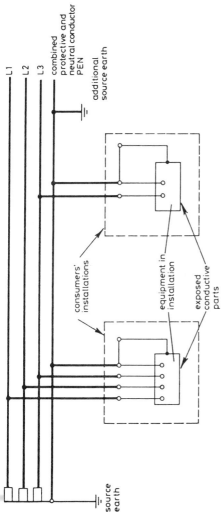

Figure 12.3 TN-C-S system. Neutral and protective functions combined in a single conductor in a part of the system. The usual form of a TN-C-S system is as shown, where the supply is TN-C and the arrangement in the installations is TN-S. This type of distribution is known also as Protective Multiple Earthing and the PEN conductor is referred to as the combined neutral and earth (CNE) conductor. The supply system PEN conductor is earthed at several points and an earth electrode may be necessary at or near a consumer's installation. All exposed conductive parts of an installation are connected to the PEN conductor via the main earthing terminal and the neutral terminal, these terminals being linked together. (Reproduced, by permission, from the 16th edition of the IEE Wiring Regulations).

233

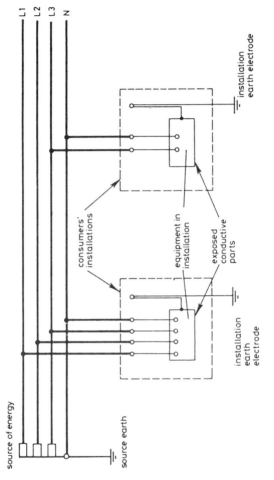

Figure 12.4 *TT system. All exposed conductive parts of an installation are connected to an earth electrode which is electrically independent of the source earth. (Reproduced, by permission, from the*

234

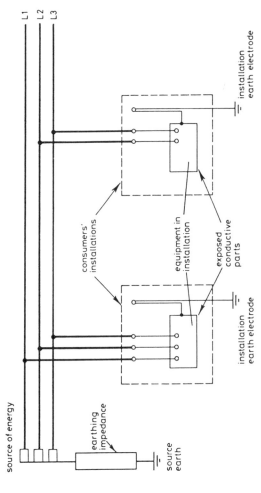

Figure 12.5 *IT system. All exposed conductive parts of an installation are connected to an earth electrode. The source is either connected to earth through a deliberately introduced earthing impedance or is isolated from earth. (Reproduced, by permission, from the 16th edition of the IEE Wiring Regulations).*

235

effective and the reliability of equipment for proper functioning must
be appropriate (Reg. 341–01–01).

PART 4. PROTECTION FOR SAFETY

Every installation, either as a whole or in its several parts, has to
comply with the requirements of Part 4 by applying the protective
measures outlined in Chapters 41 to 46 in the manner described in
Chapter 47.

Protection against direct and indirect contact shall be provided by
one of the following methods;

(i) Protection by safety extra-low voltage (Regulations 411–02 and
471–02)

(ii) Protection by limitation of discharge of energy (Regulation
411–04 and 471–03)

Functional extra-low voltage alone shall not be used as a
protective measure (see Regulation 471–14).

The source to provide the safety extra-low voltage shall be one of
the following: (a) a safety isolating transformer complying with
BS 3535, in which there shall be no connection between the
output winding and the body or the protective earthing circuit
if any; or (b) a source of current such as a motor generator
with windings providing electrical separation equivalent to that
of the safety isolating transformer specified in (a) above or (c)
an electrochemical source, i.e. a battery, or another source
independent of a higher circuit voltage, i.e. a diesel-driven
generator; or (d) electronic devices complying with appropriate
standards where measures have been taken so that even in the
case of an internal fault the voltages at the outgoing terminals
cannot exceed the value specified by Regulations 411–02–01. A
higher voltage at the output terminals is permitted under
specified circumstances.

Live parts of safety extra-low voltage equipment, other than
cables, shall be electrically separate from those of higher voltage,
neither should they be connected to earth or to a live part or a
protective conductor forming part of another system. (Reg. 411–
02–05) Circuit conductors in these installations shall preferably be
physically separated from those of any other circuit. There are
provisions made where this is impracticable (Reg. 411–02–06). Plugs
shall not be capable of being mated with socket-outlets of other
voltage systems in use in the same premises; similarly socket-outlets
must exclude plugs of other voltage systems. Socket-outlets shall not
have a protective conductor contact (Reg. 411–02–10).

If the nominal voltage of a SELV system exceeds 25 V a.c. r.m.s. 50 Hz or 60 V ripple-free d.c. additional protection against direct contact is specified (Reg. 411–02–09). Where reduced or low body resistance is to be expected or where the risk of electric shock is increased by contact with earth potential, (Regs 471–15, etc., as above) Regs 471–15 and 471–16 prescribe additional requirements. These conditions may be expected in situations where the hands and/or feet are likely to be wet or where the shock current path may not be through the extremities, or where a person is immersed in water or working in confined conductive locations.

Where extra-low voltage is used, but not all the requirements regarding safety extra-low voltage can be met, appropriate measures are outlined (Reg. 471–14) to provide protection but the systems employing these measures are termed 'functional extra-low voltage systems' or PELV systems (Reg. 471–14–02).

If the extra-low voltage system complies with the requirements of Reg. 411–02 for SELV except that the circuits are earthed at one point only, protection against direct contact shall be provided by either, (i) barriers or enclosures affording a degree of protection of at least IP2X or IPXXB, or (ii) insulation capable of withstanding a test voltage of 500 V d.c. for 60 seconds (Reg. 471–14–02–02–02).

If the extra-low voltage system does not generally comply with the safety requirements protection against direct contact shall be provided by barriers or enclosures or by insulation corresponding to the minimum test voltage for the primary circuit. Protection against indirect contact is also required (Regulation 471–14–03).

Circuits relying on protection against electric shock by limiting the discharge of energy, method (ii), shall be separated from other circuits in a manner similar to that specified in relation to safety extra-low voltage circuits (Reg. 411–02–05 and 411–02–06).

Direct contact. One or more of the following basic protective measures for protection against direct contact shall be used:

(i) Protection by insulation of live parts (Regulation 412–02 and Regulation 471–04).

(ii) Protection by a barrier or an enclosure (Regulation 412–03 and Regulation 471–05).

(iii) Protection by obstacles (Regulation 412–04 and Regulation 471–06).

(iv) Protection by placing out of reach (Regulation 412–05 and Regulation 471–07).

When protecting live parts from direct contact by insulation it must be such that it can only be removed by destruction and it must be able to withstand the electrical, mechanical, thermal and chemical stresses to which it may be subjected in service (Reg. 412–02–01). If a barrier or enclosure is employed to prevent direct contact and the opening is larger than that allowed for by IP2X (i.e. for replacement of part or for functioning purposes) suitable precautions shall be taken to prevent persons or livestock from unintentionally touching live parts. As far as is practicable persons should be made aware that live parts can be touched through the opening and should not be touched (see Reg. 471–05–02).

Protection can also be provided by obstacles, but they shall be secured in such a way as to prevent unintentional removal but may be removable without using a key or tool.

Indirect contact.   There are five basic protective measures against indirect contact specified in the Regulations and one or more should be adopted.

(i)   Earthed equipotential bonding and automatic disconnection of supply (Regulation 413–02 and Regulation 471–08).

(ii) Use of Class II equipment or equivalent insulation (Regulation 413–03 and Regulation 471–09).

(iii) Non-conducting location (Regulation 413–04 and Regulation 471–10).

(iv) Earth-free local equipotential bonding (Regulation 413–05 and Regulation 471–11).

(v)   Electrical separation (Regulation 413–06 and Regulation 471–12).

When employing method (i) the bonding conductors shall connect to the main earthing terminal of the installation extraneous conductive parts including: main water and gas pipes, other service pipes and ducting, central heating and air conditioning systems, exposed metallic parts of the building structure and the lightning protective system. Bonding to the metalwork of other services may require the permission of the undertakings responsible. Bonding to any metallic sheath of a telecommunication cable is necessary, however the consent of the owner shall be obtained. When an installation serves more than one building the requirements for earthed equipotential bonding apply to each building (Regulation 413–02–02).

The protective devices, earthing and relevant impedances of the circuits concerned shall be such that during an earth fault the voltages between simultaneously accessible exposed and extraneous conductive parts occurring anywhere in the installation shall be of such a magnitude and duration as not to cause danger (Reg. 413–02–04). Conventional means of compliance with this Regulation are given in Regulations 413–02–06 to 413–02–26 according to the type of system earthing, but other equally effective means shall not be excluded.

The Regulations provide comprehensive details of the methods of protection by earthed equipotential bonding and automatic disconnection of supply for the three types of earthing, i.e. TN, TT, and IT systems. In the UK the general practice is to employ the TN system of earthing in one of the three forms: (a) TN-C where the neutral and protective functions are combined in a single conductor (PEN) throughout the system; (b) TN-S where the neutral and protective conductors are separated throughout the system; and (c) TN-C-S where the two conductors functions are combined in a single conductor in a part of the system. System (c) is very common as it allows single-phase loads to be supplied by live and neutral with a completely separate earth system connecting together all the exposed conductive parts before connecting them to the PEN conductor via a main earthing terminal which is also connected to the neutral terminal. In this chapter therefore we will confine our remarks to the TN system of earthing only.

Section 413 is the relevant part of the Regulations and it contains a number of tables related to protection to which the reader is referred. Reproduction of any of them is not permitted by the IEE.

Two types of protective device are allowed, overcurrent or residual current (rcd) (Reg. 413–02–07). If a rcd is employed in a TN-C-S system, a PEN conductor must not be used on the load side. Connection of the protective conductor to the PEN conductor must be made on the source side of the rcd.

Regulation 413–02–04 (see above) is considered to be satisfied if the characteristic of each protective device and the earth fault loop impedance of each circuit protected by it are such that automatic disconnection of the supply occurs within a specified time when a fault of negligible impedance occurs between phase conductor and a protective conductor or an exposed conductive part anywhere in the installation. The requirement is met where $Z_s \leqslant U_o/I_a$,

and Z_s is the earth fault loop impedance;

I_a is the current causing the automatic operation of the disconnecting protective device within the time stated in table 41A (of the Regulations) as a function of the nominal voltage U_o or, under the conditions stated in Regulations 413–02–12 and 413–02–13, within a time not exceeding 5 s. U_o is the nominal voltage to earth.

The maximum disconnection times of Table 41A apply to a circuit supplying socket-outlets and to other final circuits which supply portable equipment intended for manual movement during use, or hand-held Class I equipment (Reg. 413–02–09). The requirement does not apply to a final circuit supplying an item of stationary equipment connected by means of a plug and socket-outlet where precautions are taken to prevent the use of the socket-outlet for supplying hand-held equipment, nor to the reduced low-voltage circuits described in Regulation 471–15.

Where a fuse is used to meet Regulation 413–02–09, maximum values of earth fault loop impedance (Z_s) corresponding to a disconnection time of 0.4 s are stated in Table 41 B1 (in the Regulations) for a nominal voltage to earth (U_o) of 230 V. For types and rated currents of general purpose (gG) fuses, other than those mentioned in Table 41 B1 (of the Regulations), and for motor circuit fuses (gM) the reader is referred to the appropriate British Standard, to determine the value of I_a for compliance with Regulation 413–02–08.

Table 41 B1 is in four parts and details the maximum earth fault loop impedance Z_s for fuses, for 0.4 s disconnection time with U_o of 230 V: (a) for fuse ratings from 6–50 A (gG fuses to BS 88 Parts 2 and 6); (b) fuse ratings from 5–45 A (fuses to BS 1361); (c) fuse ratings 5–45 A (fuses to BS 3036); and (d) 13 A fuses to BS 1362. Table 41 B2 gives similar details covering miniature circuit-breakers.

Irrespective of the value of U_o, for a final circuit which supplies a socket-outlet or portable equipment intended for manual moving during use, or hand-held Class I equipment, the disconnection time can be increased to a value not exceeding 5 s for the types and ratings of the overcurrent protective devices and associated maximum impedance of the circuit protective conductors shown in Table 41 C (of the Regulations). This table is in nine parts, the first four relating to the same types of fuses detailed in Table 41 B1; the other five relate to miniature circuit-breakers to EN 60898, i.e. types 1, 2, B3 and D (Reg. 413–02–12).

For a distribution circuit the disconnection time must not exceed 5 s (Reg. 413–02–13). When a fuse is used to satisfy this Regulation, Table 41D (in the Regulations) specifies the maximum values of earth fault loop impedance (Z_s) corresponding to a disconnection time of 5 s for a nominal voltage to earth of 230 V. This table again has four parts similar to Table 41 B1.

Regulation 413–02–16 covers the situation whereby a rcd is used; the following condition has to be fulfilled:

$$Z_s I_{\Delta n} \leqslant 50 \text{ V}$$

where Z_s is the earth fault loop impedance and $I_{\Delta n}$ is the residual operating current of the protective device in amperes. If the circuit protected by the rcd extends beyond the earthed equipotential zone, exposed conductive parts need not be connected to the TN system protective conductors if they are connected to an earth electrode of the correct resistance value appropriate to the rcd's operating current (Reg. 413–02–17).

Class II or equivalent.   Double or reinforced insulated equipment or low-voltage switchgear assemblies having total insulation (see EN 60439) provide suitable protection (Reg. 413–03–01). Two other systems are mentioned relating to supplementary insulation applied to electrical equipment with only basic insulation and reinforced insulation applied to uninsulated live parts. When the electrical equipment is ready for operation, all conductive parts separated from live parts by basic insulation only shall be contained in an insulating enclosure affording at least the degree of protection IP2X or IPXXB (Reg. 413–03–04).

Where a circuit supplies items of Class II equipment a circuit protective conductor shall be run to and terminated at each point in the wiring and at each accessory except a lampholder having no exposed conductive parts and suspended from such a point. (Reg. 471–09–02) This need not be observed where Regulation 471–09–03 applies. Exposed metalwork of Class II equipment should be mounted so that it is not in electrical contact with any part of the installation connected to a protective conductor. Such a contact might impair the Class II protection provided by the equipment specification.

Where a whole installation or circuit is intended to consist of Class II equipment (or equivalent) it must be verified that it is under effective supervision to ensure no change is made that would impair the effectiveness of the Class II (or equivalent) insulation. This measure cannot be applied to any circuits which include socket-outlets or where a user may change items of equipment without authorisation (Reg. 471–09–03). Certain cables cannot be described as being of Class II construction. If these cables (they are specified) are installed in accordance with Chapter 52 they are considered to provide satisfactory protection (Reg. 471–09–04).

Protection against overcurrent. Live conductors shall be protected by one or more devices for automatic interruption of the supply in the event of an overload or short-circuit except where the source is incapable of supplying a current exceeding the current-carrying capacity of the conductors (Reg. 431–01–01).

A device protecting a circuit against overload has to satisfy a number of conditions:

(i)   Its nominal current setting (I_n) shall not be less than the design current (I_b) of the circuit.

(ii)   Its I_n shall not exceed the lowest of the current carrying capacities (I_z) of any of the circuit conductors.

(iii)   The current assuring effective operation of the protective device (I_2) shall not exceed 1.45 times the lowest of the current-carrying capacities (I_z) of any of the conductors of the circuit.

When the same protective device is used with conductors in parallel the value of I_z is the sum of the current-carrying capacities of those conductors. The conductors have to be of the same construction, cross-sectional area, length and disposition, having no branch circuits throughout their length and be arranged so as to carry substantially equal currents unless the suitability of the particular arrangement is verified. (Reg. 433–03–01).

Protection against fault current. Devices for this duty shall break any fault current flowing in the conductors of each circuit before this current causes danger due to thermal and mechanical effects produced in the conductors and connections (Reg. 434–01–01). Except under the conditions outlined in the next sentence, the breaking capacity rating of each device must not be less than the prospective short-circuit current or earth fault current at the point at which the device is installed. A lower breaking capacity than that at the point at which the device(s) installed is permissible if it is backed up by another protective device(s), on the supply side, with the necessary breaking capacity. Co-ordination between the devices is important to ensure that no damage occurs to the protective device(s) on the load side of the conductors protected (Reg. 434–03–01).

Except for a ring final circuit where a single device protects two or more conductors in parallel against fault current the operating characteristics of the device and the characteristics of the parallel conductors as installed shall be suitably co-ordinated (Reg. 434–04–01).

Isolation and switching. Means shall be provided for non-automatic isolation and switching to prevent or remove hazards associated with the installation of electrically-powered equipment and machines (Reg. 460–01–01). A main linked switch or linked circuit-breaker must be provided as near as practicable to the origin of every installation as a means of switching the supply on load and as a means of isolation. For d.c. systems all poles shall be provided with a means of isolation. There are further provisions if an installation is supplied from more than one source (Reg. 460–01–02).

Means of switching off for mechanical maintenance are necessary where such maintenance may involve a risk of physical injury (Reg. 462–01–01). Precautions must also be taken to prevent any equipment being unintentionally or inadvertently re-activated (Reg. 462–01–03).

Where it is necessary to disconnect rapidly an installation from the supply emergency switching facilities shall be provided (Reg. 463–01–01). The means adopted shall be capable of cutting off the full load current of the relevant part of the installation, due account being taken of stalled motor conditions where relevant (Reg. 537–04–01).

Where isolating devices for particular circuits are placed remotely from the equipment to be isolated, provision shall be made so that the means of isolation can be secured in the open position. Where this provision takes the form of a lock or removable handle, the key or handle shall be non-interchangeable with any others used for a similar purpose within the installation (Reg. 476–02–02).

Every motor circuit shall be provided with a disconnector which disconnects the motor and all equipment including any automatic circuit-breaker (Reg. 476–02–03). There are also stringent requirements relating to electric discharge lighting installations using an open-circuit voltage exceeding low voltage (Reg. 476–02–04).

Circuits outside an equipotential bonding zone. When a circuit in an equipotential bonding zone is specifically intended to supply fixed equipment outside the zone, and that equipment may be touched by a person in contact directly with the general mass of earth, the earth fault loop impedance must be such that disconnection occurs within the time stated in Table 41 A (of the Regulations) (Reg. 471–08–03).

Socket-outlet circuits which can be expected to supply portable equipment for use outdoors shall comply with Reg. 471–16–01. This regulation refers to socket-outlets rated at 32 A or less and lays down specific requirements with some exemptions.

Where automatic disconnection as protection against indirect contact is used in an installation forming part of a TT system, every socket-outlet circuit shall be protected by a residual current device (Reg. 471–08–06 see also Reg. 413–02–16).

Special installations or locations. A completely new section has been introduced into the 16th edition as Part 6 with the title shown as the bold sidehead. It covers locations containing a bath tub or shower basin; swimming pools; locations containing a hot air sauna heater; construction site installations; electrical installations of agricultural and horticultural premises; restrictive conductive locations; earthing requirements – high earth leakage currents; electrical installations in caravans and motor caravans (Division one) and in caravan parks (Division two); and highway power supplies and street furniture. A section is reserved for marinas and there is a further unnamed section reserved for future use. Because we are highlighting protection in this part of the chapter we have introduced the relevant extracts from Part 6. More is said about this Part later on.

The particular requirements for these special installations or locations supplement or modify the general requirements contained in other Parts of the Regulations (Reg. 600–01). The absence of references to the exclusion of a Chapter, Section or Clause means that the corresponding general Regulations are applicable.

In this section we only highlight some of the Regulations relating to bath tubs, shower basins and their surroundings where the risk of electric shock is increased by a reduction in body resistance and contact of the body with earth potential. Such conditions apply to most domestic dwellings. Special requirements may be necessary for a location containing a bath for medical treatment.

Protection against electric shock. Installation of electrical equipment in a bath tub or shower basin is forbidden (Reg. 601–02–01). Where SELV (Regs 411–02 and 471–02) is used, the safety source must be installed out of reach of a person using the bath or shower and, notwithstanding the provision of the second paragraph of Regulation 411–02–09 the equipment shall incorporate protection

244

against direct contact by insulation (Reg. 412–02), capable of withstanding a test voltage of 500 V d.c. for one minute, or by barriers or enclosures (Reg. 601–03–01 and Reg. 412–03).

Protection against indirect contact. Except for SELV, for a circuit supplying equipment in a room containing a fixed bath or shower, where the equipment is simultaneously accessible with exposed conductive parts of other equipment or with extraneous conductive parts, the characteristics of the protective devices and the earthing arrangements shall be such that, in the event of a fault to earth, disconnection occurs within 0.4 s (Reg. 601–04–01). For these locations (except SELV) supplementary equipotential bonding has to be provided between simultaneously accessible exposed conductive parts of equipment, between exposed conductive parts and simultaneously accessible extraneous conductive parts, and between simultaneously accessible extraneous conductive parts (Reg. 601–04–02). Where electrical equipment is installed in the space below a bath, that space shall be accessible only by the use of a tool and, nevertheless, the requirement of Regulation 601–04–02 shall extend to the interior of that space.

There is also a Regulation against using certain measures to provide protection against direct contact which include obstacles and placing out of reach (Reg. 601–05–01). Similarly there are prohibitions against indirect contact, i.e. one cannot use a non-conducting location or earth-free local equipotential bonding (Reg. 601–06–01).

Neutral conductor. In TN or TT systems where the neutral conductor has a cross-section area at least equal or equivalent to that of the phase conductor it is not necessary to provide overcurrent detection or a disconnecting device for this conductor. Where the cross-sectional area is less than that of the phase conductors overcurrent detection for the neutral conductor shall be provided unless (i) it is protected against fault current by the protective device for the phase conductors of the circuit and (ii) the maximum current likely to be carried by the neutral conductor in normal service, is significantly less than the value of the current-carrying capacity of that conductor.

Where either or both of the conditions are not met overcurrent detection shall be provided for the neutral conductor, appropriate to its cross-sectional area. Operation of the protective device causes disconnection of the phase conductors but not necessarily the neutral conductor (Regs 473–03–03 and 473–03–04).

The cross-sectional area of the neutral conductor may be smaller than the phase conductors depending on the degree of sustained imbalance except in circuits where the load is predominantly due to discharge lighting (Regs 524–02–01, 524–02–02 and 524–02–03).

PART 5. SELECTION AND ERECTION OF EQUIPMENT

Wiring systems. Non-flexible or flexible cable or flexible cord on l.v. systems must comply with the appropriate British or harmonized Standard. Non-flexible cable sheathed with lead, PVC or an elastomeric material for aerial use may incorporate a catenary wire or include hard-drawn copper conductors. But the Regulation does not apply to flexible cord forming part of a portable appliance or luminaire where these are the subject of and comply with a British Standard, or to special flexible cables/cords for combined power and telecommunications wiring (Reg. 521–01–01). The same Regulation permits flexible cable/cord to incorporate metallic armour, braid or screen. A busbar trunking system must comply with EN 60570 (Reg. 521–01–02).

Due to electromagnetic effects single-core cables armoured with steel wire or tape must not be used in a.c. circuits. Conductors of a.c. circuits installed in ferromagnetic enclosures have to be arranged so that all phase conductors and the neutral conductor (if any) together with the appropriate protective conductor of each circuit are in the same enclosure. Where such conductors enter a ferrous enclosure they have to be arranged so that the conductors are not individually surrounded by a ferrous material, or some other provision has to be made to prevent eddy currents (Reg. 521–02–01).

Enclosures. Conduits and conduit fittings, trunking, ducting and fittings must comply with their appropriate British Standards, which are specified in the Regulations. Where BS 4678 does not apply, non-metallic trunking, ducting and their fittings shall be of insulating material complying with the ignitability characteristic 'p' of BS 476 Part 5. (Regs 521–04–01 and 521–05–01).

Cable derating. Provision for increasing the section of a cable shall be made if a run is to be installed where thermal insulation may cover it. If installed in a thermally insulating wall or above a

246

thermally insulated ceiling, with the cable being in contact on one side the current-carrying capacities are tabulated in Appendix 4 of the Regulations. If the cable is totally surrounded by thermally insulating material, derating factors have to be applied (Reg. 523–04–01).

Voltage drop.   Under normal service conditions the voltage drop in a consumer's installation must be such that the voltage at the terminals of any fixed current-using equipment must be greater than the lower limit corresponding to the British Standard relevant to the equipment. (Reg. 525–01–01). This Regulation is satisfied for a supply given in accordance with the Electricity Supply Regulations 1988 (as amended) if the voltage drop between the supply terminals and the fixed current-using equipment does not exceed 4% of the nominal supply voltage.

Environmental conditions.   The 16th edition contains a classification system relating to external influences developed for IEC Publication 364. This is at present insufficiently advanced for adoption as a basis for national regulations (see Chapter 32 of the Regulations). It is similar to, but not as extensive as, Appendix 6 in the 15th edition. The concise list of external influences now occupies a page as compared with six pages in the previous edition, but it now introduces the characteristic letters like AA, AD, etc. into the body of the text in Chapter 52. See also Appendix 5 of the Regulations.

Ambient temperature.   A wiring system has to be suitable for the highest and lowest temperature likely to be encountered (Reg. 522–01–01). Similarly when handling or installing components of a wiring system they must be only at temperatures within the limits specified (Reg. 522–01–02). Wiring systems should be protected from the effects of heat from external sources, including solar gain by one of the following means, or an equally effective method; shielding, placing sufficiently far from the source of heat, selecting a system with due regard for the additional temperature rise which may occur, reductions in current-carrying capacity or local reinforcement or substitution of insulating material (Reg. 522–02–01).

Presence of water or high humidity.   A wiring system has to be selected so that no damage can be caused by these two conditions. Where water may collect or condensation form in a wiring system, provision must be made for it to escape harmlessly. Where it may be subjected to waves protection has to be provided (Section 522–03).

Corrosive/polluting substances. Where the presence of these is likely to affect any parts of a wiring system, these parts must be suitably protected or made from materials resistant to such substances (Reg. 522–05–01).

Identification. The colour combination green and yellow is reserved exclusively for the identification of protective conductors. Where electrical conduits are required to be distinguished from pipelines of other services, orange (to BS 1710) shall be used (Regs 514–02–01 and 514–03–01).

Every single-core non-flexible cable and every core of non-flexible cable for use as fixed wiring shall be identifiable at its terminations and preferably throughout its length, by the appropriate method described in items (i) to (v) below:

(i)   For rubber- and PVC-insulated cables, the use of core colours is specified in Table 12.1, or the application at terminations of tapes, sleeves or discs of the appropriate colours in the table.

(ii)   For armoured PVC-insulated auxiliary cables, either (i) above or the use of numbered cores in accordance with BS 6346.

(iii) For paper-insulated cables, the use of numbered cores in accordance with BS 6480, with Nos. 1, 2, 3 signifying phase conductors, number 0 the neutral conductor and No. 4 for the fifth (special purpose) core if any.

(iv)   For cables with thermosetting insulation, the use of core colours as in Table 12.1 or numbered cores in accordance with BS 5467 with phase and neutral conductors numbered as in (iii) above.

(v)   For m.i. cables, the application at terminations of tapes, sleeves or discs of the appropriate colours specified in Table 12.1. Identification sleeves for cables should comply with BS 3858 where appropriate (Reg. 514–06–01).

Bare conductors should be identified in a similar manner by tapes, sleeves and discs or by painting with the appropriate colours specified in Table 12.1 (Reg. 514–06–03).

Flexible cable and cord shall be identifiable throughout its length as appropriate to its functions as indicated in Table 12.2. Such cables shall not use the following core colours: green alone; yellow alone; or any bi-colour other than the colour combination green and yellow (Regs 514–07–01 and 514–07–02).

Table 12.1   Colour identification of cores of non-flexible cables and bare conductors for fixed wiring

Function	Colour identification
Protective (including earthing) conductor	green-and-yellow
Phase of a.c. single-phase circuit	red†
Neutral of a.c. single- or three-phase circuit	black
Phase R of 3-phase a.c. circuit	red
Phase Y of 3-phase a.c. circuit	yellow
Phase B of 3-phase a.c. circuit	blue
Positive of d.c. 2-wire circuit	red
Negative of d.c. 2-wire circuit	black
Outer (positive or negative) of d.c. 2-wire circuit derived from 3-wire system	red
Positive of 3-wire d.c. circuit	red
Middle wire of 3-wire d.c. circuit	black*
Negative of 3-wire d.c. circuit	blue
Functional earth	cream

For armoured PVC insulated cables and paper-insulated cables, see Regulation 514-06-01 (ii) and (iii)

*Only the middle wire of three-wire circuits may be earthed
† As alternatives to the use of red, if desired, in large installations, on the supply side of the final distribution board, yellow and blue may also be used.

Table 12.2   Colour identification of cores of flexible cables and flexible cords

Number of cores	Function of core	Colour(s) of core
1	Phase	Brown (Note 1)
	Neutral	Blue
	Protective	Green-and-yellow
2	Phase	Brown
	Neutral	Blue (Note 2)
3	Phase	Brown (Note 3)
	Neutral	Blue (Note 2)
	Protective	Green-and-yellow
4 or 5	Phase	Brown or black (Note 4)
	Neutral	Blue (Note 2)
	Protective	Green-and-yellow

Notes to Table 12.2.
(1) Or any other colour not prohibited by Regulations 514–03–01 and 514–07–02, except blue.
(2) The blue core can be used for functions other than the neutral in circuits which do not incorporate a neutral conductor, in which case its function must be appropriately identified during installation; provided that the blue core is not in any event used as a protective conductor. If the blue core is used for other functions, the coding L1, L2, L3, or other coding where appropriate should be used.
(3) In three-core flexible cables or cords not incorporating a green and yellow core, a brown core and a black core may be used as phase conductors.
(4) Where an indication of phase rotation is desired, or it is desired to distinguish the functions or more than one phase core of the same colour, this must be by use of numbered or lettered (not coloured) sleeves to the core, preferably using the coding L1, L2, L3 or other coding where appropriate. (Reg. 514–07–01).

Detrimental influences.   Cables of Category 1 circuits shall not be drawn into the same conduit, duct or ducting as cables of Category 2 circuits unless the latter cables are insulated in accordance with the highest voltage present in the Category 1 circuits or the cables of Category 1 circuits are effectively partitioned from those of Category 2 (Reg. 528–01–05).

Fire alarm and emergency lighting circuits (Category 3 circuits) must be segregated from all other cables and from each other in accordance with BS 5266 and BS 5839. Telecommunication circuits have to be segregated in accordance with BS 6701 as appropriate. (Reg. 528–01–04)

250

Where Category 3 circuits are installed in a channel or trunking containing circuits of any other category, these circuits shall be segregated from the latter by continuous partitions so that the specified integrity of the Category 3 circuits is not reduced. These partitions shall also be provided at any common outlets in a trunking system accommodating Category 3 circuits and circuits of other categories. If m.i. cables, to BS 6387 are used for the Category 3 circuits these partitions are not normally required. (Reg. 528–01–06)

Bends. The internal radius of every bend in a non-flexible cable must be such as not to cause damage to the conductors and cable. (Reg. 522–08–03).

Switchgear. No fuse or, excepting where linked, switch or circuit-breaker shall be connected in the neutral conductor of TN or TT systems (Reg. 530–01–02).

When using a residual current device its operating current shall comply with Section 413 as appropriate to the type of system earthing.

In a TN system where for certain equipment in a certain part of the installation, one or more of the conditions in Regulation 413–02–08 cannot be satisfied, an rcd can be used as a protective device. There are certain other conditions which have to be fulfilled. (Reg. 531–03–01) where a single rcd is used to protect part of a TT installation, it must be placed at the origin of the installation, unless the part between the origin and the device complies with the requirements for protection by the use of Class II equipment or equivalent insulation. The Regulation applies to each origin if there is more than one (Reg. 531–04–01).

For an IT system where an rcd is used for protection and disconnection following a first fault is not envisaged, the non-operating residual current of the device has to be at least equal to the current which circulates on the first fault to earth of negligible impedance affecting a phase conductor (Reg. 531–05–01).

Earthing arrangements. The types of earth electrode recognised by the Regulations are: earth rods or pipes; earth tapes or wires; earth plates; underground structural metalwork embedded in foundations; welded metal reinforcement of concrete (except pre-stressed concrete) embedded in the earth; lead sheaths and other metallic coverings of cables not precluded by the Regulations; other suitable underground metalwork (Reg. 542–02–01). Certain precautions are specified when using some of these systems (Regs. 542–02–02 to 02–05).

Cross-sectional areas of earthing conductors where buried in the ground shall not be less than the figures in Table 54 A of the Regulations (Reg. 542–03–01). Every earthing conductor also has to comply with Section 543.

Regulations 542–03–02 and 547–01–01 banning the use of aluminium and copperclad aluminium for final connections to earth electrodes or for bonding connections have been deleted by Amendment No. 1.

Protective conductors. The cross-sectional area of every protective conductor, other than an equipotential bonding conductor shall be calculated in accordance with Reg. 543–01–03 or selected in accordance with Reg. 543–01–04. If this conductor is not an integral part of a cable, and is not formed by conduit, ducting or trunking, and is not contained in an enclosure formed by a wiring system, the cross-sectional area shall not be less than 2.5 mm² copper equivalent if protection against mechanical damage is provided and 4 mm² copper equivalent if mechanical protection is not provided. (Reg. 543–01–01). The reader is also referred to Regulation 543–03–01. For an earthing conductor buried in the ground Regulation 542–03–01 applies.

The formula for calculating the cross-sectional area is $S = \sqrt{(I^2 t)}/k$ mm², where S is cross-sectional area in mm², I is the value (a.c., r.m.s.) of the fault current for a fault of negligible impedance that can flow through the protective device in amperes (with certain provisions), t is the operating time of the disconnecting device in seconds corresponding to the fault current I amperes, and k is a factor taking account of the resistivity, temperature coefficient and heat capacity of the conductor material, and the appropriate initial and final temperatures. The values of k are tabulated in the Regulations in Tables 54B–F (Reg. 543–01–03).

When it is desired not to calculate the minimum cross-sectional area of a protective conductor the value may be selected in accordance with Table 12.3. Where the application of Table 12.3 produces a non-standard size, a conductor having the nearest larger standard cross-sectional area shall be used (Reg. 543–01–04).

Types of protective conductors. Flexible or pliable conduit cannot be used as a protective conductor; neither can a gas or oil pipe (Reg. 543–02–01). It may consist of one or more of the following: a single-core cable; a conductor in a cable; an insulated or bare conductor in a common enclosure with insulated live conductors; a

252

Tble 12.3 **Minimum cross-sectional area of protective conductor in relation to the cross-sectional area of associated phase conductor**

Cross-sectional area of phase conductor (S)	Minimum cross-sectional area of the corresponding protective conductor	
	If the protective conductor is of the same material as the phase conductor	If the protective conductor is not the same material as the phase conductor
mm²	mm²	mm²
S ≤ 16	S	$k_1 S$ / k_2
16 < S ≤ 35	16	$k_1 16$ / k_2
S > 35	S / 2	$k_1 S$ / $k_2 2$

fixed bare or insulated conductor; a metal covering, for example, the sheath, screen or armouring of a cable; a metal conduit or other enclosure or electrically continuous support system for conductors; and an extraneous conductive part complying with Regulation 543–02–06. (Reg. 543–02–02).

A protective conductor of the types described above and of cross-sectional area 10 mm² or less, shall be of copper (Reg. 543–02–03). Where a metal enclosure or frame of a l.v. switchgear or control gear assembly or busbar trunking system is employed as a protective conductor it has to satisfy three requirements which are laid down (Reg. 543–02–04).

Where the protective conductor is formed by conduit, trunking, ducting, or the metal sheath and/or armour of cables, the earthing terminal of each accessory shall be connected by a separate protective conductor to an earthing terminal incorporated in the associated box or other enclosure (Reg. 543–02–07).

PEN conductors. The provisions on combined protective and neutral conductors forming the section of Reg. 546 are applicable only under certain conditions including that of privately owned generating plant, or where the installation is supplied by a privately owned transformer or converter in such a way that there is no metallic connection with the general public supply, or where special authorisation has been granted (Reg. 546–02–01).

Conductors of the following types may serve as PEN conductors provided that the part of the installation concerned is not supplied through a residual curent device:

(i) For fixed installations, conductors of cables not subject to flexing and having a cross-sectional area not less than 10 mm² for copper, or 16 mm² for aluminium.

(ii) The outer conductor of concentric cables where that conductor has a cross-sectional area not less than 4 mm² in a cable complying with an appropriate British Standard and selected and erected in accordance with Regs. 546–02–03 to 02–08 (Reg. 546–02–02).

The conductance of the outer conductor of a concentric cable is specified in Reg. 546–02–04.

Dimensions of equipotential bonding conductors. Except where PME conditions apply cross-sectional area of main equipotential bonding conductors shall not be less than half that of the earthing conductor of the installation, subject to a minimum of 6 mm². The cross-sectional area need not exceed 25 mm² if the bonding conductor is of copper or a cross-sectional area affording equivalent conductance in other materials. Where PME conditions apply the main equipotential bonding conductor must be selected in accordance with the neutral conductor of the supply and Table 54 H of the Regulations (Reg. 547–02–01). The position of the main bonding connections to water or gas services are specified in Reg. 547–02–02.

Plugs and socket-outlets. In low-voltage circuits these must conform to British Standards listed in Table 12.4 (Reg. 553–01–02 to 553–02–02).

GENERAL CHARACTERISTICS OF NETWORKS

Part 3 of the Regulations 'Assessment of general characteristics', includes the purpose for which the installation is intended, supplies and structure (Chapter 31); the external influences to which it is exposed (Chapter 32); compatibility of the equipment (Chapter 33); and its maintainability (Chapter 34).

Table 12.4 Plugs and socket-outlets for low voltage circuits

Type of plug and socket-outlet	Rating, amperes	British Standard
Fused plugs and shuttered socket-outlets, 2-pole and earth for a.c.	13	BS 1363 (fuses to BS 1362)
Plugs, fused or non-fused and socket-outlets, 2-pole and earth	2, 5, 15, 30	BS 546 (fuses, if any to BS 646)
Plugs, fused or non-fused, and socket-outlets, protected type, 2-pole with earthing contact	5, 15, 30	BS 196
Plugs and socket-outlets (theatre type)	15	BS 5550 subsection 7.3.1
Plugs and socket-outlets (industrial type)	16, 32, 63, 125	EN 60309–2

Maximum demand. The maximum demand of an installation, expressed in amperes, has to be assessed. In determining this diversity may be taken into account (Reg. 311–01–01).

Live conductors and earthing. The number and type of live conductors and the method of earthing (depending on the type of system, i.e. TN-C, TN-S, TN-C-S, TT and IT) govern the method of protection for safety adopted in order to comply with Part 4 of the Regulations (Reg. 312–01–01).

Nature of supply. This relates to the nominal voltage; current; frequency; prospective short-circuit current at the origin of the installation; the earth fault loop impedance (Z_e) of that part of the system external to the installation; suitability of the installation to meet requirements, including the maximum demand; and the nature of the protective device acting at the origin of the installation (Reg. 313–01–01). Section 313–02 details requirements for safety services and standby purposes.

Circuit arrangement. Every installation must be divided into circuits as necessary to avoid danger and minimize inconvenience in the event of a fault and to facilitate safe operation, inspection, testing and maintenance. (Reg. 314–01–01). See also other 314 Regulations.

Compatibility. The installer must check whether any equipment will have harmful effects upon other electrical equipment or other services. For an installation supplied from an external source of supply the supplier should be consulted about any equipment in the installation whose operation may have a significant influence on the supply.

PART 6. SPECIAL INSTALLATIONS OR LOCATIONS

Earlier on in this chapter details were given of the special requirements for locations containing a bath tub or shower basin. Here we provide some information from the sections dealing with other special installations. It is important to appreciate that requirements for special installations supplement or modify the general requirements contained in other Parts of the Regulations. Absence of reference to the exclusion of a Chapter, Section or Clause means that the corresponding general Regulations are applicable.

Swimming pools. This covers swimming and paddling pools and their surrounding zones; special requirements may be necessary for swimming pools for medical use. Three zones are identified in Regulation 602–02. Zone A is the interior of the basin, chute or flume and includes the portions of essential appertures in its walls and floor which are accessible to persons in the basin.

Zone B is limited (a) by the vertical plane 2 m from the rim of the basin, and (b) by the floor or surface expected to be accessible to persons, and (c) by the horizontal plane 2.5 m above that floor or surface, except where the basin is above ground, when it shall be 2.5 m above the level of the rim of the basin. This Zone also includes other areas where applicable like diving and spring boards, starting blocks or a chute and the immediate vicinity around such areas.

Zone C is limited by (a) the vertical plane circumscribing Zone B and the parallel vertical plane 1.5 m external to Zone B, and (b) by the floor or surface expected to be occupied by persons and the horizontal plane 2.5 m above that floor or surface.

Regulations cover protection for safety and selection and erection of equipment. For example where SELV is used as protection against electric shock, irrespective of the nominal voltage, protection against direct contact has to be provided either by barriers and enclosures affording at least IP2X or IPXXB degree of protection, or by insulation able to stand a test voltage of 500 V d.c. for 1 minute.

(Reg. 602–03–01). In Zones A and B only the protective measure against electric shock by SELV (Reg. 411–02) at a nominal voltage not exceeding 12 V a.c. rms or 30 V d.c. can be used, the safety source being installed outside the Zones A, B and C with certain specified exceptions (Reg. 602–04–01).

Some protective measures cannot be used in any zone: they include protection by means of obstacles (Reg. 412–05); by means of placing out of reach (Reg. 412–05); by means of a non-conducting location (Reg. 413–04); and by means of earth-free local equipotential bonding (Reg. 413–05). These prohibitions form Regulation 602–04–02.

In Zones A and B surface wiring systems must not employ metallic conduit or metal trunking or an exposed metallic cable sheath or an exposed earthing or bonding conductor (Reg. 602–06–01). Zones A and B shall contain only wiring necessary to supply equipment located in those Zones (Reg. 602–06–02). Accessible metal junction boxes must not be installed in Zones A and B (Reg. 602–06–03).

A socket-outlet, switch or accessory is permitted in Zone C only under certain conditions which are specified in Regulations 602–07–02. Socket-outlets must comply with EN 60309–2.

Hot air saunas. The particular requirements of this section apply to locations in which hot air sauna heating equipment to EN 60335–2–53 is installed (Reg. 603–01–01). Four temperature classification zones are illustrated in Figure 603A (in the Regulations).

Where SELV is used to protect against electric shock, irrespective of the nominal voltage, one has to provide protection against electric shock by insulation able to withstand a test voltage of 500 V d.c. for one minute or barriers or enclosures affording at least an IP24 or IPX4B degree of protection (Reg. 603–03–01). Two methods of protection against direct contact are prohibited, by using obstacles or placing out of reach (Reg. 603–04–01). Two methods of protection against indirect contact are also prohibited, non-conducting location and earth-free local equipotential bonding (Reg. 603–05–01). All equipment used must have at least an IP24 degree of protection (Reg. 603–06–01). Equipment characteristics for the four temperature zones A, B, C and D are indicated (Reg. 603–06–02). As regards wiring systems only flexible cords complying with BS 6141 having 150°C rubber insulation can be employed and it must be mechanically protected with material which complies with Regulation 413–03–01 (Reg. 603–07–01).

Construction site. This section applies to installations provided for the purpose of electricity supply for the following works: new building construction, repair, alterations, extensions or demolition of existing buildings; engineering construction; earth-works; and similar works (Reg. 604–01–01). These requirements do not apply to installations in construction site offices, cloakrooms, meeting rooms, canteens, restaurants, dormitories, toilets, etc. where the general requirements of the Regulations apply. Construction site fixed installations are limited to the assembly of the main switchgear and principal protective devices. An installation on the load side is considered a movable one except for parts which are designed according to Chapter 52 of the Regulations (Reg. 604–01–01).

There is a limit to the voltages one can use which are as follows: portable hand lamps in confined or damp locations – 25 V single-phase SELV or 50 V single-phase centre point earthed; portable hand lamps for general use and portable hand-held tools and local lighting up to 2 kW–110 V single-phase centre point earthed; small mobile plant up to 3.75 kW–110 V three-phase star point earthed; fixed floodlighting – 230 V single-phase; and fixed and movable equipment above 3.75 kW–400 V three-phase. High voltage supplies can be used for large equipment where necessary. (Reg. 604–02–02).

Considerable space is devoted to the requirements for protection for safety against indirect contact for TN, TT and IT systems. An important Regulation 604–03–01 states that where an alternative system is available an IT system shall not be used. If it is then permanent earth monitoring must be provided. Where the protection against indirect contact is by earthed equipotential bonding and automatic disconnection of supply, (Regs 413–02 and 471–08 as appropriate to the type of earthing system) then Regulations 604–06 to 604–08 apply.

For a TN system many of the Regulations in Section 413 of the fifteenth edition are replaced by new Regulations contained in Section 604. For example Regulation 413–02–08 Table 41 A is replace by a new table, Table 604 A where the maximum disconnection times are appreciably reduced. For example for U_o 120 V the time drops from 0.8 to 0.35 s. (Reg. 604–04–01). For a TT system the formula in Regulation 413–02–20 is replaced by $R_a I_a \leqslant 25$ V (Reg. 604–05–01). This is the only change for such a system. For an IT system (see the preceding paragraph), the formula in Regulation 413–02–03 is replaced by $R_b I_a \leqslant 25$ V

(Reg. 604–06–01). Also Table 41 E in Regulation 413–02–26 is replaced by Table 604 E and the definition of I_a is altered to relate to the new table (Reg. 604–06–02).

As regards wiring systems there is a requirement to protect cables that run across site roads or walkways (Reg. 604–10–02). All assemblies used for site distribution networks must comply with BS 4363 and EN 60439–4 (Reg. 604–09–01). Except for assemblies covered by Regulation 604–09–01 equipment must have at least IP44 degree of protection; every socket-outlet must also form part of an assembly complying with this Regulation. (Regs 604–09–02 and 604–12–01). All plugs and socket-outlets and cable couplers must comply with EN 60309–2 (Regs 604–12–02 and 604–13–01).

Agricultural and horticultural premises. The requirements laid down in Section 605 apply to all parts of fixed installations of these types of premises both outdoors and indoors, and to where livestock is kept. Where the premises include dwellings solely for human habitation they are excluded from the scope of this section.

Where SELV is used, irrespective of the nominal voltage, protection against direct contact must be provided either by barriers or enclosures to at least IP2X or IPXXB insulation capable of withstanding a test voltage of 500 V d.c. for one minute (Reg. 605–02–02). All socket-outlet circuits, except those supplied from a SELV supply, must be protected by an rcd complying with the appropriate British Standard and having the characteristics specified in Reg. 412–06–02 (ii). (Reg. 605–03–01).

Where protection against indirect contact for livestock is provided by earthed equipotential bonding and automatic disconnection Regulations 605–05 to 605–09 apply and these cover TN systems (Reg. 605–04–01). In these Regulations many of the requirements of the Regulations in Section 413 are replaced including provision of new tables. For TT and IT systems the same conditions apply about the formulas $R_a I_a$ and $R_b I_d$ as for construction sites (Reg. 605–06–01 and 605–07–01). For an IT system Table 41 E is replaced by Table 605 E and the definition of I_a is altered similar to that for construction sites (Reg. 605–07–02).

Restrictive conductive locations. Section 606 deals only with installations of this nature and not to locations in which freedom of movement is not physically constrained. Regulation 606–02–01 specifies the means of protection against direct and indirect contact by SELV or FELV (Reg. 411–02–02 to 411–03 and 471–02) regardless of the voltage, protection against direct contact must be provided by a barrier or enclosure affording at least IP2X or IPXXB degree of protection or insulation able to withstand a test voltage of 500 V d.c. for one minute. On cannot use obstacles or placing out of reach as a means of protection against direct contact (Reg. 606–03–01). Protection against indirect contact can be provided by (i) SELV (Reg. 606–2); (ii) electrical separation; (iii) automatic disconnection in which case a supplementary equipotential bonding conductor shall be provided; (iv) use of Class II equipment. Certain other requirements are specified for some of these methods (Reg. 606–04–01).

Equipment with high earth leakage currents. The requirements of Section 607 apply to installations where a piece of equipment has high leakage current, usually exceeding 3.5 mA. It includes information technology (IT) equipment and industrial control equipment (Reg. 607–01–01). Where more than one item of stationary equipment with a leakage current exceeding 3.5 mA in normal service is supplied from an rcd protected supply, the total leakage current must not exceed 25% of the nominal tripping current of the rcd (Reg. 607–02–03). If this cannot be achieved the items of equipment should be supplied through a double-wound transformer or equivalent. There are other Regulations covering the situation where the earth leakage current exceeds 10 mA for a piece of equipment. One requires the provision of a high integrity protective connection complying with one of six arrangements (Reg. 607–02–07). Equipment with a high earth leakage current must not be connected directly to an IT system (Reg. 607–04–01).

Caravans, motor caravans and caravan parks. Section 608 dealing with caravans and their parks is divided into two divisions, the first dealing with caravans themselves and the second with parks. Division one restricts the Regulations to caravans and motor caravans where the supply does not exceed 250/440 V. Electrical circuits and equipment covered by Road Vehicles Lighting Regulations 1989 and installations covered by BS 6765 Part 3 are not covered (Reg. 608–01–01). Certain requirements of Section 601 (bath tubs and shower basins) also apply to caravans and motor caravans. They do not apply to certain installations which are specified in BS 7671.

Protection against direct contact cannot be provided by obstacles or by placing out of reach (Reg. 608–02–01); and similarly for protection against indirect contact, non-conducting locations, earth-free equipotential bonding or electrical separation are prohibited (Reg. 608–03–01). If automatic disconnection of supply is employed a double pole rcd must be provided and the wiring system must include a circuit protective conductor which must be connected to (i) the protective contact of the inlet; and (ii) the exposed conductive parts of the electrical equipment; and (iii) the protective contacts of the socket-outlets. (Reg. 608–03–02). A final circuit must be provided with overcurrent protection capable of disconnecting all live conductors of that circuit (Reg. 608–04–01).

Wiring systems can use (i) flexible single-core insulated conductors in non-metallic conduits; (ii) stranded insulated conductors with a minimum of seven strands in non-metallic conduits; or (iii) sheathed flexible cables (Reg. 608-06-01). Pliable polyethylene conduits shall not be used. Minimum conductor size is 1.5 mm². The limit of 6 mm² in Regulation 543-03-02 does not apply and all protective conductors, irrespective of cross-sectional area, have to be insulated (Reg. 608-06-03). No electrical equipment must be installed in a compartment intended for fuel storage. A notice of durable material has to be fixed near the main isolating switch inside the caravan or motor caravan and carry specified text headed 'Instructions for electricity supply' (Reg. 608–07–05). There are a number of Regulations dealing with socket-outlets and luminaires.

The means of connection to the caravan or motor caravan switch socket-outlet are specified (Reg. 608–08–08). This Regulation also includes Table 608 A giving details of the cross-sectional areas of flexible cords and cables used for caravan connectors from 16 A up to 100 A.

Division Two of Section 608 dealing with caravan parks also prohibits the use of protection by obstacles, placing out of reach, non-conducting locations, earth-free equipotential bonding and electrical separation (Regs 608–10–01 and 608–11–01). Under-ground cables should, as far as possible be used to connect caravan pitch supply equipment. Overhead conductors can be employed subject to certain conditions (Reg. 608–12–03).

Supplies for highways and street furniture. These Regulations do not apply to supplier's works as defined by the Electricity Supply Regulations 1988 (as amended) in accordance with Regulation 110–02. This comparatively short Section 611 deals with protection against electric shock; devices for isolation and switching; identification of cables; external influences: and temporary supplies. There are a number of protective measures which cannot be used with highway power supplies and street furniture.

CONVENTIONAL CIRCUIT ARRANGEMENTS

This section gives details of conventional circuit arrangements which satisfy the requirements of Chapter 46 for isolation and switching, together with the requirements as regards current-carrying capacities of conductors prescribed in Chapter 52 and Appendix 4. It is the responsibility of the designer and installer when adopting them to take the appropriate measures to comply with the other chapters or sections of the Regulations that are appropriate. The following information has been extracted from Appendix E of Guidance Note No. 1.

Circuit arrangements other than those detailed below are not precluded where they are specified by a suitably qualified electrical engineer, in accordance with the general requirements of the Regulation 314–01–03. The conventional circuit arrangements are:

A. Final circuits using socket-outlets complying with BS 1363.
B. Final circuits using socket-outlets complying with BS 196.
C. Final radial circuits using socket-outlets complying with BS 4343.
D. Cooker final circuits in household premises.

Circuits under A. A ring or radial circuit, with spurs if any, feeds permanently connected equipment and an unlimited number of socket-outlets. The floor area served by the circuit is determined by the known or estimated load but does not exceed the value given in Table 12.5.

For household installations a single 30 A ring circuit may serve a floor area of up to $100\,\text{m}^2$ but consideration should be given to the loading in kitchens which may require a separate circuit. For other types of premises, final circuits complying with Table 12.5 may be used subject to conditions laid down in the Appendix.

The number of socket-outlets shall be such as to ensure compliance with the Regulation which states that every portable appliance must be able to be fed from an adjacent and conveniently accessible socket-outlet. Account should be taken of the length of flexible cord normally fitted to the majority of appliances and luminaires. Each socket outlet of a twin or multiple socket-outlet unit is regarded as one socket-outlet.

Conductor size *(Circuits under A).* The minimum size of conductor in the circuit and in non-fused spurs is given in Table 12.5. If cables of more than two circuits are bunched together or the ambient temperature exceeds 30°C, the size of the conductor shall be increased by applying the appropriate correction factors from Appendix 4 of the Regulations such that the size then corresponds to a current-carrying capacity not less than: 20 A for circuit A1; 30 A or 32 A for circuit A2; 20 A for circuit A3.

The conductor size for a fused spur is determined from the total current demand served by that spur, which is limited to a maximum of 13 A. The minimum conductor size for a spur serving socket-outlets is: 1.5 mm^2 for rubber or PVC-insulated cables with copper conductors; 2.5 mm^2 for rubber or PVC-insulated cables with copperclad aluminium conductors; 1 mm^2 for mineral insulated cables with copper conductors.

Spurs *(Circuits under A)*. The total number of fused spurs is unlimited but the number of non-fused spurs shall not exceed the total number of socket-outlets and stationary equipment connected directly in the circuit.

A non-fused spur feeds only one single or one twin socket-outlet or one permanently connected item of equipment. Such a spur is connected to a ring circuit at the terminals of socket-outlets, or at joint boxes, or at the origin of the circuit in the distribution board.

A fused spur is connected to a ring circuit through a fused connection unit the rating of the fuse in the unit not exceeding that of the cable forming the spur, and in any event, not exceeding 13 A.

Circuits under B. A ring or radial circuit, with fused spurs if any, feeds equipment the maximum demand of which, allowing for diversity, is known or estimated not to exceed the rating of the overcurrent protective device and in any event does not exceed 32 A. No diversity is allowed for permanently connected equipment. The number of socket-outlets is unlimited and the total current demands of points served by a fused spur shall not exceed 16 A. The overcurrent protective device rating shall not exceed 32 A.

The size of conductor is obtained by applying the appropriate correction factors from Appendix 4 of the Regulations and is such that it then corresponds to a current-carrying capacity of:
(i)   For ring circuits – not less than 0.67 times the rating of the overcurrent device.

263

Table 12.5 Final circuits using BS 1363 socket-outlets

| Type of circuit | Overcurrent protective device | | Minimum conductor size*, mm² | | Maximum floor area served, m² |
	Rating, A	Type	Copper conductor rubber- or PVC-insulated cables mm²	Copper conductor mineral insulated cables mm²	
A1 Ring	30 or 32	Any	2.5	1.5	100
A2 Radial	30 or 32	Cartridge fuse or circuit-breaker	4	2.5	50
A3 Radial	20	Any	2.5	1.5	20

*The tabulated values of conductor size may be reduced for fused spurs.

(ii) For radial circuits – not less than the rating of the overcurrent protective device.

The conductor size for a fused spur is determined from the total demand served by that spur which is limited to a maximum of 16 A.

Spurs *(Circuits under B).*   A fused spur is connected to a ring circuit through a fuse connection unit, the rating of the fuse in the unit not exceeding that of the cable forming the spur and in any event not exceeding 16 A. Non-fused spurs are not used.

Circuits under C.   A radial circuit feeds equipment the maximum demand of which, having allowed for diversity, is known or estimated not to exceed the rating of the overcurrent protective device, and in any event does not exceed 20 A. The number of socket-outlets is unlimited. The overcurrent protective device has a rating not exceeding 20 A.

Conductor size is determined by applying appropriate correction factors from Appendix 4 of the Regulations and is such that it then corresponds to a current-carrying capacity not less than the rating of the overcurrent protective device. Socket-outlets have a rated current of 16 A.

Circuits under D.   The circuit supplies a control switch or a cooker control unit which may incorporate a socket-outlet. The rating of the circuit is determined by the assessment of the current demand of the cooking appliance(s) and control unit socket-outlet, if any, in accordance with Table J1 of Appendix J.

A circuit of rating exceeding 15 A but not exceeding 50 A may supply two or more cooking appliances where these are installed in one room, subject to certain conditions.

CLASSIFICATION OF EXTERNAL INFLUENCES

Appendix 5 gives the classification and codification of external influences developed for IEC Publication 364-3 Second Edition. Each condition of external influences is designated by a code comprising a group of two capital letters and number as follows. The first letter relates to the general category of external influences:

A. Environment
B. Utilisation
C. Construction of buildings.

The second letter relates to the nature of the external influence:

A.

B.

C.

The number relates to the class within each external influences:

1.

2.

3.

For example, AA4 signifies:

A = Environment; AA = Environment – ambient temperature; AA4 = Environment – ambient temperature in the range −5°C to 40°C.

Because of the importance of the table contained in this Appendix the reader is referred to the Wiring Regulations themselves. The contents covers; ambient temperatures within lower and upper limits of −60°C and +5°C (AA1) to −50°C and +40°C (AA8); atmospheric humidity under consideration; altitude (AC1 and AC2); presence of water from negligible (AD1) to submersion (AD8); presence of foreign solid bodies from negligible (AE1) to heavy dust (AE6); presence of corrosive or polluting substances from negligible (AF1) to continuous (AF4); impact from low severity (AG1) to high severity (AG3); vibration from low severity (AH1) to high severity (AH3); other mechanical stresses (AJ) under consideration; presence of flora and/or mould growth from no hazard (AK1) to hazard (AK2); presence of fauna from no hazard (AL1) to hazard (AL2); radiation (AM1) to induction (AM6); solar radiation from low (AN1) to high (AN3); seismic effects from negligible (AP1) to high severity (AP4); lightning from negligible (AQ1) to direct exposure (AQ3); movement of air from low (AR1) to high (AR3); wind from low (AS1) to high (AS3). Similarly there are two other sections of the table dealing with utilisation (B) and construction of buildings (C).

LIMITATION OF EARTH FAULT LOOP IMPEDANCE

Guidance Note No. 1 details how to calculate the earth fault loop impedance Z_s in order to comply with the Regulation dealing with a.c. circuits. For cables having conductors of cross-sectional area not exceeding 35 mm² the inductance can be ignored so that Z_s is given by:

(i) For radial circuits $Z_s = Z_E + R_1 + R_2$ ohms

Where Z_E is that part of the earth fault loop impedance

external to the circuit concerned; R_1 is the resistance of the phase conductor from the origin of the circuit to the most distant socket-outlet or other point of utilisation; R_2 is the resistance of the protective conductor from the origin of the circuit to the most distant socket-outlet or other point of utilisation.

(ii) For ring circuits without spurs $Z_s = Z_E + 0.25R_1 + 0.25R_2$ ohms

where Z_E is as described under (i) above; R_1 is now the total resistance of the phase conductor between its ends prior to them being connected together to complete the ring; R_2 is similarly the total resistance of the protective conductor.

Having determined Z_s, the earth fault current I_F is given by

$$I_F = \frac{U_o}{Z_s} \text{ amperes}$$

where U_o is the nominal voltage to earth (phase to neutral voltage). From the relevant time/current characteristics the time for disconnection (t) corresponding to this fault current is obtained. Substituting for I_F, t and the appropriate k value in the equation given in Regulation 543-01-03 gives the minimum cross-sectional area of the protective conductor and this must be equal to or less than the size chosen.

When the cables are to Table 5 of BS 6004 or are other PVC-insulated cables to that standard, the Guidance Note gives in tabular form the maximum earth loop impedance for circuits having protective conductors of copper ranging from 1 to 16 mm² cross-section area and the overcurrent protective device is a fuse to BS 88 Part 2, BS 1361 or BS 3036. These tables also apply if the protective conductor is bare copper and in contact with cable insulated with PVC. For each type of fuse, two tables are given:

(i) Where the circuit concerned feeds socket-outlets and the disconnection time for compliance is 0.4 seconds, and

(ii) Where the circuit concerned feeds fixed equipment and the disconnection time for compliance is 5 seconds.

The graphs and tables are not reproduced here, being too numerous (totalling 6); the reader is referred to the Guidance Note for details.

CABLE CURRENT-CARRYING CAPACITIES

Current-carrying capacities and voltage drops for cables and flexible cords are contained in Appendix 4. Typical methods of installation of cables and conductors are contained in a table ranging from single and multicore sheathed cables clipped direct to or lying on a non-metallic surface (No. 1 method) to single and multicore cables in an enclosed trench 600 mm wide by 760 mm deep (minimum dimensions) including 100 mm cover (No. 20 method). Correction factors for groups of more than one circuit of single-core cables or more than one multicore cable, and for cables in trenches and for mineral insulated cables installed on perforated tray are specified in other tables. These correction factors are applied as necessary to the current-carrying capacities and associated voltage drops of the types of cables covered.

There are many tables for copper conductor cables and somewhat fewer for cables with aluminium conductors. Cables with copper conductors include PVC-insulated, thermosetting, 85°C rubber insulated, flexible cables and cords and mineral-insulated cables. Cables with aluminium conductors include PVC insulated and thermosetting cables.

The current-carrying capacities take account of IEC Publication 364.5.523 (1983) so far as the latter is applicable. For types of cable not treated in the IEC Publication (e.g. armoured cables) the current-carrying capacities of the Appendix are based on data provided by ERA Technology and the British Cable Makers Confederation. Readers are referred to the Regulations for fuller details.

To comply with the requirements of Chapter 52 of the Regulations for the selection and erection of wiring systems in relation to risks of mechanical damage and corrosion, The IEE On-site Guide to BS 7671 tabulates types of cable and flexible cord for particular uses and external influences. The tables are not intended to be exhaustive and other limitations may be imposed by the relevant regulations, in particular those concerning maximum permissible operating temperature.

An example of the type of information contained in these tables is as follows: PVC- or rubber-insulated non-sheathed cables for fixed wiring can be used in conduits, cable ducting or trunking, but not when the enclosures are buried underground. Mineral-insulated cables are suitable for general use but additional precautions in the form of a PVC sheath are necessary when such cables are exposed to the weather or risk of corrosion, or where installed underground, or in concrete ducts.

METHODS OF CABLE SUPPORT

Examples of methods of support for cables, conductors and wiring systems are described in Guidance Note No. 1 but other methods are not precluded when specified by a suitably qualified electrical engineer.

Cables. For cables supported on structures which are only subject to vibration of low severity and a low risk of mechanical impact the following conditions should be observed:

(i) For cables of any construction, installation in conduit, trunking or ducting without further fixing of the cables, precautions being taken against undue compression or other mechanical stressing of the insulation at the top of any vertical runs exceeding 5 m in length.

(ii) For sheathed and/or armoured cables installed in accessible positions, support by clips at spacings not exceeding the appropriate value stated in Table 12.6.

(iii) For sheathed and/or armoured cables in vertical runs which are inaccessible and unlikely to be disturbed (e.g. installations in an inaccessible cavity), supported at the top of a run by a clip and a rounded support of a radius not less than the appropriate value stated in Appendix I, Table I1 (of Guidance Note No. 1). For unarmoured cables, the length of run without intermediate support should not exceed 2 m for a lead sheathed cable or 5 m for a rubber of plastics sheathed cable. Where these figures are intended to be exceeded the advice of the manufacturer of the cable should be obtained. Care should be taken that the cable does not bridge a cavity between external and internal faces in a manner which would transmit moisture.

In this Note there are recommendations for the support of cables in caravans, for flexible cords used as pendants and overhead wiring. For overhead wiring cables sheathed with rubber or PVC, support

Table 12.6 Spacing of supports for cable in accessible positions the entire support derived from the clips

Overall diameter of cable*, mm	Maximum spacing of clips, mm							
	Non-armoured rubber-, plastics or lead-sheathed cables				Armoured cables		Mineral-insulated copper-sheathed or aluminium-sheathed cables	
	Generally		In caravans					
	Horizontal†	Vertical†	Horizontal†	Vertical†	Horizontal†	Vertical†	Horizontal†	Vertical†
≤ 9	250	400	150 (for all sizes)	250 (for all sizes)	–	–	600	800
> 9 ≤ 15	300	400			350	450	900	1200
> 15 ≤ 20	350	450			400	550	1500	2000
> 20 ≤ 40	400	550			450	600	–	–

Note: For the spacing of supports for cables of overall diameter exceeding 40 mm, and for single-core cables having conductors of cross-sectional area 300 mm² and larger, the manufacturer's recommendations should be observed.

*For flat cables taken as the measurement of the major axis.

†The spacings stated for horizontal runs may be applied also to runs at an angle of more than 30° from the vertical. For runs at an angle of 30° or less from the vertical, the vertical spacings are applicable.

Table 12.7 Spacings of supports for conduits and cable trunking

(a) *Conduits*

Nominal size of conduit, mm	Maximum distance between supports, m					
	Rigid metal		*Rigid insulation*		*Pliable*	
	Horizontal	*Vertical*	*Horizontal*	*Vertical*	*Horizontal*	*Vertical*
Not exceeding 16	0.75	1.0	0.75	1.0	0.3	0.5
Exceeding 16 and not exceeding 25	1.75	2.0	1.5	1.75	0.4	0.6
Exceeding 25 and not exceeding 40	2.0	2.25	1.75	2.0	0.6	0.8
Exceeding 40	2.25	2.5	2.0	2.0	0.8	1.0

(b) *Cable trunking*

Cross-sectional area, mm^2	Maximum distance between supports, m			
	Metal		*Insulating*	
	Horizontal	*Vertical*	*Horizontal*	*Vertical*
Up to 700	0.75	1.0	0.5	0.5
Exceeding 700 and not exceeding 1500	1.25	1.5	0.5	0.5
Exceeding 1500 and not exceeding 2500	1.75	2.0	1.25	1.25
Exceeding 2500 and not exceeding 5000	3.0	3.0	1.5	2.0
Exceeding 5000	3.0	3.0	1.75	2.0

Note 1. The spacings tabulated allow for maximum fill of cables to the Wiring Regulations and the thermal limits specified in the relevant British Standards. They assume that the conduit or trunking is not exposed to other mechanical stress.

2. The above figures do not apply to lighting suspension trunking, or where special strengthening couplers are used. A flexible conduit is not normally required to be supported in its run. Supports should be positioned within 300 mm of bends or fittings.

271

by a catenary wire is required either continuously bound up with the cable or attached thereto at intervals, these intervals not to exceed those given in Appendix I Table 13 (of Guidance Note No. 1). Three other overhead wiring systems are also contained in the same Table.

Conduit and cable trunking.   Table 12.7 gives supporting spacings for rigid conduit and cable trunking. Conduit, embedded in the material of the building and pliable conduit embedded in the material of the building or in the ground need no further support. Otherwise the support of pliable conduit must be in acordance with Table 12.7. Cable trays should be supported in accordance with the installation designer's requirements.

Cables installed in conduit and trunking.   This subject is dealt with in Appendix A of Guidance Note No. 1.

METHODS OF TESTING

The standard methods of testing described in a Guidance Note are suitable for the corresponding testing prescribed in Part 7 of the Regulations. They are given as examples, and the use of other methods giving no less effective results is not precluded.

   Each leg of the ring circuit is identified. The phase conductor of one leg and the neutral conductor of the other leg are temporarily bridged. The resistance is measured between the remaining phase and neutral conductors; a finite reading confirms that there is no open-circuit on the ring conductors under test. These remaining conductors are then temporarily bridged together. The connections and instrument are indicated in the Guidance Note. The resistance between phase and neutral contacts at each socket-outlet around the ring is measured and noted. The readings obtained should be substantially the same, provided that no multiple loops exist over the length of the ring. Where the protective conductor is in the form of a ring, the test is repeated, transposing the circuit protective conductor with the phase conductors. The phase conductor from one leg of the ring is temporarily bridged with the circuit protective conductor of the other leg of the ring. The resistance is measured between the remaining phase conductor and the remaining unconnected circuit protective conductor at the origin of the circuit. A finite reading confirms that there is no open-circuit on the ring conductors under test. The remaining circuit protective

272

conductor and phase conductor are then temporarily bridged together. The resistance is measured between the circuit protective conductor and phase conductor contacts at each socket-outlet around the ring. The readings obtained should be substantially the same, provided that no multiple loops exist, and the readings at the centre point of the ring are approximately equal to $(R_1 + R_2)$ for the circuit. This value should be recorded and, when corrected for temperature, may be used to calculate the earth fault loop impedance of the circuit Z_s, to verify compliance with the requirements of Table 41B1 and Table 41B2 of the Wiring Regulations.

Where single-core cables are used, special precautions should be taken and these are outlined in the Guidance Note.

Continuity of protective conductors main and supplementary bonding. Every protective conductor including any bonding conductor must be tested to ensure it is electrically sound and correctly connected.

For cables having conductors of cross-sectional area not exceeding 35 mm^2 their inductance can be ignored. Above that figure the inductance becomes significant and an appropriate a.c. instrument should be used for the measurement. The test methods detailed below, as well as checking the continuity of the protective conductor, also measure $(R_1 + R_2)$ which, when corrected for temperature, enables the designer to verify the calculated earth-fault loop impedance Z_s. Use a low-resistance ohmmeter for these tests.

Test method 1. Strap the phase conductor to the protective conductor at the distribution switchboard to include all the circuit. Then test between phase and earth terminals at each outlet in the circuit. The measurement $(R_1 + R_2)$ at the circuit's extremity should be recorded to verify compliance with the Wiring Regulations.

When the testing of ring circuit continuity including protective conductors, is required, the test should be made prior to connecting supplementary bonds to the protective conductors.

Test method 2. Connect one terminal of the continuity tester to one test lead and connect this to the consumer's earth terminal. Then connect the other terminal of the continuity tester to another test lead and use this to make contact with the protective conductors at various points on the circuit, such as light fittings, switches, spur outlets, etc.

The resistance reading obtained by the above method includes the resistance of the test leads. The resistance of the test leads should be measured and deducted from any resistance reading obtained using this method. To test the bonding conductors' continuity use Test method 2.

Earth electrode resistance. After an earth electrode has been installed it is necessary to verify that the resistance meets the conditions of the Wiring Regulations for TT and IT installations. Use an earth electrode resistance tester for this test. Refer to Section 16 of Guidance Note No. 1.

Two test methods are described but we only cover one in this book. The test requires the use of two test spikes (electrodes), and is carried out in the following manner.

Connection to the earth electrode is made using terminals C1 and P1 of a four-terminal earth tester. Details are given to exclude test lead resistance. Connection to the temporary spikes is made as shown in Fig. 12.6. The distance between the test spikes is important. In general, reliable results may be expected if the distance between the electrode under test and the current spike is at least ten times the maximum dimension of the electrode system, *e.g.* 30 m for a 3 m long electrode.

Three readings are taken: with the potential spike initially midway between the electrode and current spike, secondly at a position 10 per cent of the electrode-to-current spike distance

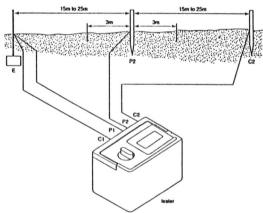

Figure 12.6   Measurement of earth electrode resistance

274

back towards the electrode, and finally at a position 10 per cent of the distance towards the current spike. By comparing the three readings, a percentage deviation can be determined. This is calculated by taking the average of the three readings, finding the maximum deviation of the readings from this average in ohms, and expressing this as a percentage of the average. The accuracy of the measurement using this technique is typically 1.2 times the percentage deviation of the readings. It is difficult to aim for a measurement accuracy better than 2 per cent, and inadvisable to accept readings that differ by more than 5 per cent. To improve the accuracy of the measurement to acceptable levels, the test must be repeated with a larger separation between the electrode and the current spike.

The instrument output current may be a.c. or reversed d.c. to overcome electrolytic effects. Because these testers employ phase-sensitive detectors the errors associated with stray currents are eliminated.

If the temporary spike resistances are too high, measures to reduce these will be necessary, such as driving the spikes deeper into the ground or watering with brine to improve contact resistance.

Earth fault loop impedance. The earth fault current loop (phase to earth loop) comprises the following parts, starting at the point of fault:

● the circuit protective conductor
● the main earthing terminal and earthing conductor
● for TN systems the metallic return path, _or_ for TT and IT systems the earth return path
● the path through the earth and neutral point of the transformer
● the transformer winding and the phase conductor from the transformer to the point of fault.

The earth fault impedance Z_s should be determined at the farthest point of each circuit including socket-outlets, lighting points, sub-main cables and any other fixed equipment. The value obtained, after adjustment to take into account the effects of fault current, should not exceed that detailed in Table 41B or Table 41D, or should not exceed a value which might prevent conformity with the Regulations for r.c.d. protected circuits.

Z_e is measured using a phase earth loop impedance tester at the source of the installation supply.

The impedance measurement is made between the main phase supply and the main means of earthing with the main switch open or with all the circuits isolated. The means of earthing will be isolated from the installation earthed equipotential bonding for the duration of the test. Care should be taken to avoid any shock hazard to the testing personnel and other persons on the site whilst both establishing contact, and performing the test. Use a loop impedance tester for this test. Refer to Section 16 of this Guidance Note.

While testing the continuity of protective conductors of radial circuits, or whilst testing the continuity of ring final circuits, a value of $(R_1 + R_2)$ will have been measured at ambient temperature. The measured values of $(R_1 + R_2)$ for the final circuit should be added to the appropriate values of $(R_1 + R_2)$ for any distribution circuits.

An alternative means to determine $(R_1 + R_2)$ is to measure the loop impedance value at the extremity of the final circuit, taking care to use the correct phase supply and protective conductor return for the circuit. This loop impedance less the Z_e value measured earlier can be taken as the value of $(R_1 + R_2)$. These $(R_1 + R_2)$ values will be used to determine the value of Z_s of the circuit under fault conditions, but must be first corrected for conductor temperature. This requires the use of correction factors dependent on the type of cable used in the circuit, and ambient temperature at the time of continuity testing. Details of how to obtain them are given.

Residual current operated general purpose (non-delayed) devices to BS 4293. Prior to these r.c.d. tests it is essential, for safety reasons that the earth loop impedance is tested to check the requirements have been met. Use an r.c.d. tester for these tests.

Tests are made on the load side of the r.c.d. between the phase conductor of the circuit protected and the associated c.p.c. The load supplied should be disconnected during the test. Since r.c.d. testers require a few milliamperes to operate the instrument, these are normally obtained from the phase and neutral of the circuit under test. When testing a three-phase r.c.d. protecting a three-wire circuit, its neutral is required to be connected to earth. This means that the test current will be increased by the instrument supply current and will cause some devices to operate during the 50 per cent test at a time when they should not operate. It is necessary to check the operating parameters of the r.c.d. with the manufacturers before failing the device.

Under certain conditions (which are given) these tests can result in a potentially dangerous voltage on exposed and extraneous-conductive-parts. Precautions must therefore be taken to prevent contact of persons or livestock with such parts.

Tests are as follows: (i) with a fault current flowing equivalent to 50 per cent of the rated tripping current of the r.c.d. for a period of 2 s, the device should not open. (ii) with a fault current flowing equivalent to 100 per cent of the rated tripping current of the r.c.d., the device will open in less than 200 ms. Where the r.c.d. incorporates an intentional time delay it should trip within the time range of 50 per cent of the rated time delay plus 200 ms and 100 per cent of the rated time range plus 200 ms. Because of the variability of the time delay it is not possible to specify a maximum test time. It is therefore imperative that the circuit protective conductor does not rise more than 50 V above earth potential. It is suggested that in practice a 2 s maximum test time is sufficient. (iii) where the r.c.d. is used to provide supplementary protection against direct contact, with a test current of 150 mA the device should open in less than 40 ms. The maximum test time must not be longer than 50 ms. An integral test device is incorporated in each r.c.d. This device enables the mechanical parts of the r.c.d. to be checked. Tripping the r.c.d. by means of the above electrical tests and the integral test device establishes the following: (i) that the r.c.d is operating with the correct sensitivity. (ii) the integrity of the electrical and mechanical elements of the tripping device. Operation of the integral test device does not provide a means of checking: (i) the continuity of the earthing conductor or the associated circuit protective conductors, or (ii) any earth electrode or other means of earthing, or (iii) any other part of the associated installation earthing.

Battery insulation resistance testers. In the insulation resistance test covered by a Guidance Note the testing voltage may be derived from a hand-driven generator. There are now a number of battery-powered insulation testers available in which the high voltage is derived from special circuitry inside the unit.

The BM 101, figure 12.7 from Avo Megger Instruments Ltd has four scales, one reading insulation resistance values in megohms from 0–200 and infinity; the other three reading resistance from 0–1 MΩ, 0–200Ω and 0–2Ω respectively. The latter scale is suitable for continuity testing. A taut-band suspension indicator is used which is inherently robust so that the instrument is ideal for field use. A battery condition indicator is incorporated and a pushbutton is used to energise the circuit. A single 9 V dry battery provides the power source and the insulation testing voltage is 500 V d.c.

Intended for use where high values of insulation resistance may be met, such as h.v. cable, or where tests may be of long duration, is the BM8/2 battery-operated multi-voltage tester. It has four testing voltages selected by a rotary switch 100 V, 250 V, 500 V and 1,000 V. At 1,000 V the range is 0–20,000 megohms and infinity. At 100 V the range is 0–2,000 megohms and infinity. A black and white band on the scale indicates the battery condition. Like the BM101 it is pushbutton operated and has a power source provided by six 1.5 V cells.

Guidance Notes
1. Selection and Erection
2. Isolation and Switching
3. Inspection and Testing
4. Protection against Fire
5. Protection against Electric Shock
6. Protection against Overcurrent
7. Special Locations (not yet published)

Figure 12.7   Type BM101 insulation and continuity tester (Avo Megger Instruments Ltd)

13

Lighting

Virtually all buildings have electric lighting which serves two purposes. It helps us to recognise objects quickly and in sufficient detail to learn all we need to know about them, and it contributes to making buildings safe and pleasant places in which to work or take part in other activities. There must always be enough light to make objects visible but other factors are no less important. The directions from which light comes, the brightness and colour contrasts created between details of interest and their background, the presence or absence of bright reflections in the part of the object being looked at, and changes in colour resulting from the type of lamp used can all affect ease of recognition.

CIBSE. The Chartered Institute of Building Services Engineers, Lighting Division (formerly the Illuminating Engineering Society) has produced a code for interior lighting and some of the information contained in this chapter is taken from this code which was last published in January 1984. Some of the more common terms used in lighting design and the associated units are given below.

Luminous flux. The light emitted by a source, or received by a surface. It is expressed in lumens. Symbol: ϕ.

Lumen. This is the SI unit of luminous flux. An ordinary 100 W lamp for example emits about 1200 lumens. One lumen is the luminous flux emitted within unit solid angle (one steradian) by a point source having a uniform luminous intensity of one candela. Symbol: lm.

Luminous intensity. The quantity which describes the power of a source or illuminated surface to emit light in a given direction. It is the luminous flux emitted in a very narow cone containing the given direction divided by the solid angle of the cone. The result is expressed in candelas. Symbol: I.

Candela. The SI unit of intensity. It is 1 lumen per steradian. Symbol: cd.

Illuminance. The luminous flux density at a surface, i.e. the luminous flux incident per unit area. The quantity was formerly known as the illumination value or illumination level. It is expressed in lux (lumens/m^2 or lm/m^2). Symbol: E.

Standard service illuminance. The service illuminance throughout the life of an installation and averaged over the relevant area. This area may be the whole area of the working plane in an interior, or the area of the visual task and its immediate surround. It is expressed in lux. The tables that are contained in this chapter specifying the illuminance for specific areas is based on the standard service illuminance.

Lux. SI unit of illuminance. It is equal to one lumen per square metre.

Foot-candle. An obsolete unit of illuminance. It has the same value as the lumen per square foot.

Mean spherical illuminance: scalar illuminance. The average illuminance over the whole surface of a very small sphere located at a given point. It is expressed in lux. Symbol: E_s.

Mean cylindrical illuminance. The average illuminance over the curved surface of a very small cylinder located at a given point. Unless otherwise stated, the axis of the cylinder is taken to be vertical. It is expressed in lux. Symbol: E_c.

Luminance. A term which expresses the intensity of the light emitted in a given direction by unit area of a luminous or reflecting surface. It is the luminous flux emitted in the given direction from a surface element, divided by the product of the projected area of that element perpendicular to the prescribed direction and the solid angle containing the direction. It is expressed in lumens per square metre per steradian which is equivalent to candelas per square metre. In interior lighting design it is the product of the illuminance and the luminance factor (q.v.) for the particular conditions of illumination and viewing. If the surface can be assumed without too much error to be perfectly matt, its luminance in any direction is the product of the surface illuminance and its reflectance, and can be expressed in lm/m² (apostilbs). Symbols: L.

Candela per square metre.   SI unit of luminance in lumens per square metre per steradian. Unit luminance in this system is that of a uniform plane diffuser emitting π lumens per square metre. Symbol: cd/m^2. To convert the luminance from apostilbs to the SI unit divide apostilbs by π.

Apostilb.   A metric unit of luminance. Unit luminance is expressed in this system as that of a uniform diffuser emitting 1 lm/m². Symbol: asb. 1 asb = $1/\pi$ cd/m².

Luminosity.   Visual sensation associated with the amount of light emitted from a given area. The term brightness is used coloquially.

Glare.   The discomfort or impairment of vision experienced when parts of the visual field are excessively bright in relation to the general surroundings. There are a number of terms to express the amount of glare. For example, disability glare which prevents seeing detail; discomfort glare which causes visual discomfort but might not impair ability to see detail; direct glare caused when excessively bright parts of the visual field are seen directly, i.e. unshielded light sources.

I.E.S. glare index.   A numerical index calculated according to the method described in I.E.S. Technical Report No. 10. It enables the discomfort glare to be ranked in order of severity and the permissible limit of this from an installation to be prescribed quantitatively. This Technical Report has been updated and is now available as Technical Memorandum No. 10 of the CIBSE.

Luminous efficacy.   The ratio of the luminous flux emitted by a lamp to the power consumed by it. Expressed in lm/W.

Reflectance.   The ratio of the flux reflected from a surface to the flux incident upon it. The value is always less than unity. Symbol: ρ.

British Zonal system (BZ). A system for classifying luminaires according to their downward light distribution. The BZ class number denotes the classification of a luminaire in terms of the flux from a conventional installation directly incident on the working plane, relative to the total flux emitted below the horizontal. Although sometimes still quoted it has now been replaced, for the purpose of flux calculations by the methods outlined in CIBSE Technical Memorandum No. 5.

Room index. An index related to the dimensions of a room, and used when calculating the utilization factor and the characteristics of the lighting installation. It is given by:

$$\frac{lw}{h_m (l+w)}$$

where l is the length and w the width of the room and h_m the height of the luminaires above the working plane.

Utilisation factor. The total flux reaching the working plane divided by the total lamp flux.

PSALI. An abbreviation of 'permanent supplementary artificial lighting of interiors'.

PAL. An abbreviation of 'permanent artificial lighting'.

Reflection. Whenever light falls onto surface, some of it is absorbed and the remainder is either reflected or transmitted. If the surface is opaque and smoothly polished, the *specularly* reflected light leaves the surface at the same angle as it arrived (as a billiard-ball striking a cushion), and by suitably shaping the surface it is possible to redirect the light in any desired direction (e.g. a motor-car headlight, with lamp placed at the focal point of a polished parabolic mirror directing most of the light forward).

Diffuse reflection occurs from matt surfaces. The light is reflected most strongly at right-angles to the surface (whatever the direction from which the light arrives) and progressively more weakly at other angles. Matt surfaces show no highlights. Most painted and many other surfaces are partly specular and partly diffuse reflectors of light and are classified according to which type of reflection predominates.

282

Diffusion. Light passes straight through a transparent material, but is scattered or diffused to a greater or lesser extent in a translucent material. Flashed opal glass or its plastics equivalent scatters it completely so that it emerges in all directions, and complete concealment of lamps behind a panel of this material is easily achieved. Frosted glass diffuses the light less perfectly, so that it emerges mainly in the same general direction as when it entered the glass; in effect, it is usually possible to see vaguely the positions of lighted lamps behind frosted panels. Hammered and rolled glasses and clear plastics with a similar finish generally have less diffusing and concealing power than frosted glass but have a sparkle that may be preferred in many cases.

Refraction. If light passes through a transparent material which does not have parallel sides, it will be bent away from its original direction by a process known as *refraction*. Ribbed glass or plastic fittings in which each rib is a carefully designed prism can therefore be made to control light very accurately in a required direction, and this principle is very widely used in electric street lighting fittings.

Shadows. When an obstruction completely masks the *only* source of light from a point on, say, the floor, the shadow at that point will be complete, but where the source is only partially masked there will be only a partial shadow. It follows, then, that a concentrated light source tends to promote deep shadows with hard edges, while physically large sources promote soft and faint shadows; also that the greater the number of sources in a room, and the more light that is reflected from the ceilings and walls, the softer and fainter shadows become.

Fluorescent lighting is often called shadowless. This does not apply to single lamps, for though the length of the lamp tends to reduce shadows of linear objects at right-angles to its main axis, its small width tends to promote those of objects parallel to it. Thus the shadows cast depend partly on the shape of the obstruction and partly on the orientation of the fluorescent lamp with regard to it. When reduction of shadow from a particular fitting is desired, the aim should be to ensure that every important point on the working area can 'see' at least part of the fitting; thus, for lighting a kitchen sink, a fluorescent lamp should be placed above a parallel to the front edge of the sink, for in this position the housewife's head and shoulders obstructs less light than if the lamp were placed at right angles to this position.

Glare. Glare has been described as 'light out of place'. Properly applied light makes it easy and comfortable to see, glare has the reverse effect and means that money is being wasted reducing visual effectiveness. Broadly, glare is caused either by too much light entering the eye from the wrong directions, or by some things being too bright in relation to other surfaces in the normal field of view, and may be prevented or at least minimized by applying the following rules:

1. Use lighting fittings which put the downward light mainly where it is wanted – on the work – with relatively little escaping in the direction of the worker's eye. It is just as necessary to screen fluorescent lamps from normal angle of view as any other.

2. Make the actual detail being looked at a little brighter than anything else seen at the same time (e.g. white paper against a light-coloured desk-top). If dark cloth were viewed against a light-coloured surface, a local light would be required to make the cloth look bright.

3. Use light-coloured decorations and ensure that sufficient light from the fittings goes upwards and sideways to illuminate the ceiling and walls to make them fairly bright, and thus reduce the contrast of brightness between the fittings and their background.

4. Avoid if possible the use of glossy working surfaces (e.g. polished wood or glass table-tops) which mirror the lighting fittings.

5. The more extensive the installation, and the higher the illumination, the more carefully should these rules be followed.

The latest CIBSE Code for interior lighting published in 1984 giving recommendations for good interior lighting makes reference to glare index. This factor should be applied in all illumination calculations. Due to the complicated nature of the subject the reader should refer to the Code for a full explanation of its use.

ELECTRIC LAMPS

Filament lamps. Almost all filament lamps for general lighting service are made to last an average of at least 1000 hours. This does not imply that every individual lamp will do so, but that the short-life ones will be balanced by the long-life ones; with British lamps the precision and uniformity of manufacture now ensures that the spread of life is small, most individual lamps in service lasting more or just less than 1000 hours when used as they are intended to be used.

In general, vacuum lamps, which are mainly of the tubular and fancy shapes, can be used in any position without affecting their performance. The ordinary pear-shaped gasfilled lamps are designed to be used in the cap-up position in which little or no blackening of the bulb becomes apparent in late life. The smaller sizes, up to 150 W, may be mounted horizontally or upside-down, but as the lamp ages in these positions the bulb becomes blackened immediately above the filament and absorbs some of the light. Also vibration, may have a more serious effect on lamp life in these positions. Over the 150 W size, burning in the wrong position leads to serious shortening of life.

Coiled-coil lamps.   By double coiling of the filament in a lamp of given wattage a longer and thicker filament can be employed, and additional light output is obtained from the greater surface area of the coil, which is maintained at the same temperature thus avoiding sacrificing life. The extra light obtained varies from 20% in the 40 W size to 10% in the 100 W size.

Effect of voltage variation.   Filament lamps are very sensitive to voltage variation. A 5% over-voltage halves lamp life due to over-running of the filament. A 5% under-voltage prolongs lamp life but leads to the lamp giving much less than its proper light output while still consuming nearly its rated wattage. The rated lamp voltage should correspond with the supply voltage. Complaints of short lamp life very often arise directly from the fact that mains voltage is on the high side of the declared value, possibly because the complainant happens to live near a sub-station.

Bulb finish.   In general, the most appropriate use for clear bulbs is in wattages of 200 and above in fittings where accurate control of light is required. Clear lamps afford a view of the intensely bright filament and are very glaring, besides giving rise to hard and sharp shadows. In domestic sizes, from 150 W downwards, the pearl lamp – which gives equal light output – is greatly to be preferred on account of the softness of the light produced. Even better in this respect are silica lamps; these are pearl lamps with an interior coating of silica powder which completely diffuses the light so that the whole bulb surface appears equally bright, with a loss of 5% of light compared with pearl or clear lamps. Silica lamps are available in sizes from 40–200 W. Doublelife lamps compromise slightly in lumen output to provide a rated life of 2,000 hours.

Table 13.1 Average lighting design lumens
throughout of G.L.S. lamps

Watts	Doublelife lamps* 240 V	Neta-bulb* 240 V	g.l.s. lamps 240 V	g.l.s. lamps 110 V
25	–	–	200	–
40	350	365	390	460
60	595	625	665	770
100	1140	1185	1260	1450
150	1800	1925	2037	2160
200	–	–	2730	2980
300	–	–	4300	4710
500	–	–	7700	8270
750	–	–	12 400	–
1000	–	–	17 300	–

*These lamps have coiled-coil filaments

Reflector lamps. For display purposes reflector lamps are available in sizes of 25 W to 150 W. They have an internally mirrored bulb of parabolic section with the filament at its focus, and a lightly or strongly diffusing front glass, so that the beam of light emitted is either wide or fairly narrow according to type. The pressed-glass (PAR) type of reflector lamp gives a good light output with longer life than a blown glass lamp. Since it is made of borosylicate glass, it can be used out-of-doors without protection.

Tungsten halogen lamps. The life of an incandescent lamp depends on the rate of evaporation of the filament, which is partly a function of its temperature and partly of the pressure exerted on it by the gas filling. Increasing the pressure slows the rate of evaporation and allows the filament to be run at a higher temperature thus producing more light for the same life.

If a smaller bulb is used, the gas pressure can be increased, but blackening of the bulb by tungsten atoms carried from the filament to it by the gas rapidly reduces light output. The addition of a very small quantity of a haline, iodine or bromine, to the gas filling overcomes this difficulty, as near the bulb wall at a temperature of about 300°C this combines with the free tungsten atoms to form a gas. The tungsten and the haline separate again when the gas is carried back to the filament by convection currents, so that the haline is freed to repeat the cycle.

286

Tungsten halogen lamps have a longer life, give more light and are much smaller than their conventional equivalents, and, since there is no bulb blackening, maintain their colour throughout their lives. Mains-voltage lamps of the tubular type should be operated within 5 degrees of the horizontal. A 1000 W tungsten halogen lamp gives 21 000 lumens and has a life of 2000 hours. These lamps have all but replaced the largest sizes of g.l.s. lamps for floodlighting, etc. They are used extensively in the automotive industry. They are also making inroads into shop display and similar areas in the form of l.v. (12 V) single-ended dichroic lamps.

Discharge lamps. In discharge lamps the light output is obtained, not from incandescence, by by 'exciting' the gas or vapour content of the discharge tube. The excited gas emits energy of a characteristic wavelength and this may appear as several disconnected spectral lines. Light emitted in such a discontinuous manner may result in serious distortion of the colour appearance of objects seen in it. The main advantage of discharge lamps is their high luminous efficacy with very long life.

In any discharge lamp, once started, the current tends to increase instantaneously to a destructive value so to protect the lamp and wiring a device such as a choke has to be incorporated in the circuit to limit the current to a designed safe value. On d.c. supplies a resistance may be used which, in most cases, consumes about as much power as the lamp itself and thus lowers the luminous efficacy compared with operation on a.c.

The choke is connected in series with the lamp, and a power-factor capacitor is placed across the mains on the mains side of the choke. The lamp takes a few minutes, according to type, to reach full brightness. If switched off when hot, it will not re-start until it has cooled down, but since, in general, it will withstand a sudden voltage drop of about 30 V, ordinary fluctuations of mains voltage do not affect it seriously.

Fluorescent tubes. A tube is a mercury discharge lamp in which the inside of the discharge tube is coated with fluorescent phosphors. Because the vapour-pressure is low a higher proportion of u.v. radiation than of visible light is emitted and this radiation excites the fluorescent powders which then give off visible light. By choosing the phosphors correctly any colour can be produced. Because the light comes from the phosphors, the proportion of visible light to radiant heat is much greater in fluorescent tubes than in incandescent lamps. Consequently the lamps have a high efficacy and also 'run cool' so that they can be used in situations where filaments lamps would generate too much heat.

It used to be that the efficacies of the 'white' lamps in the range were closely related to their colour rendering properties. This is because the eye is less sensitive to the red and blue light at the ends of the spectrum than to the green and yellow in the middle and so more electrical energy must be used to produce the sensation of red light than is needed for the same apparent value of green and yellow. Consequently lamps with a high efficacy (white or warm white) have poor colour rendering properties while lamps which show objects in their true colours, i.e. Natural, Kolor-rite or Northlight, have a lower efficacy. In about 1980 narrow band or so-called 'rare earth' phosphors were introduced into fluorescent tubes. These enable good colour rendering lamps to be produced with high efficacy by having three or more narrow bands of energy in the red, green and blue wavelengths, which the eye interprets as 'white light'. These phosphors can operate at higher temperatures and have permitted the development of a new generation of compact lamps.

A range of higher efficacy narrower diameter (26 mm instead of 38 mm) lamps have been introduced. A change of gas fill has enabled the tubes to have similar electrical characteristics to conventional fluorescent tubes but with a 5% reduction in power consumption. Combined with the rare earth phosphors, lamps of higher efficacy and better colour rendering are now available. These tubes can be operated on the existing circuits with a reduction in power.

Several compact fluorescent lamps have been introduced recently, aimed at eventually replacing the g.l.s. design. The colour rendering of these is nearly as good as the g.l.s. lamps and typically they are five times more efficient. The Thorn 2D is an example and its characteristics have been included in Table 13.5.

Nearly all present day fluorescent tubes are of the 'hot cathode' type. This means that the electrodes at the ends of the lamp must be heated by passing an electric current through them before an arc will strike between them. These cathodes consist of tungsten wire or braid covered with or encasing a slug of electron emitting material. When the cathode is heated, a cloud of electrons forms around the cathodes at either end of the tube, ionising the argon within it. A choke is normally place in series with the tube to control the current in the arc and the heating current passes through it and both the cathodes.

The various circuits used to start fluorescent tubes are described below.

Glow starter circuit. As soon as the main circuit switch is closed, the full mains voltage is applied across the the electrodes of the glow starting switch. The voltage is sufficient to cause a glow discharge in the starting switch bulb. This has the effect of warming up the bi-metallic strips on which the switch contacts are mounted. The heating of these bi-metallic strips causes them to bend towards each other until the contacts touch. The glow discharge in the starter switch then disappears. The heater

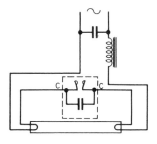

Figure 13.1   Glow starter circuit

elements which form the electrodes in the fluorescent tube are heated by the current which now passes through them. In the meantime the bi-metallic strips not being heated by the glow discharge cool down and spring away from each other. This sudden interruption of the circuit, which contains a choke, causes a voltage surge across the fluorescent lamp electrodes which starts the discharge in the fluorescent lamp.

Electronic ignitors. A number of electronic ignitors for fluorescent tubes are now on the market. These may take the form of a simple replacement for a conventional starter switch or be incorporated in a complete ballast circuit. Most of them are solid state devices that provide a cathode-heating current for a controlled period, followed by a starting pulse voltage which may utilise the choke winding as does the conventional glow starter. They operate at temperatures down to −5°C, require only a simple choke ballast, and have the same low watts loss as a glow switch. They are a 'first time' device and eliminate the blinking which sometimes occurs with conventional starters and shortens lamp life. The 'stuck starter' condition cannot occur and they are permanent devices and do not need to be replaced. They are available for all sizes of fluorescent tubes.

The series circuit. Fluorescent lamps more than 600mm long require one set of control gear each; lamps of 600mm length and less can either be run singly on 100–130 V or on 200–260 V (except 40 W 600mm) or, as is generally the case, two in series with a single choke on 200–260 V. The circuit is shown in Figure 13.2 and it will be seen that the lamps will start one after the other. Both

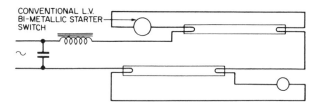

Figure 13.2 Series operation of short fluorescent lamps with switched starting

must be of the same wattage but not necessarily of the same colour. Series running is economical since only one choke is necessary per pair of lamps, and choke losses are small, but if one lamp or starter misbehaves it may also affect the other lamp; therefore trouble in this circuit should be investigated as soon as pssible. If series-operated lamps are to be run on a Quickstart circuit a double-wound transformer is needed as shown in Figure 13.3 and both lamps must have the earthed metallic strip.

Quickstart circuit. A fluorescent tube circuit designed to give a rapid start without flickering is shown in Figure 13.4. The unit consists of an auto-transformer, the primary winding of which is connected across the fluorescent tube, with the secondary winding in two separate sections, one across each cathode. This method is not normally used today but is included because it may be found in old installations.

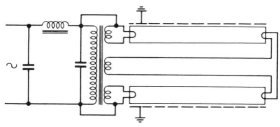

Figure 13.3 Series operation of short fluorescent lamps with Quickstart circuits

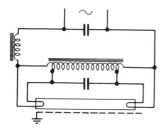

Figure 13.4 Quickstart circuit for single-lamp operation

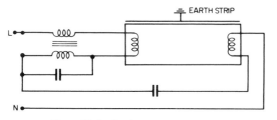

Figure 13.5 Semi-resonant start circuit

Semi-resonant start circuit. In this circuit the place of the choke is taken by a specially wound transformer. Current flows through the primary coil to one cathode of the lamp, and thence through the secondary coil which is wound in opposition to the primary. The other end of the secondary is connected to a fairly large capacitor, and thence through the second cathode of the lamp to neutral (Figure 13.5). This method is not normally used today but is included because it may be found in old installations.

Table 13.2 Resistors for D.C. operation

Mains voltage	Resistor values (ohms)					
	1–80 W	1–40 W 1200 mm	1–30 W	2–40 W 600 mm	2–20 W	2–15 W
200	103	208	264	116	182	235
210	116	235	293	128	208	264
220	128	264	330	147	235	293
230	147	293	380	147	264	330
240	166	330	420	166	293	380
250	166	330	420	166	330	380

Electronic ballast circuit. The recently introduced high frequency (HF) electronic ballast provides silent, instant flicker-free starting for single and twin standard fluorescent lamps up to 1800 mm in length, both 26 mm and 38 mm in diameter. The new ballasts eliminate the need for a separate starter, choke and capacitor. The basic construction of a HF ballast involves a harmonic filter, rectifier and inverter, similar to those used in emergency lighting. The principle of operation is to convert 50 or 60 Hz mains input into a d.c. voltage and then convert this to a high frequency, around 32 kHz, for operating fluorescent lamps. The HF ballast takes advantage of a characteristic of fluorescent lamps whereby greater efficiency is obtained at high frequency. The overall lighting system efficacy can be increased by 20 to 30% due to:

(i) Improved lamp efficacy at high frequency operation.
(ii) Reduced circuit power losses.
(iii) Lamp operates closer to optimum performance in most enclosed luminaires.

For instance an 1,800 mm lamp normally rated at 70 W with standard control gear can now be run at 62 W for the same light output. Furthermore ballast losses are reduced – in the case of a twin 1,800 mm circut from 26 W (starter switch circuit) to only 8 W with a HF ballast. The overall achievement then for a twin 70 W 1,800 mm circuit is some 20% being 132 W (that is 2×62 W plus 8 W) compared with 166 W (2×70 W plus 26 W). Together these energy saving features enable lighting levels to be maintained with a dramatic cut in electricity costs.

D.C. working. If desired, fluorescent lamps may be used on d.c. supplies by omitting the normal power-factor capacitor and inserting a suitable resistor in series with the choke. This resistor consumes a wattage generally roughly equal to that of the lamp, so that luminous efficacy of the circuit is about half of that on a.c. Suitable resistance values are given in Table 13.2.

After a period of working the positive end of the tube may darken owing to migration of the mercury to the negative end, and to counteract this tendency it is usual to fit a polarity-reversing switch to the sub-circuit, preferably of the unidirectional rotary type so that polarity is changed at every switching. A thermal starter is normally used.

Colour appearance and colour rendering. The light output of a fluorescent tube is not necessarily the only consideration. Good colour-rendering can be most important and there are many situations where the use of a de luxe colour such as Polylux can produce a stimulating atmosphere which far outweighs the small additional cost. In some cases where critical colour matching is important the old Kolor-rite or Natural tubes may be appropriate.

Operating characteristics. Fluorescent lamps emit about one-third as much total heat as filament lamps giving the same amount of light, but only about one-fifth as much radiated heat. Their light output varies by about 1% for each 1% change in mains voltage (compared with 4% variation in light of filament lamps). In general, light output after 5000 hours' burning is about 85% of that at 100 hours.

Effect of operating temperature. Fluorescent lamps are designed to operate at ambients of about 25°C. If they are used at elevated temperatures, a drop in light output occurs due to reduction in u.v. emission and enhancement of the visible mercury lines caused by an increase in vapour pressure. One way to

Table 13.3 Tubular fluorescent lamps

Lamp rating W	Nominal lamp length, mm	Actual length, mm	Diameter, mm	Nominal tube voltage	Nominal lamp current, A
125	2400	2389.1	38	152	0.94
100	2400	2389.1	38	128	0.96
85	2400	2389.1	38	185	0.54
70	1800	1778.0	26	128	0.70
75/85	1800	1778.0	38	123	0.77
80/65	1500	1514.3	38	100/110	0.87/0.67
58	1500	1514.3	26	113	0.63
50	1500	1514.3	26	160	0.38
40	1200	1213.6	38	102	0.44
36	1200	1213.6	26	104	0.42
30	900	908.8	26	101	0.36
40	600	604.0	38	104	0.42
20	600	604.0	38	58	0.38
18	600	604.0	26	58	0.38
15	450	451.6	26	57	0.34
13	525	531.0	16	92	0.17
8	300	302.4	16	55	0.17
6	225	226.2	16	43	0.16
4	150	150.0	16	30	0.15
2D lamps	Lamp width and length, mm		Depth mm		
38	205		35	110	0.49
28	205		35	107	0.32
16	140		27	97	0.20

overcome this is to use 'Amalgam' lamps in which a ring of indium near the end of the lamps absorbs mercury and thus reduces pressure. The other is to use ventilated or ventilating fittings.

Low ambient temperatures also cause a fall-off in light output due to the condensation of mercury on the tube walls. Enclosing lamps in a clear acrylic sheath usually maintains them at the correct working temperature in low temperature surroundings.

Fault finding.   A fault in some part of the circuit may often be traced by observing the following symptoms:

Lamp glows continuously at both ends, or one end, but makes no effort to start.   In the first case, faulty starter; in the second, an earth in some part of the circuit and possibly a faulty starter also. With the instant-start circuit, lamp has no earthing strip, or

Table 13.4 Light output of standard fluorescent tubes

	38 / 2400 / 125	38 / 2400 / 100	38 / 2400 / 85	38 / 1800 / 75	26 / 1800 / 70	38 / 1500 / 65	26 / 1500 / 58	26 / 1500 / 50	38 / 1200 / 40	26 / 1200 / 36	26 / 1050 / 40	26 / 900 / 30	38 / 600 / 40	38 / 600 / 20	26 / 600 / 18	26 / 525 / 15	16 / 525 / 13	16 / 300 / 8	16 / 225 / 6	16 / 150 / 4
Nominal diameter mm	38	38	38	38	26	38	26	26	38	26	26	26	38	38	26	26	16	16	16	16
Nominal length mm	2400	2400	2400	1800	1800	1500	1500	1500	1200	1200	1050	900	600	600	600	525	525	300	225	150
Rated wattage	125	100	85	75	70	65	58	50	40	36	40	30	40	20	18	15	13	8	6	4
Lighting design lumens 2000 hours																				
Polylux 2700		8900		6300		5100			3200											
Polylux 3500		8900		6300		5100			3200											
Polylux 4000		8900		6300		5100			3200											
Pluslux 3000 (warm white)		7900		5500		4400			2750											
Pluslux 3500 (white)		8000		5600		4500			2800											
Pluslux 4000 (cool white)		7600		5400		4400			2750											
White	8800	8000	6850	5750		4750	3600	2800	2800	2150	1700	1100			1300	800	750	420	250	100
Pluswhite	8400	7600	6500	5500		4500		2700							1300					
Warmwhite	8700	7900	6750	5650		4600	3550	2700	2700	2150	1700	1100			1300	800	750	420	250	
Coolwhite	8500	7800	6500	5450		4450		2650		2050	1600	1050			1300	750	700	360	240	
Natural	6500	5900	5000	4000		3400	2400	2100		1600						800	600			
Kolor-rite	5700	5200	4400	3500		3000		1800								750				
Northlight colour matching	5600	5000	4100	3200		2700		1700		1250						700	500			
Deluxe natural	4800	4400	3800	2900		2500		1500		1100						600	500			
Artificial daylight	3800			2400		2100		1200								500				
Gro-lux						1300		810		530						340	200			

Table 13.5 Colour appearance and colour rendering properties of fluorescent lamps

Tube colour	Colour rendering quality	Colour appearance	Applications
Polylux 2700 Pluslus 3000	Very good Fair	Warm Warm	Tubes of various efficacies for use in social residential and domestic situations
Polylux 3500 Pluslux 3500 Plus White	Very good Fair Good	Intermediate Intermediate Intermediate	Tubes of various efficacies for general illumination of work areas – shops, factories, warehouses, etc.
Polylux 4000 Pluslux 4000 Natural	Very good Fair Very good	Cool Cool Cool	Tubes of various efficacies for work areas requiring illumination to blend with natural daylight – offices, shops, etc.
Deluxe natural	Good	Intermediate	Butchers, fishmongers, supermarkets. Enhances the appearance of red objects
Kolor-rite	Excellent	Cool	Complies with DHSS requirements for hospital lighting
Northlight/Colour matching	Very good	Cool	Areas for matching materials, etc. Any application where a wintry effect or an impression of coolness is required
Artificial daylight	Excellent	Cool	Areas for exact colour matching. Best colour rendering with cool appearance. Meets BS950 part 1
Gro-lux	–	–	For vivid colour effects on tropical fish or plant displays
Colours	Poor	–	Saturated colours for display, floodlighting, stage lighting

earth is poor, or low mains voltage; if glowing at one end only, faulty transformer or leads to one end of the lamp short- or open-circuited.

Lamp flashes repeatedly but cannot start. Faulty starter giving insufficient pre-heating time; or an old lamp. In the latter case one or both ends of the lamp may have become blackened and the lamp may light normally for a few moments, then die away with a shimmering effect.

Lamp lights normally but extinguishes after a few seconds, then repeats. Probably abnormally low mains voltage. May also be due to faulty starter.

Swirling effect of light in lamp. Probably disappears after a few switching operations on a new lamp. If it persists, change starter. If it still persists, renew lamp.

Cold cathode fluorescent tubes. Another type of fluorescent tube – the cold cathode type is usually made of 20 mm tubing coated with fluorescent powder and with either a mercury and argon or neon filling. It is of the high-voltage type, operated from a step-up transformer and there is no delay period in switching on. A life of 10 000 hours or more may be expected and is unaffected by the frequency of switching. Mercury-filled tubes show the usual drop in efficiency throughout life, but that of neon-filled lamps remains constant. Both types remain alight with severe reductions in mains voltage, and can be dimmed by suitable apparatus.

Cold cathode lamps are manufactured in a wide range of standard colours including daylight, warm white, blue, green, gold and red. The colours can be used separately or mixed to give any desired result, the gold and red lamps being particularly useful for providing a warm-toned light. Dimming of one of mixed colours will provide a colour change. As the tubes can be manufactured in a variety of shapes and curves, they are suitable for decorative illumination in restaurants, etc. in situations where the higher voltage required for operating them will not be objectionable.

The main problems are their relative inefficiency as compared to hot cathode lamps and the high voltages needed for starting and operation.

Mercury and metal halide lamps. The mercury spectrum has four well-defined lines in the visible area and two in the invisible ultra violet region. This u.v. radiation is used to excite fluorescence in certain phosphors, by which means some of the missing colours can be restored to the spectrum. The proportion of visible light to u.v. increases as the vapour pressure in the discharge

Table 13.6 Electric discharge lamps

Type	Watts	Bulb shape	Length, mm	Dia., mm	Design lm	Cap
Mercury MBF	50	Oval	129	56	1900	E27 (ES)
	80	Oval	154	71	3650	E27 (ES)
	125	Oval	175	76	5800	E27 (ES)
	250	Oval	227	91	12 500	E40 (GES)
	400	Oval	286	122	21 500	E40 (GES)
	700	Oval	328	143	38 000	E40 (GES)
	1000	Oval	410	167	58 000	E40 (GES)
MBF-DL	80	Oval	154	71	3650	E27 (ES)
	125	Oval	175	76	6200	E27 (ES)
	250	Oval	227	91	13 300	E40 (GES)
	400	Oval	286	122	22 800	E40 (GES)
Mercury-reflector MBFR	250	Parabolic	260	166	10 500	E40 (GES)
	400	Parabolic	300	181	18 000	E40 (GES)
	700	Parabolic	328	202	32 500	E40 (GES)
	1000	Parabolic	380	221	48 000	E40 (GES)
Mercury-tungsten MBFT	160	Oval	175	76	2560	B22 (BC)
	250	Oval	227	91	4840	E40 (GES)
	500	Oval	286	122	11 500	E40 (GES)
Metal-halide MBIL	750	Linear	254	21	58 500	R × 75
	1000	Linear	254	21	110 000	R × 75
MBI-T	150	Linear (SE)	80	22	12 000	G12
CSI	1000	Parabolic	175	205	67 000	G38
MBIF	250	Oval	227	91	16 000	E40 (GES)
	400	Oval	286	122	24 000	E40 (GES)
	1000	Oval	410	167	85 000	E40 (GES)
High-pressure sodium SON	50	Oval	154	71	3100	E27 (ES)
	70	Oval	154	71	5300	E27 (ES)
	150	Oval	227	91	15 000	E40 (GES)
	250	Oval	227	91	25 500	E40 (GES)
	400	Oval	286	122	45 000	E40 (GES)
	1000	Oval	410	167	110 000	E40 (GES)
SONDL	150	Oval	227	91	11 000	E40 (GES)
	250	Oval	227	91	19 000	E40 (GES)
	400	Oval	286	122	33 000	E40 (GES)
SON-T	50	Cylindrical	154	39	3 100	E27 (ES)
	70	Cylindrical	154	39	5500	E27 (ES)
	150	Cylindrical	210	47	15 500	E40 (GES)
	250	Cylindrical	257	47	27 000	E40 (GES)
	400	Cylindrical	285	47	47 000	E40 (GES)
	1000	Cylindrical	380	67	120 000	E40 (GES)
SONDL-T	150	Cylindrical	210	47	11 500	E40 (GES)
	250	Cylindrical	257	47	20 500	E40 (GES)
	400	Cylindrical	285	47	34 000	E40 (GES)
SON.TD	250	Linear	189	24	25 000	2R × 7s
	400	Linear	254	24	46 000	2R × 7s
Low-pressure sodium SOX	18	Linear (SE)	210	53	1750	B22 (BC)
	35	Linear (SE)	311	53	4500	B22 (BC)
	55	Linear (SE)	425	53	7500	B22 (BC)
	90	Linear (SE)	528	67	12 500	B22 (BC)
	135	Linear (SE)	775	67	21 500	B22 (BC)

tube, so that colour correction is less effective in a high-pressure mercury lamp than in a low-pressure (fluorescent) tube.

High pressure mercury lamps are designated MBF and the outer bulb is coated with a fluorescent powder. MBF lamps are now commonly used in offices, shops and indoor situations where previously they were considered unsuitable. Better colour rendering lamps have recently been introduced with a slight increase in efficacy. They are designated MBF de-luxe or MBF-DL lamps and are at present slightly more expensive than ordinary MBF lamps.

A more fundamental solution to the problem of colour rendering is to add the halides of various metals to mercury in the discharge tube. In metal halide lamps (designated MBI) the number of spectral lines is so much increased that a virtually continuous emission of light is achieved, and colour rendering is thus much improved. The addition of fluorescent powders to the outer jacket (MBIF) still further improves the colour rendering properties of the lamp, which is similar to that of a de luxe natural fluorescent tube.

Metal halide lamps are also made in a compact linear form for floodlighting (MBIL) in which case the enclosed floodlighting projector takes the place of the outer jacket and in a very compact form (CSI) with a short arc length which is used for projectors, and encapsulated in a pressed glass reflector, for long range floodlighting of sports arenas, etc. In addition, single-ended low wattage (typically 150 W) metal halide lamps (MBI-T) have been developed offering excellent colour rendering for display lighting, floodlighting and uplighting of commercial interiors.

No attempt should ever be made to keep an MB and MBF lamp in operation if the outer bulb becomes accidentally broken, for in these types the inner discharge tube of quartz does not absorb potentially dangerous radiations which are normally blocked by the outer glass bulb.

Sodium lamps.   Low pressure sodium lamps give light which is virtually monochromatic; that is, they emit yellow light at one wavelength only, all other colours of light being absent. Thus white and yellow objects look yellow, and other colours appear in varying shades of grey and black.

However, they have a very high efficacy and are widely used for streets where the primary aim is to provide light for visibility at minimum cost; also for floodlighting where a yellow light is acceptable or preferred.

The discharge U-tube is contained within a vacuum glass jacket which conserves the heat and enables the metallic sodium in the tube to become sufficiently vaporized. The arc is initially struck in neon, giving a characteristic red glow; the sodium then becomes vaporised and takes over the discharge.

Sometimes leakage transformers are used to provide the relatively high voltage required for starting, and the lower voltage required as the lamp runs up to full brightness – a process taking up to about 15 minutes. Modern practice is to use electronic ignitors to start the lamp which then continues to operate on a conventional choke ballast. A power-factor correction capacitor should be used on the mains side of the transformer primary.

A linear sodium lamp (SLI/H) with an efficacy of 150lm/W is available and in the past was used for motorway lighting. The outer tube is similar to that of a fluorescent lamp and has an internal coating of indium to conserve heat in the arc. Mainly because of its size the SLI/H lamp has been replaced with the bigger versions of SOX lamps as described above.

Metallic sodium may burn if brought into contact with moisture, therefore care is necessary when disposing of discarded sodium lamps; a sound plan is to break the lamps in a bucket in the open and pour water on them, then after a short while the residue can be disposed of in the ordinary way. The normal life of all sodium lamps has recently been increased to 4 000 hours with a objective average of 6 000 hours.

SON high-pressure sodium lamps. In this type of lamp, the vapour pressure in the discharge tube is raised resulting in a widening of the spectral distribution of the light, with consequent improvement in its colour-rendering qualities. Although still biassed towards the yellow, the light is quite acceptable for most general lighting purposes and allows colours to be readily distinguished. The luminous efficacy of these lamps is high, in the region of 100lm per watt, and they consequently find a considerable application in industrial situations, for street lighting in city centres and for floodlighting.

Three types of lamp are available; an elliptical type (SON) in which the outer bulb is coated with a fine diffusing powder, intended for general lighting; a single-ended cylindrical type with a clear glass outer bulb, used for flood-lighting, (SON.T); and a double-ended tubular lamp (SON.TD) also designed for flood-lighting and dimensioned so that it can be used in linear parabolic reflectors designed for tungsten halogen lamps. This type must always be used in an enclosed fitting.

The critical feature of the SON lamp is the discharge tube. This is made of sintered aluminum oxide to withstand the chemical action of hot ionised sodium vapour, a material that is very difficult to work. Recent research in this country has resulted in improved methods of sealing the electrodes into the tubes, leading to the production of lower lamp ratings, down to 50 W, much extending the usefulness of the lamps.

Most types of lamps require some form of starting device which can take the form of an external electric pulse ignitor or an internal starter. At least one manufacturer offers a range of EPS lamps with internal starters and another range that can be used as direct replacements for MBF lamps of similar rating. They may require small changes in respect of ballast tapping, values of p.f. correction capacitor and upgrading of the wiring insulation to withstand the starting pulse voltage. Lamps with internal starters may take up to 20 minutes to restart where lamps with electronic ignition allow hot restart in about 1 minute.

Considerable research is being made into the efficacy and colour rendering properties of these lamps and improvements continue to be introduced.

Recent developments have led to the introduction of SON de luxe or DL lamps. At the expense of some efficacy and a small reduction in life far better colour rendering has been obtained. They are increasingly being used in offices and shops as well as for industrial applications.

ULTRA-VIOLET LAMPS

The invisible ultra-violet portion of the spectrum extends for an appreciable distance beyond the limit of the visible spectrum. That part of the u.v. spectrum which is near the visible spectrum is referred to as the near u.v. region. The next portion is known as the middle u.v. region and the third portion as the far u.v. region. 'Near' u.v. rays are used for exciting fluorescence on the stage, in discos, etc.

'Middle' u.v. rays are those which are most effective in therapeutics. 'Far' u.v. rays are applied chiefly in the destruction of germs, though they also have other applications in biology and medicine, and to excite the phosphors in fluorescent tubes.

Apart from their use in the lamps themselves fluorescent phosphors are used in paints and dyes to produce brighter colours than can be obtained by normal reflection of light from a coloured surface. These paints and dyes can be excited by the use of fluorescent tubes coated with phosphors that emit near ultra violet to reinforce that from the discharge. They may be made of clear glass in which case some of the visible radiation from the arc is also visible, or of black 'Woods' glass which absorbs almost all of it. When more powerful and concentrated sources of u.v. are required, as for example, on stage, 125 W and 175 W MB lamps with 'Woods' glass outer envelopes are used.

Since the 'black light' excites fluorescence in the vitreous humour of the human eye, it becomes a little difficult to see clearly, and objects are seen through a slight haze. The effect is quite harmless and disappears as soon as the observer's eyes are no longer irradiated.

Although long wave u.v. is harmless, that which occurs at about 3000 nm is not, and it can cause severe burning of the skin and 'snow blindness'. Wavelengths in this region, which are present in all mercury discharges are completely absorbed by the ordinary soda lime glass of which the outer bulbs of high pressure lamps and fluorescent tubes are made, but they can penetrate quartz glass. A germicidal tube is made in the 30 W size and various types of high pressure mercury discharge lamps are made for scientific purposes. It cannot be too strongly emphasised that these short-wave sources of light should not be looked at with the naked eye. Ordinary glass spectacles (although not always those with plastics lenses) afford sufficient protection.

Note that if the outer jacket of an MBF or MBI lamp is accidentally broken, the discharge tube may continue to function for a considerable time. Since short-wave u.v. as well as the other characteristic radiations will be produced these lamps can be injurious to health and should not be left in circuit.

'Life' and light output of lamps. In an incandescent lamp failure occurs long before its light output has fallen a substantial amount, indeed in a tungsten halogen lamp there is virtually no difference between the light output of a new lamp and that of one at the point of failure. This is by no means the case with discharge lamps, including fluorescent tubes, which will often 'run' for many

302

hours longer than their rated 'life' and may finally prove quite uneconomical in use, since, although they still consume about the same amount of power, their light output may be reduced to a fraction of its original value.

In consequence, all British and most European manufacturers no longer publish 'life' figures for these types of lamp, but issue lumens maintenance and lamp mortality curves to show the rate at which light-output may be expected to diminish during use and the percentage of lamp failures which may occur within a representative batch. These factors taken in conjunction with the likely

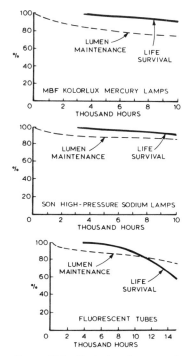

Figure 13.6 Typical performance in laboratory conditions

deterioration of reflecting surfaces of ceilings, walls and floors and the accumulation of dirt on the reflectors and diffusers of luminaires, make it possible to determine the point at which lamps should be replaced. Typical curves for various types of discharge lamps are shown in Figure 13.6.

Failure of a fluorescent tube is not normally related to fracture of the filament or cathode. It generally occurs when the electron emitting material on the cathodes ceases to produce electrons in sifficient quantity to permit the lamp to strike. A symptom of the end of life is a strong yellow glow at one end and severe blackening of both ends of the tube.

Before this occurs, however, the reduction of light-output of the phosphors will have brought the total light-output down to an uneconomic figure. For this reason tubes are normally replaced before actual failure.

INTERIOR LIGHTING TECHNIQUES

Types of fittings. Fittings should be chosen according to their qualities in the following respects: appearance; light distribution; brightness; ease of erection, relamping and cleaning; cost; and luminous efficacy. The order of importance changes with different applications, but light distribution should always be considered first.

Fittings classed as 'mainly direct' give most of their light in one direction. With any given type of lamp they will give the highest illuminance per watt of electrical input, with some likelihood of glare either direct from the lamp or by reflection from polished working surfaces, with a tendency to hard and deep shadows, but with the possibility of ensuring that at any particular working position most of the light comes from a desired direction – as may be required in say a drawing office.

On the other hand, 'mainly indirect' fittings, which give all or nearly all their light upwards, have the opposite characteristics. Indirect lighting can be an efficient system measured in illuminance per watt, and there is no likelihood of glare except from the ceiling at high illuminance; shadows are very soft and weak, and the light at every working point is received more or less equally from all directions. Midway between these extremes is the 'General Diffusing' class of fitting which, of course, has intermediate characteristics.

304

'Direct' lighting is almost always used for industry and for display purposes in shops; it gives a brisk, lively effect which emphasizes light and shade and reveals shape well. Indirect lighting is usually considered more restful and is used mainly for restaurants, hotels and other 'social' interiors, and in combination with direct lighting for many classes of shops although it is becoming increasingly popular for office lighting using so-called 'Uplights' fitted with SON-DL or MBIF lamps. 'General diffusing' lighting, or 'direct', is generally used in offices and schools.

Spacing of fittings. In most interiors which contain several lighting fittings the aim is usually to provide 'general lighting', i.e. a comparatively even illumination at working level all over the room. This allows furniture and plant to be moved or added to without altering the lighting installation Too wide a spacing of fittings would give the effect of comparative darkness between pools of light, whereas an unnecessarily close spacing would be uneconomic. As a general rule, applying both to fluorescent lamp fittings and other types normally used for general lighting, a satisfactorily even illumination results when the fittings are not spaced more than 1½ times their distance above the plane of work, with end fittings half this distance from the wall (one-third, if work is done next to the wall).

Interior lighting design. It is not the purpose of this section to provide enough information to design a complete lighting installation but to give general guidance on the matter. One design method is based on calculating the number of lamp lumens required and from this, based on the light source, the number of fittings and the mounting conditions, the layout of the installation.

Multiplying the area to be lit by the illuminance required gives the total lumens needed at the working place, taken at a point 0.85 m above floor level. For example if a room 10 m by 25 m is to have an illuminance of 1000 lux, the total lumens required at the working plane is $10 \times 25 \times 1000 = 250\,000$ lumens. Because of the absorption of light by the walls, ceiling and floor and by the lighting fittings themselves the number of lumens emitted by the lamps must be in excess of 250 000. The ratio of the actual number of lumens reaching the working plane in a particular area to the number of lumens emitted by the lamp is called the utilance or coefficient of utilisation. In addition allowance in the calculation must be made for the effect of dirt or dust on the fittings themselves, the maintenance factor. The importance of maintenance is stressed on page 308.

The formula for calculating the total lamp lumens is:

$$F = \frac{E_{av} \times A}{UF \times M \times Abs}$$

where F is the total lamp lumens; E_{av} the average illuminance (in lux); A the area to be lighted in square metres; UF the utilance factor; M the maintenance factor; Abs the absorption factor, which can be ignored except in very dusty atmospheres.

The next step is to calculate the number of lamps and fittings required. There is a close relationship between the spacing and mounting height of the lighting fittings. Too wide a spacing results in a fall off in illuminance between the fittings. Illumination anywhere on the working plane should not drop below 70% of that directly below a fitting. A space/mounting height ratio of $1\frac{1}{2}$:1 above the working plane is usually considered adequate in most cases. It is normal practice to space fittings half the distance from the walls that they are spaced apart. In some cases, e.g. where the wall must be lighted this space is reduced. The actual maximum spacing to mounting height ratio that a fitting can be used at is normally incorporated with the UF tables for the fitting. This value should never be exceeded in normal circumstances.

The majority of lighting engineers arrive at the minimum number of fittings required by scaling them off on the plan. Alternatively it can be derived from the formula:

$$\frac{L}{MS} \times \frac{W}{MS}$$

where L and W are the length and width of the room (m) and MS is the maximum space between fittings (m). Each part of the equation is worked out separately to the nearest whole number *above* the actual answer.

The lamp lumens per fitting having been calculated, the number and rating of lamps or tubes is determined from the lamp characteristic tables (see pages 286, 294, 298).

A practical example illustrates the use of the information above. It is proposed to light an office 15 m long by 10 m wide to an illuminance of 400 lux using fluorescent tubes. The ceiling height is 3 m and its reflectance 75%. Wall reflectance is 30% but owing to windows running along both long sides the average reflectance is dropped to 10%. The ceiling is unobstructed by beams. Maximum spacing allowed is 4.5 m. This gives a requirement of 12/9 fittings from the formula but the best arrangement which fits into a room of that shape is the three rows of fluorescent fittings, making a total of nine. The room index is 3 if the height of the working plane above floor level is taken as 1 m. (*Note:* Room index is calculated from the formula on page 282.)

If Thorn 'Clipper' fittings with prismatic diffusers are employed and mounted on the ceiling a utilisation factor of 0.61 obtains for a single lamp and 0.68 for a twin lamp fitting.

From the formula $F = E_{av} \times A/UF \times M \times Abs$, we see that the total lumens required is 60000 (floor area 150 m^2 $\times$ 400 lux) divided by the utilance (0.68 for a twin-lamp fitting) and a maintenance factor of 0.8. (We can ignore *Abs* as this is a clean atmosphere.) This gives us a figure of 108000 lumens which, divided by the twelve fittings, shows 9000 lumens per fitting or 4500 lumens per lamp. A 1500 mm 65 W white tube gives 4750 lumens, so we can use twelve fittings, each carrying two lamps of this rating. Note that occasionally the number of fittings installed may have to be higher than the calculated minimum. This can happen where fittings of the required lumen output are not available or where a certain rating of lamp or tube is specified to allow standardisation of replacement lamps throughout an installation.

Maintenance factor. Lighting levels within a building decrease progressively due to the accumulation of dirt on windows, fittings and room surfaces, and to the fall in lumen output of the lamps themselves. Illuminance values, therefore, vary continuously, decreasing because of depreciation but being restored by cleaning redecorating and relamping.

When natural and artificial lighting installations are designed, allowance is made for depreciation by incorporating appropriate factors in the design formulae as indicated earlier. The tables in this chapter giving recommended illuminances are for average service conditions and so the factors adopted to take account of the lamp, fitting and room surface depreciation relate to these average conditions. The maintenance factor M is defined as the ratio of the illuminance provided by an installation in the average condition of dirtiness expected in service, to the illuminance from the same installation when clean. Selection of maintenance factor for an installation is rather complicated and the reader is referred to IES technical Report No. 9, 1967. A modified and shortened version of this is reprinted in Appendix 7 of the 1984 CIBSE code.

Cleaning of fittings. The fall in light output due to dirt collecting on the light controlling surfaces of a fitting can usually be almost completely recovered by cleaning. The optimum economic cleaning interval (T) is that at which the cost of the light loss by dirt on the fitting equals the cost of cleaning the fitting. It is given by:

$$T = \frac{-C_c}{C_a} + \sqrt{\frac{2C_c}{C_a\Delta}}$$

where C_c is the cost of cleaning the fittings once; C_a is the annual cost of operating the fittings without cleaning; Δ is the notional average rate of decrease of luminous flux caused by dirt deposition and diminution of light from the tube, Δ can be calculated from the formula:

$$\Delta = (E_0 - E_1)/(E_0 \times T)$$

where E_0 is the initial illumination; E_1 is the minimum illumination after time T and T is time in years.

Maintenance techniques. Techniques for lighting maintenance fall into three categories, relamping, cleaning of fittings and cleaning and redecoration of room surfaces. Spot replacement of filament lamps calls for no comment but discharge lamps should be changed as quickly as possible after failure to prevent damage to associated control gear. If bulk replacement of lamps is adopted then this should take place when failures have reached about 20% of the total. Alternatively bulk replacement can take place when the illuminance falls below an agreed level which should not be less than the minimum value as defined in IES Technical Report No. 9.

All inside walls, partitions and ceilings should be washed at least once every 14 months and either whitewashed every 14 months or repainted every 7 years. Surfaces decorated with a washable water paint should be repainted every 3 years.

FLOODLIGHTING TECHNIQUES

This form of outdoor illumination has three main applications, (*a*) for industrial purposes, i.e. the lighting of railway sidings and

Table 13.7 Recommended values of illuminance

The recommendations in this table are taken from the CIBSE Code 1984. For more detailed recommendations the reader is advised to consult the Code itself.

Task group and typical task or interior	Standard service illuminance (lux)
Storage areas and plant rooms with not continuous work	150
Casual work	200
Rough work Rough machining and assembly	300
Routine work Offices, control rooms, medium machining and assembly	500
Demanding work Deep-plan drawing or business machine offices Inspection of medium machinery	750
Fine work Colour descrimination, textile processing, fine machining and assembly	1000
Very fine work Hand engraving, inspection of fine machining and assembly	1500
Minute work Inspection of very fine assembly	3000

goods yards and for outdoor constructional work, (b) for decorative purposes, i.e. the illumination of buildings, monuments and gardens on special occasions, and (c) outdoor sports.

Industrial floodlighting.   All the high pressure discharge lamps described in the preceding pages are suitable for this type of application. The SON lamp for example, is commonly used for area floodlighting and SON.TD or MBIL/H or lamps used in accurately designed parabolic reflectors are also used for long distance industrial floodlighting.

Quarries and surface workings of mines.   The advantages of floodlighting for this class of work are so obvious as to need little emphasis. The projectors can be located at stations remote from the working face, thus avoiding all risk of accidental damage to lighting apparatus, wiring, etc. The portable tripod type of projector base enables the apparatus to be moved easily from place to place, following the work as it progresses.

Docks loading wharves.   For illumination of both docks and wharves floodlight projectors have many advantages over the old-fashioned 'cluster' fittings. Their flexibility and much greater efficiency ensures that adequate illumination is readily available over a working area. As a general rule, the projectors should be mounted high up on the crane platform or runway, and facilities provided for the spreading and training of the beams to meet the exigencies of the work.

Railway shunting yards, sidings, etc.   The introduction of floodlighting has enabled a much more even distribution of light to be achieved than was possible with fittings mounted on short posts and has to a large extent eliminated the accident hazard. General practice is to mount the projectors on masts no less than 13 m high.

Shipyards, constructional work.   A comparatively low level of illumination is required over extensive areas and a high intensity as certain points. Projectors should be mounted high to avoid glare and be located on the berth structure.

High mounting of a similar character should be employed for the lighting of buildings under construction. Skilfully planned floodlighting is a valuable aid to the modern building contractor in his race against time.

Industrial floodlighting design.   The amount of light required for any of these industrial applications will depend upon the nature of the work carried out. The illuminance required for manufacturing operations is given in Table 13.7 at the beginning of this section. The size of lamp required can be calculated by considering the area to be lighted, the illuminance required, depreciation factor, and the beam factor of the particular projector to be employed. In such cases it is, not always, the horizontal area that must be taken as the surface to be lighted.

Decorative floodlighting.   It is necessary in this application first to decide upon the average illumination required. This naturally depends on the reflection factor, which depends on the texture of

310

the surface and whether it is clean or dirty. For clean Portland stone the factor is 60%, which is reduced to 20% if the surface is dirty. On the other end of the scale clean red brick has a reflection factor of 25%, dropping to 8% if dirty.

The question of arranging the projectors requires careful consideration. Lighting directly from the front is seldom satisfactory, while lighting from an angle can be most pleasing. Lighting of polished surfaces is seldom successful using floodlights, as they tend to act as plane mirrors, producing an image of the floodlight. One technique is to light objects in front of the polished surface so that it stands out in silhouette.

Football and other sports. Practice or casual training areas can be lighted from one side by low-wattage projectors housing tungsten halogen lamps (e.g. Thorn 'Sunfloods') mounted on 4–5 m poles but tournament areas require a higher luminance and more rigorous glare control. Minimum mounting heights of 10 m are recommended and the area should be lighted from both sides.

Stadia are lighted to higher standards, especially if colour television of matches is likely. The most popular method is to light the area by means of 1000 W CSI, MBIL or SON deluxe lamps mounted on towers up to 20 m high at the corners of the stadium. Alternatively, the lamps may be mounted on the roofs of the stands or, as at Wembley on stub towers mounted upon them. Full descriptions of this method and of other floodlighting techniques are given in the Thorn Outdoor Lighting Handbook.

Escape lighting and emergency lighting. In accordance with the Fire Precautions Act 1971 and the Health and Safety at Work Act 1974, BS 5266 'Code of practice for the emergency lighting of premises' lays down minimum standards. Interpretations of these may vary from one Authority to another. Installations and equipment are defined as follows:

Definitions of terms. The basic aim of legislation and supporting documents such as British Standards, Codes of Practice and local authority regulations, is to encourage uniformity of application of emergency lighting on a national scale. A number of terms and their meanings associated with emergency lighting systems are indicated below.

Emergency lighting. Lighting from a separate source independent of the mains supply, which continues after failure of the normal lighting of premises. Such lighting may be for standby or escape purposes.

Standby lighting. Standby lighting is emergency lighting, fed from a separate source from the mains, which comes into operation during a power failure to enable essential activities to continue.

Escape lighting. Lighting fed from a separate source, which is brought into immediate operation in the event of a power failure, to enable a building to be evacuated quickly and safely.

Sustained luminaire. A lighting fitting containing at least two lamps, one of which is energised from the normal supply and the other from an emergency lighting source.

Self-contained luminaire or sign. A luminaire or sign in which all the associated control units are housed and which only requires connection to the normal supply.

Slave luminaire or sign. One operated from a central control system.

Maintained system. In a maintained emergency lighting system the same lamps remain alight regardless of whether the normal lighting supply has failed. The maintained system may be provided with power from a floating battery system, which means the battery charger and power rectifier are connected in parallel with the battery and load, Figure 13.7. The floating system is now rarely used as the high float voltage results in short lamp life and low end of discharge light output. Alternatively, an automatic transfer switch may be used to connect the battery to the load terminals at the instant of normal supply failure, Figure 13.8. The taper type charger is now replaced by the constant voltage type but is still to be found in use.

The floating battery maintained system provides d.c. to the load terminals. The transfer switch circuit is normally provided with power by a step-down isolating transformer, and provides low voltage a.c. until at the instant of supply failure the automatic transfer switch connects the battery to the load terminals thus providing d.c. during supply failure.

312

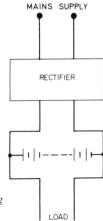

Figure 13.7 Circuit for a single floating battery system

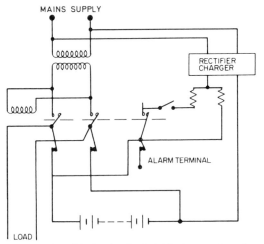

Figure 13.8 Simplified maintained trickle charge automatic switch circuit

313

Non-maintained system. A system in which lamps are normally off but are automatically illuminated on failure of the mains supply, Figure 13.9. Sustained emergency lighting is a category of a non-maintained system whereby a lighting fitting contains lamps illuminated from normal supply and separate lamp(s) illuminated from the battery only when normal supply fails. In this case the luminance of the fitting may differ during normal and emergency periods.

Central system. A number of lamps (luminaires) fed from a central secondary power supply is known as a central system.

Equipment available. Battery equipment available for emergency lighting systems may be classified as follows:

Category 1. Central storage batteries supplying tungsten filament lamps.

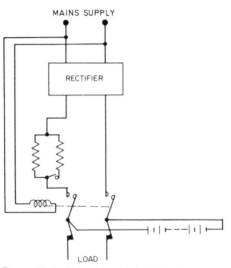

Figure 13.9   Non-maintained trickle charge automatic switch circuit

Category 2. Central storage batteries supplying fluorescent tubes, each fluorescent tube fitting being equipped with its own individual in-built static inverter to give high frequency a.c. to the tube. Various frequencies are used, 10, 20 and 40 kHz for example.

Category 3. Central storage batteries supplying bulk inverters which can produce several kilowatts of a.c. at normal mains frequency i.e. 50 Hz.

Category 4. Self-contained luminaires, each containing its own battery, usually of the sealed nickel cadmium type, complete with charging facility. Self-contained luminaires are available in a wide range of designs and are usually equipped with low power filament bulbs or transistor-inverter fluorescent tubes of 8 W and 13 W rating.

Chloride Keepalite systems. Operation of maintained and non-maintained Chloride Keepalite systems are described below and reference should be made to the accompanying electrical diagrams.

Figure 13.10 shows a Keepalite maintained system. Only one control switch is provided this being of an On-off nature in the mains a.c. supply. After installation this is switched on and operation thereafter is entirely automatic. While the mains supply is healthy the maintained lights circuit is supplied with a.c.

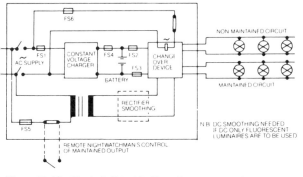

Figure 13.10   Typical Chloride Keepalite maintained system.

through the transformer, and the battery is kept in a fully charged condition by the constant voltage charger. On failure of the mains the maintained lights circuit and the non-maintained lights circuit are instantly and automatically connected to the battery by the changeover contactor. When the mains supply is restored the contactor takes up its normal position which disconnects the emergency non-maintained lights circuit and re-connects the maintained lights circuit to the transformer. The charger then takes over and automatically recharges the battery. When the fully charged condition is reached the battery voltage reaches the constant voltage setting, reducing the charge current to the low level needed to balance open-circuit cell losses.

Figure 13.11 shows a Keepalite non-maintained system. This also has only one switch in the mains supply circuit. This system

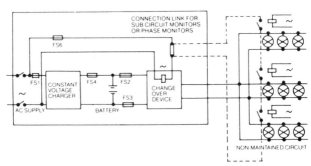

Figure 13.11 Typical Chloride Keepalite 'Non-maintained' system.

energises the emergency circuits only when the mains supply fails. While the mains is healthy the battery is kept in a fully charged condition by the charger. Following a mains failure the emergency lights are connected to the battery virtually instantaneously. Upon restoration of the mains supply the operation is similar to that described above for the maintained system.

Statutory requirements are as follows:

A minimum illuminance of 0.2 lux on the centre line of the escape route with a maximum diversity of 40:1. Since photometric

316

data for spacing is based upon luminaire performance alone, the usual average illuminance is likely to be in the order of 0.5 lux. Exit doors, stairs and other hazards may be lighted to 40 times the minimum, i.e. to 8.0 lux by luminaires positioned over them.

Positioning of luminaires must be such as to indicate exit and change of direction signs and hazards such as steps or ramps on the exit route, which includes the area outside the final door. They must be placed so that they indicate the lines of escape.

Legal responsibility. Fire officers are responsible for the issue of fire certificates for most premises, except Government offices, schools and local council property which come under the aegis of H.M. Factory Inspectorate. The inspector will demand a written guarantee from the contractor to the effect that the installation conforms to local standards and requirements, usually to BS 5266. The final legal responsibility rests with the owner or occupier.

Choice of luminaire. The choice of the type of luminaire will be governed partly by economic and partly by engineering considerations. Centrally controlled (slave) luminaires are less expensive than the self-contained type, but the necessity of providing separate wiring from the central control and power position to each fitting largely offsets this, especially as the provision of heat-resisting wiring is essential. In addition a plant room, housing the central batteries and control switchgear is necessary. There is no certainty that in the event of a sub-circuit failing the system will come into operation unless an expensive relay system is provided.

Self-contained luminaires require little or no special wiring as they can be operated from the same distribution boxes as the main system. They can easily be added to an existing installation and, since they automatically come into action when the supply to the part of the installation in which they occur is interrupted can easily be tested by removing a fuse or operating an isolating switch. Each contains its own power supply, usually consisting of nickel cadmium cells located in the luminaire and operating miniature florescent tubes. They are maintained at full charge by a mains-powered solid state module and operate for a minimum period of three hours from the interruption of supply. Consequently this type of emergency lighting system is rapidly gaining popularity compared to the centrally operated type.

Planning the installation. Escape routes should be marked on a floor plan. They should be as direct as possible but avoid congested areas or those where rapid spread of flame is likely. Danger points, doors, staircases and places where the route changes direction should be specially emphasised and an emergency lighting fitting placed at each of them. The isolux diagram provided by the manufacturer should then be centred on each of these points and positions on the centre line of the escape route where the illuminance falls below 0.2 lux marked and extra luminaires positioned at them. It is essential to ensure that each luminaire is so placed as to be visible from the position of the preceding one, as in smokey conditions they may have to act as guides on the escape route.

Exit signs should be mounted between 2 and 2.5 m above floor level and positioned close to the points they indicate. Where no direct sightline to the exit exists subsidiary signs should be used to point the way to it. Exit signs should be of the 'maintained' type; i.e. they should be lighted all the time that the building is occupied.

Maintenance and testing. Each self-contained sign should be energised once a month from its internal battery and every six months it should be left on for an hour. (BS 5266 Part 1) All testing should take place at times of least risk and personnel operating in the area be notified.

Recommendations for further reading. Detailed instructions for designing interior lighting installations are contained in 'Interior Lighting Design' published by the Electricity Council, and for floodlighting in the Thorn Outdoor Lighting Handbook. A great deal of information on lamps and lighting equipment is to be found in the Thorn Technical Handbook, and the student is advised to obtain this and also the CIBSE Code, without which no serious study of lighting can be attempted.

This section edited by Thorn Lighting Ltd to whom acknowledgement is made.

14
Motors and control gear

D.C. MOTORS

D.C. motors are divided into three classes, as follows:

1. *The series-wound motor.* In this type (Figure 14.1) the field is in series with the armature. This type of motor is only used for direct coupling and other work where the load (or part of the load) is permanently coupled to the motor. This will be seen from the speed-torque characteristic, which shows that on no load or light load the speed will be very high and therefore dangerous.

2. *The shunt-wound motor.* In this case the field is in parallel with the armature, as shown in Figure 14.2, and the shunt motor is the standard type of d.c. motor for ordinary purposes. Its speed is nearly constant, falling off as the load increases due to resistance drop and armature reaction.

3. *The compound-wound motor.* This is a combination of the above two types. There is a field winding in series with the armature and a field winding in parallel with it (Figure 14.3). The relative proportions of the shunt and series winding can be varied in order to make the characteristics nearer those of the series motor or those of the shunt-wound motor. The typical speed-torque curve is shown in the diagram.

Compound-wound motors are used for cranes and other heavy duty where an overload may have to be carried and a heavy starting torque is required.

Speed control. Speed control is obtained as follows:

Series motors. By series resistance in parallel with the field winding of the motor. The resistance is then known as a diverter resistance. Another method used in traction consists of starting up

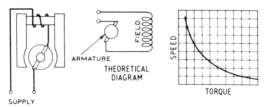

Figure 14.1   Series wound motor

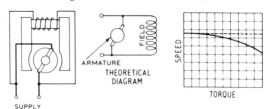

Figure 14.2   Shunt wound motor

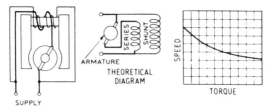

Figure 14.3   Compound wound motor

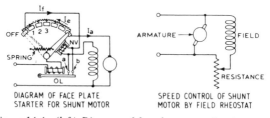

*Figure 14.4   (left) Diagram of faceplate starter for shunt motor.
(right) Speed control of shunt motor by field rheostat*

two motors in series and then connecting them in parallel when a certain speed has been reached. Series resistances are used to limit the current in this case.

Shunt- and compound-wound motors. Speed regulation on shunt- and compound-wound motors is obtained by resistance in series with the shunt-field winding only. This is shown diagrammatically for a shunt motor in Figure 14.4.

Starting. The principle of starting a shunt motor will be seen from Figure 14.4 which shows the face-plate type starter, the starting resistance being in between the segments marked 1, 2, 3, etc.

The starting-handle is held in position by the no-volt coil, marked NV, which automatically allows the starter to return to the off position if the supply fails. Overload protection is obtained by means of the overload coil, marked OL, which on overload short circuits the no-volt coil by means of the contacts marked *a* and *b*.

When starting a shunt-wound motor it is most important to see that the shunt rheostat (for speed control) is in the slow-speed position. This is because the starting torque is proportional to the field current and this field current must be at its maximum value for starting purposes. Many starters have the speed regulator interlocked with the starting handle so that the motor cannot be started with a weak field.

These methods of starting are not used much today but are left in because many installations still exist. Modern methods of control employ static devices described on the next page.

Ward-Leonard control. One of the most important methods of speed control is that involving the Ward-Leonard principle which comprises a d.c. motor fed from its own motor-generator set. The diagram of connections is shown in Figure 14.5. The usual components are an a.c. induction or synchronous motor, driving a d.c. generator, and a constant voltage exciter; a shunt-wound d.c. driving motor and a field rheostat. The speed of the driving motor is controlled by varying the voltage applied to the armature, by means of the rheostat in the shunt winding circuit of the generator. The d.c. supply to the field windings of the generator and driving motor is obtained by means of an exciter driven from the generator shaft.

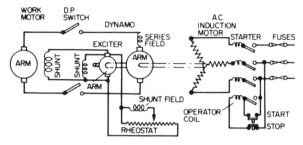

Figure 14.5   Ward-Leonard control

With the equipment it is possible to obtain 10 to 1 speed range by regulation of the generator shunt field and these sets have been used for outputs of 360 W and upwards. On the smaller sizes speed ranges up to 15 to 1 have been obtained, but for general purposes the safe limit can be taken as 10 to 1. Speed control obtained in this way is extremely stable and the speed regulation between no load and full load at any particular setting is from 7 to 10%, depending on the size and design of the equipment.

This type of drive has been used for variety of industrial applications and has been particularly successful in the case of electric planers and certain types of lifts, with outputs varying from 15 kW to 112 kW, also in the case of grinders in outputs of 360 W, ¾ kW and 1½ kW with speed ranges from 6:1 to 10:1.

Thyristor regulators.   The development of thyristors with high current carrying capacity and reliability has enabled thyristor regulators to be designed to provide a d.c. variable drive system that can match and even better the many a.c. variable speed drive systems on the market. This has meant a redesign of the d.c. motor to cater for the characteristics of the thyristor regulator. Machines have laminated poles and smaller machines may also have laminated yokes. This is to improve commutation by allowing the magnetic circuit to respond more quickly to flux changes caused by the thyristor regulators. Square frame designs of d.c. machines have also been developed with much improved power/weight ratios together with other advantages. A typical motor is shown in Figure 14.6.

Figure 14.6   A Mawdsley's square frame industrial d.c. motor rated at 250 kW with forced ventilation

An electrical schematic, Figure 14.7, shows a three-phase six-pulse thyristor bridge, d.c. motor and feedback loop. It is usual for a thyristor-controlled d.c. motor to be fitted with a tachogenerator which provides a speed related signal for comparison with the reference signal. It is the error or difference between these two signals which advances or retards the firing angle of the thyristors in the regulator to correct the d.c. voltage and hence the speed of the motor. In addition to, or instead of using the motor speed, other parameters such as current, load sharing, web or strip tension and level may be employed.

Switched reluctance drives.   A fairly recent development in the d.c. variable speed drive is that of the switched reluctance motor. The machine has salient poles on the stator and as in the case of other d.c. motors the number of poles does not determine the motor speed. The rotor is built up of shaped laminations but carries no winding at all.

Uni-directional current pulses are applied sequentially to the field poles at a rate determined by the required speed and rotor

323

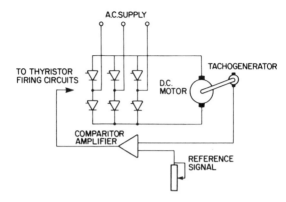

Figure 14.7   Three-phase six-pulse thyristor bridge, d.c. motor and feedback loop

position. Control of the pulse timing is derived from a position transducer mounted on the motor shaft. By appropriate pulse timing either positive or negative torques can be achieved and full four quadrant control is available over a wide speed range and both constant torque or constant power characteristics can be provided.

A.C. MOTORS

Alternating current motors can be grouped as follows:

- (a) Induction motors.
- (b) Synchronous motors.
- (c) Variable-speed commutator motors including the Schrage motor.
- (d) Series motors.
- (e) Single phase repulsion, capacitor and shunt motors.
- (f) Pole-changing and other special motors.

The first three are used in all sizes, and for all general purposes induction motors are employed on account of their simplicity, reliability and low first cost. Synchronous motors are generally installed where it is desirable to obtain power-factor improvement or where a constant speed is required. They are only economical in the case of loads of 75 kW and over, although there are instances where smaller machines are in use for special purposes.

The three-phase commutator motor until recently has been the only a.c. motor for large outputs which gives full speed control, and although expensive it is being used for duties where variable speed is required. Pole-changing and other special motors with speed control characteristics are now available and details of them are given later on in this chapter. Static inverters for use with cage induction motors are becoming very popular as variable speed-drives and are also described later.

Groups (d) and (e) represent the types used for small power motors, which also include induction motors, These small motors have been developed to a great extent because of the number of small machines incorporating individual drives. Normally small power motors include machines developing from 20 W to about 1½ kW. The reason for inclusion of the 1½ kW motor is that a different technique is used for manufacturing small motors which are turned out in large quantities by mass-production methods. Most of the manufacturers of these small motors can supply them with gearing incorporated giving final shaft speed of any value down to one revolution in 24 hours or even longer.

The induction motor, which can be termed the standard motor, is now made on mass production lines, and as a result of the standardization of voltage and frequency, the cost of standard-sized new motors is exceptionally low. The absence of a commutator and, in the case of the cage motor, of any connection whatever to the rotor, combined with the simplicity of starting, make it the most reliable and cheapest form of power-drive available.

There are a number of specialized motors which are used in a few unusual applications, but these will not be described as they are really rather of academic interest than of general use in industry. The linear induction motor falls within this category although this has been applied to overhead cranes and as a means of actuation where force without movement is required. The most recent example of a linear induction motor application is the rail transport car operating between Birmingham Airport and the nearby Birmingham International Railway Station.

The use of synchronous motors for improvement of power factor is referred to in the section dealing with power factor correction, but it should be realised that the essential points of a synchronous motor are its constant speed (depending on the frequency) and the fact that the power factor at which it operates can be varied at will over a certain range – usually from 0.6 leading to 0.8 lagging – by varying the excitation current.

INDUCTION MOTORS

The essential principle of an induction motor is that the current in the stator winding produces a rotating flux which because it cuts the rotor bars induces a current in the winding of the rotor. This current then produces its own field which reacts with the rotating stator flux thus producing the necessary starting and running torque.

The stator winding to produce this rotating flux is fairly simple in the case of a three-phase motor, being based on three symmetrical windings, as shown in Figure 14.8. In the case of single-phase and now vanishing two-phase it is not quite so easily understood.

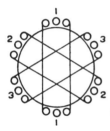

Figure 14.8   Three-phase stator winding to produce rotating field.

The stator field will revolve at synchronous speed and if no power whatever was required to turn the rotor it would catch up with the flux and would also revolve at synchronous speed. The condition for torque production would then have vanished. As, however, a certain amount of power is required to turn the rotor even if unconnected to any load, the speed is always slightly less than synchronous. As the load increases the speed falls in order to allow the additional rotor currents to be induced to give a larger torque.

The difference between the actual speed and synchronous speed is termed the *slip*, which is usually expressed as a percentage or a fraction of the synchronous speed. For standard machines the maximum slip at full load is usually about 4%.

Calculation of synchronous speed. Induction motors are made with any number of poles (in multiples of 2), but it is not usual to make motors with more than 10 poles, and for ordinary use 2, 4 and 6 poles are chosen, if possible, on account of the lower first cost and higher efficiency.

$$\text{Synchronous speed in rev/min} = \frac{\text{frequency} \times 60}{\text{number of pairs of poles}}$$

Thus a 2-pole motor on 50 Hz will have synchronous speed of 3000 rev/min, a 4-pole 1500 rev/min, and 6-pole 1000 rev/min. The suitability of 1500 rev/min for many purposes has made the 4-pole motor the more usual.

The actual rotor speed for 4% slip is given for various motors on 50 Hz in Table 14.3 at the end of this sub-section. Slip is calculated from

$$\text{Percentage slip} = \frac{(\text{syn. speed} - \text{rotor speed}) \times 100}{\text{syn. speed}}$$

and the rotor speed for any given slip will be

$$\text{Rotor speed} = \text{syn. speed} \frac{100 - \text{slip}}{100}$$

the slip being the percentage slip.

Variation of slip with torque. It can be shown that the torque of an induction motor is proportional to

$$T = \frac{kE_2sR_2}{R_2{}^2 + (sX_2)^2}$$

where T = torque
k = constant
E_2 = rotor voltage
s = fractional slip
R_2 = rotor resistance
X_2 = rotor reactance.

The variation of slip with torque can therefore be calculated and typical torque-slip curves are given in Figure 14.9. These curves are for the same motor, and curve (a) is for the rotor short-circuited, whereas (b) and (c) are for cases where additional or added resistance has been put in the rotor circuit. It will be seen from these curves and also from the formula above that the

327

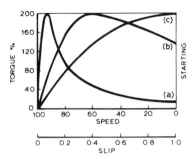

Figure 14.9
Torque-slip curves
(a) No added resistance in rotor $R = 1$.
(b) With added resistance $R = 4$
(c) With more added resistance $R = 20$

torque at starting or low speeds is greatly increased by adding resistance in the rotor circuit, and this principle is made use of in the wound-rotor induction motor which is used to start up against heavy loads.

It can also be shown that maximum torque occurs when

$$R_2 = sX_2 \text{ when the slip will be } s = R_2/X_2$$

Wound-rotor or slip-ring motor. The slip-ring induction motor is used for duties where the motor has to start up against a fairly heavy load and the slip-rings are arranged for added resistance to be inserted in the rotor circuit for starting purposes. Figure 14.10 shows how the various circuits are connected to the supply and to the variable rotor resistance.

This type of motor is referred to as a *wound-rotor* motor, because for this purpose the rotor has to be wound with insulated conductors similar to those used for the stator. In the case of

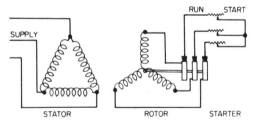

Figure 14.10 Starting wound-rotor motor.

328

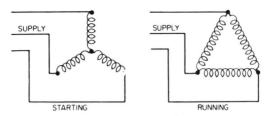

Figure 14.11   Star-delta starting for cage motors

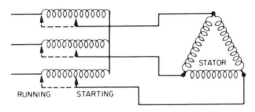

Figure 14.12   Auto-transformer starting for cage motors

larger motors, a device is fitted to the rotor shaft enabling the slip-rings to be short-circuited and the brushes lifted off the slip rings, thus reducing both electrical and friction losses. Starters for these motors are described in the sub-section on 'Motor Starters' later in this chapter.

With the external resistance in the rotor circuit, it is possible to obtain a large starting torque without drawing an excessive starting current from the line. Generally it will be found that full-load torque may be obtained when passing about 1¼ times full-load current; other values of torque, up to 2 to 2½ times full load (depending upon the characteristics of the motor), may be obtained when passing from 2½ to 3 times full-load current.

Cage motors. Cage motors should always be employed whenever possible because of their robustness, simplicity, lower cost and lower maintenance compared with slip-ring machines. They may be employed when the starting torque is sufficient to run the drive up to speed and when the starting current to do this is not too high for the conditions of the supply.

329

Figure 14.13a   Drip-proof motor to BS5000 Parts 10 and 99 and BS4999 for general clean industrial conditions. It is available in ratings from 5.5 to 650 kW with foot or flange mounting. Enclosure is to IP22 (Brook Crompton Parkinson Motors)

Figure 14.13b   Totally enclosed ventilated motor with IP54 enclosure in ratings from 0.18–425 kW foot or flange mounting. It complies with BS5000 Part 10 (Brook Crompton Parkinson Motors)

Figure 14.13c Frame size 7–D280M totally enclosed fan ventilated foot mounted motor to IP55. Ratings 18.5–425 kW (Brook Crompton Parkinson Motors)

The rotor conductors of cage rotors normally consist of aluminium bars arranged round the periphery of the rotor and connected at each end by a ring of aluminium. Frequently both conductors and endrings are die cast, although in larger sizes, of the order of 200 or 300 kW the conductors and endrings would be made of copper. The resistance of the rotor is thus low and very little starting torque is available in comparison with the current taken from the supply system. On the other hand, a very cheap and efficient motor is produced. Figures 14.13a to 13d show four different types of cage rotor machines.

Typical characteristics for an average purpose cage motor are shown in Figure 14.14. It will be seen that the torque developed at standstill by the motor *on full voltage* is 125% of full-load torque, the starting current being just over 600% full-load current, i.e. corresponding to the delta curves.

331

Figure 14.13d Flameproof totally enclosed fan-ventilated motor for Group I applications in mines as certified by the Health & Safety Executive, and for Groups IIA and IIB Ex d industrial use, certified by BASEEFA. Available as foot or flange mounting with IP54 enclosure. Electrical performance to BS 5000 Part 99, 0.37 to 150 kW and outputs to BS 4683 Part 2 (Brook Crompton Parkinson Motors)

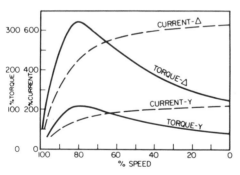

Figure 14.14 Torque/speed and current/speed curves for normal cage induction motor connected in star and delta

In view of the large starting current on full voltage, regulations imposed by supply companies protect the interests of neighbouring consumers by limiting the size of cage motors that may be started by switching direct on mains voltage. This rating of the machine that can be switched direct on line will vary depending on the electrical characteristics of the supply. In some cases the limit may be about 5 kW, in others over 75 kW on a 415 V supply.

In cases where the direct starting method cannot be employed because of the high initial peak current, the cage motor has to be started on reduced voltage. There are three methods of starting on reduced voltage, all of which involve a reduction in starting torque. This excludes static inverter starting which is discussed later. In such a method the voltage applied to the motor is also reduced. A recent development in cage rotor design, also described later, includes an integral eddy current inductor which produces a high torque at low current:

(a) Star-delta connections,
(b) The use of an auto-transformer,
(c) Primary resistance starting.

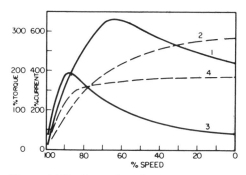

Figure 14.15 Torque/speed and current/speed curves for high resistance and high-reactance cage motors.

Torque curves: 1. High resistance; 3. High reactance;

Current curves: 2. High reactance; 4. High resistance.

Starting on reduced voltage. Star-delta starting is the most usual method of starting on reduced voltage and consists of connecting the stator winding in star until a certain speed has been reached, when it is switched over to delta, the circuit connections being as shown in Figure 14.11. Six terminals must be available on the machine for these connections. By this means the voltage across each winding is reduced to 58% of the supply voltage. The star connection means that at starting the line current and torque values are one-third of the full voltage figures. This method is simple and inexpensive, but can only be used where the starting load is small. Repeated starting can be obtained.

Wauchope type continuous torque star-delta starter. The Wauchope type star-delta starter is an important advance in the technique of the starting of three-phase induction motors of cage design, and achieves a smooth acceleration approaching that of a slip-ring motor, although sustained excess torque during starting is not claimed.

Considering standard star-delta starting, if it were possible to change from star to delta instantaneously, the current would rise to a value approximately three times the current in star obtaining at the instant of changeover. But owing to the transition pause and consequent loss of speed, the delta peak current is very much increased, and may even approximate to the current which would be drawn from the line when the motor is connected in delta direct to the supply.

The Wauchope star-delta starter improves the starting characteristics associated with a standard star-delta starter by the insertion of resistance when changing over from star to delta, and incidentally, provides three steps of acceleration, instead of two steps as in the conventional starter.

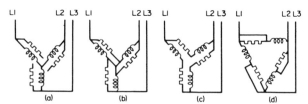

Figure 14.16   Wauchope star-delta starter – switching sequence

Figure 14.16 shows the changes on connections in the Wauchope starter and the sequence of starting comprises in detail:

(*a*) The motor is connected in star and is permitted to accelerate to a stable speed in the normal manner.

(*b*) The three resistances are then connected in parallel with the motor windings. The line current is increased by the amount taken by the resistances, but this does not affect the running of the motor. This is simply a preparatory step and is in operation for a fraction of a second only.

(*c*) The star point of the windings is now opened. It will be seen from Figure 14.16 that the resistances have been so connected that the motor windings are in delta, with a resistance in series with each winding. At this stage the voltage across the motor windings is increased, giving a corresponding increase in torque, and the motor accelerates to a higher steady speed.

(*d*) The resistances are then short-circuited and the motor windings connected in delta across full line voltage. The motor accelerates to full speed and the starting operation is complete.

Auto-transformer starting. The use of an auto-transformer is confined to those cases where a definite limit is required to the starting current, and the arrangement is shown in Figure 14.12, the transformer being disconnected from the supply in the running position. Transformer tappings enable the starting voltage to be selected to obtain the starting torque required for the load. Starting torque and current are each equal to the square of the transformer tapping (fraction) × direct switching value; e.g. 50% tapping gives $(\frac{1}{2})^2$ or 25% of the direct switching value.

Primary resistance starting. In this method of reduced voltage starting, the stator is connected through an adjustable three-phase series resistance. As the motor accelerates, the resistance is short-circuited in one or several steps. A heavier line current is required for a certain starting torque compared with other methods, e.g. 50% line voltage and current gives 25% torque and 80% line voltage and current gives 64% torque. Heavy peak currents are avoided. The number of starts is limited by the resistance rating. When high torque motors are employed, this method shows appreciable saving in first cost, and the simple yet robust construction will give very low maintenance costs. If it is not important to strictly limit the starting current, these advantages recommend its more extended use, and in an emergency or breakdown, the method allows considerable scope for ingenious improvisation.

Integral eddy current inductor.　　A recent development in cage rotor design produces high torque at low current. It involves the inclusion of an eddy current inductor as part of the rotor, Figure 14.17. Manufactured in the output range 18.5 kW to 150 kW it is being used as an alternative to wound rotor motors.

Figure 14.17　Complete pipe cage rotor for 30 kW 4-pole motor (Brook Crompton Parkinson Motors)

Known as the pipe cage motor, it has superior performance characteristics in terms of frequency of starting and driving high inertia loads whilst only requiring the cheaper direct-on-line starter.

A cage rotor is used having a core in alignment with that of the stator, but with copper rotor bars extended at the opposite drive end of the motor. This extension takes up the space normally occupied by the sliprings of wound rotor motor. Each extended rotor bar passes through a close fitting steel pipe before connection to the usual short-circuiting ring. The rotor bars are insulated in the slots and pipes by high temperature materials.

The pipe cage assembly acts as a series of individual short-circuited transformers with the rotor bars as their respective primary windings. The pipe forms both the core and secondary winding of these transformers which are short-circuited by the supporting discs. The pipes are made of steel and, therefore, exhibit a high electrical resistance which is further increased when the primary (i.e. rotor) current is at mains frequency. This is due to skin effect, thus making the effective pipe resistance dependent on rotor frequency and hence motor speed.

At start, with rotor current at mains frequency, a longitudinal current flows in the pipes and adds impedance to the rotor circuit. Both the resistance and inductive components of this impedance are dependent on rotor current and frequency and, therefore, decrease as the motor accelerates to full speed. It is this automatic adjustment of impedance that gives the pipe cage motor its torque, current, and speed characteristics. Starting current is decreased at start in comparison with a normal cage machine and the additional resistance increases the starting torque. Figures 14.18 and 14.19 show the relationship of pipe cage motors to cage and slipring motors.

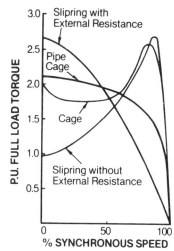

Figure 14.18 Performance of a pipe cage motor compared with cage and slipring motors – torque/speed curves

Cage versus slip-ring. A slip-ring motor may be chosen for a particular application in preference to a cage machine for one or more of several reasons:

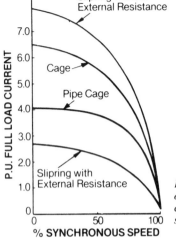

Figure 14.19 Performance of a pipe cage motor compared with cage and slipring motors – full load current/speed curves

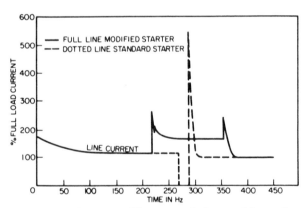

Figure 14.20 Variation of line current during star-delta starting Wauchope starter indicated by full line. Standard starter indicated by broken line (Simplex – GE Ltd)

1. To limit the starting current drawn from the line.
2. To obtain a high value of starting torque for a comparatively low current drawn from the line.
3. To obtain speed control.

If a cage motor can be used that will satisfy either (1) or (2) for a particular duty, it will be possible to dispense with the slip-ring machine on that particular score. Both the starting torque and starting current of cage motors depend on the characteristics of the rotor employed; and there is a wide choice of motor for different types of drive.

'High resistance' cage motor. As with the slip-ring (wound rotor) motor, the starting torque of the motor can be increased by using a rotor with a higher resistance; but since this resistance has to remain permanently in circuit (as in the ordinary cage motor), it cannot be too high if the motor is to retain anything like normal efficiency. A 'high resistance' rotor has been used in applications such as forge hammers and power presses. The motor is allowed to slow down with increasing load so that the stored energy of the flywheel and other moving parts of the system may be utilized to relieve the motor of very heavy overloads. By this means also the supply line is relieved of very heavy peak currents, since the additional peak torque is provided by the flywheel and not by the motor. With such machines, therefore, since the load is of an intermittent nature, efficiency is not of so great importance as overload capacity.

High-resistance rotors have been frequently used on cranes and hoists where a high starting torque is required, but generally it may be said that, unless there is some definite advantage to be gained, the inefficiency will be such as to preclude their use in any situation where consumption is of importance.

Double cage high torque motor. In order to overcome the disadvantages of the cage motor, and avoid having to use the more expensive slip-ring motor and its associated gear, increasing attention has been given to the use of the double cage in which the resistance of the cage rotor is increased temporarily while starting.

The double cage rotor in its simple form consists of two separate cages. The outer or starting cage is made of high resistance material and is arranged to have the smallest possible reactance. The inner cage is of the ordinary low-resistance type, and since it is sunk deep into the iron, has a high reactance. The four qualities – reactance and resistance of inner and outer cage – can be varied in an infinite number of combinations and many different shapes of speed-torque curve can be obtained.

At starting, the frequency of the currents in the rotor conductors is the same as the supply frequency, thus the high reactance of the inner cage produces a choking effect and reduces the current flowing in this winding. The outer cage, being of high resistance, develops a high starting torque depending largely on the value of its resistance. As the rotor accelerates and approaches synchronism, the frequency of the emfs in its conductors falls and the choking effect in the inner cage is reduced; the inner cage now carries practically all the current until finally, when near synchronism, the rotor operates with the characteristics of an ordinary low resistance rotor. The general result is to produce a machine having a high start torque and a high running efficiency, with reasonably small values of starting current.

Often the higher starting torque allows the motor to be started on reduced voltage provided the load against which it is required to accelerate is not too great.

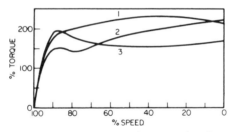

Figure 14.21   Torque-speed curves of various double cage motors

By altering the relative values of resistance and reactance, a wide variety of torque-speed characteristics can be obtained (Figure 14.21). For example, one make of standard high torque motors available up to about 55 kW for direct switching develops at least twice full-load torque at standstill and takes a starting current of about 350% full load current. In another range of standard machines for star-delta starting, 100% of full load torque is developed at starting in the star connection, the starting current being 150–175% full load current.

It should be noted that the cost of a double cage machine and its starting gear is considerably less than an equivalent slip-ring motor with resistance starter, and performance is obtained without sacrifice of running efficiency.

Extra high-speed induction motors. Certain types of wood- and metal-working machines and portable tools require very high speeds of rotation (e.g. up to 27 000 rev/min). Such speeds cannot be obtained with a direct drive from an ordinary induction motor supplied at standard frequency (50 Hz), as the maximum speed is 3000 rev/min (corresponding to two poles). In these special cases, where a direct drive is particularly advantageous, the only solution to the problem is to raise the frequency of the supply to the motor. This is accomplished by means of a frequency changer set, or frequency booster.

A frequency booster is a slip-ring induction motor. If this is excited from the a.c. mains and driven in the opposite direction to that of the rotating magnetic field of its primary winding, a high-frequency supply can be drawn from the secondary winding.

Either the stator or rotor can be made to act as primary, selection being according to whichever best suits the requirements.

The frequency booster may be driven by a motor (direct or belt coupled) or by any other source of mechanical power.

With different driving-motor speeds and excitation windings, different frequencies can be generated as indicated by the following formula:

High-frequency output

$$= \left(\text{Pairs of poles in booster} \times \frac{\text{rev/min of motor}}{60} \right) + \text{Excitation frequency}$$

Special precautions are necessary in the arrangement of the switchgear. The driving motor starting switches must be interlocked with the switches controlling the power supply of the induction machine stator and rotor, so that it is impossible to connect either the input or the output of the high-frequency unit before the machines are run up to speed. Similarly, it must be arranged so that in the event of the motor circuit being tripped the input and output circuits of the booster are tripped at the same time.

The standard high-frequency portable electric tool is designed to operate at 200 Hz. compared with the 'universal' type of motor, the speed does no drop in accordance with the load applied, and there are not commutators or brushes to wear or to be replaced.

SYNCHRONOUS MOTORS

The synchronous motor is essentially a reversed alternator and is often used for power-factor correction. As its name implies, it has a constant speed (running at synchronous speed at all loads), and its power factor can be controlled by varying the exciting current. It can thus be made to take a leading current for power-factor improvement purposes. The synchronous motor itself is not self-starting and it must be synchronised on to the supply when it has been run up to speed by a special starting motor or by some other means.

Self-contained motor. Refer to Figure 14.22 which shows the arrangement of a self-contained synchronous motor suitable for driving a steady load but does not show any method of starting. The three-phase supply is taken direct to the stator and a d.c. supply is necessary for excitation. This can either be obtained from a separate d.c. system (sometimes used where there are several motors in use) or from the individual exciter mounted direct to the motor as shown. Power-factor control is obtained by varying the excitation – this being regulated in large motors by means of a rheostat in the field circuit of the exciter.

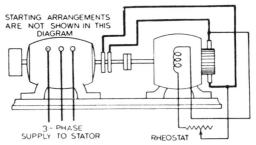

Figure 14.22   Diagrammatic arrangement of motor with exciter. Power factor is controlled by rheostat in exciting circuit of exciter

This type of motor in its simple form has no starting torque and will not therefore start up under load. Also if the overload capacity is exceeded the motor will fall 'out of step' and will shut down. It must then be started up and synchronised in the usual manner.

Table 14.1 Efficiency and power factor of 4-pole induction motors on 50 Hz – full load

The following are average values for standard motors running at 1440 to 1470 rev/min.

	Efficiency				Power factor			
	Single phase				Single phase			
kW	Split phase	Capacitor	2 phase	3 phase	Split phase	Capacitor	2 phase	3 phase
	Per cent	Per cent	Per cent	Per cent				
¾	65	70	73	74	0.80	0.90	0.79	0.81
1⅛	69	74	76	78	0.81	0.90	0.79	0.81
1½	72	76	78	80	0.82	0.91	0.82	0.84
2¼	74	78	82	83	0.82	0.92	0.82	0.84
3	76	80	83	84	0.82	0.93	0.82	0.84
3¾		82	84	85		0.93	0.84	0.86
5⅝		83	85	86		0.94	0.85	0.87
7½		84	87	88		0.94	0.86	0.88
9⅜		84	87	88		0.90	0.86	0.88
11¼		85	88	88		0.90	0.87	0.89
15		86	88	90		0.90	0.88	0.90
22½			89	90			0.88	0.90
30			90	90			0.89	0.90
37½			91	91			0.90	0.91
56¼			91	91			0.90	0.91
75			92	92			0.91	0.92

Synchronous-induction motor. The diagram for a typical self-starting synchronous-induction motor is shown in Figure 14.23. It will be seen that by means of a starting resistance the machine will start up as an induction motor. As full speed is reached the motor will pull into synchronism (against full load if required) and the starting resistances are then short-circuited.

A two-phase winding is used on the rotor and arranged so that the neutral point is used as one connection for the excitation circuit.

Table 14.2 Full load currents (amperes) of alternating current motors

The values below may vary slightly with different types of motors but can be accepted as reasonably accurate

kW	Split phase 100 V	Split 200 V	Split 230 V	Split 400 V	Split 480 V	Capacitor 100 V	Cap 200 V	Cap 230 V	Cap 400 V	Cap 480 V	Two phase 200 V	2ph 400 V	2ph 480 V	Three phase 200 V	3ph 220 V	3ph 350 V	3ph 400 V	3ph 440 V	3ph 500 V
¾	12.9	6.5	5.6	3.2	2.7	11.6	5.9	5.1	2.9	2.4	3.4	1.7	1.4	3.4	3.1	2	1.7	1.5	1.4
1½	24	12	10.4	6	5	22	11	9.5	5.4	4.5	6.6	3.3	2.9	6.6	6	3.8	3.3	3	2.6
2¼	35	17	15	9	7	31	15	13	7.8	6.5	9.8	4.9	4.2	9.8	8.9	5.7	4.9	4.5	3.9
3	45	23	20	11	9.4	40	20	18	10	8.5	13	6.5	5.6	13	12	7.5	6.5	6	5.2
3¾	56	28	25	14	12	51	25	22	13	11	16	8	6.8	16	14	9.1	7.9	7.2	6.3
4½	66	33	29	17	14	60	29	26	15	13	19	9.4	8.1	19	17	11	9.4	8.5	7.5
5⅝	82	41	36	21	17	74	37	33	19	16	24	12	10	24	21	14	12	11	9.4
6	87	43	38	22	18	78	39	34	20	16	25	13	11	25	21	15	13	12	9.8
7½	109	54	47	27	23	98	48	42	24	20	30	15	13	30	27	17	16	14	12
11¼	159	79	69	40	33	147	73	64	37	31	44	22	19	44	40	26	22	20	18
15	209	105	91	52	44	193	97	84	48	41	58	29	25	58	53	34	29	26	23
16¾	256	128	111	64	53	237	118	103	59	49	72	36	31	72	66	42	36	33	29
22½	306	152	134	77	64	283	142	124	71	59	86	43	37	86	78	50	43	39	34
30	400	200	174	100	83	370	185	161	93	77	111	56	48	111	101	64	56	51	45
37½	487	244	212	122	102	450	226	196	113	94	137	68	59	137	124	79	68	62	55
45	586	292	254	147	122	542	270	235	136	113	162	81	70	162	147	94	81	74	65
56¼	715	358	310	179	150	662	330	287	166	139	198	99	86	198	180	114	99	90	79
75	941	471	410	235	196	870	435	380	217	182	263	132	114	263	239	152	132	120	105
102½	—	—	—	—	—	—	—	—	—	—	388	194	168	388	356	225	194	176	155
150	—	—	—	—	—	—	—	—	—	—	517	258	223	517	468	299	258	235	207

The current required for any alternating current motor can be obtained from the following equations. The power factor and efficiency can be obtained from Table 14.1.

Single phase
$$\text{Current} = \frac{\text{kW} \times 1000}{\text{Voltage} \times \text{power factor} \times \text{efficiency}}$$

Three phase
$$\text{Current} = \frac{\text{kW} \times 1000}{1.732 \times \text{line voltage} \times \text{power factor} \times \text{efficiency}}$$

Two phase, four-wire supply
$$\text{Current} = \frac{\text{kW} \times 1000}{2 \times \text{line voltage} \times \text{power factor} \times \text{efficiency}}$$

Two phase, three-wire supply
Current = In outers, as above

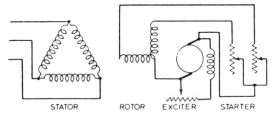

Figure 14.23 Synchronous induction motor. A two-phase rotor is used. Diagram shows method of starting as an induction motor

Hunting. One of the features of a synchronous motor is that on a fluctuating load it may *hunt*.

In modern industrial motors this is prevented by means of a damping winding in which eddy currents are induced by the variations in speed should hunting occur.

Hunting is more likely to occur with weak excitation than with strong. Temporary hunting can therefore often be cured by strengthening the field.

Table 14.3 Synchronous speed and rotor speed at 4% slip

No. of poles on stator	Synchronous speed on 50 Hz, rev/min	Rotor speed at 4% slip
2	3000	2880
4	1500	1440
6	1000	960
8	750	720
10	600	576

SINGLE-PHASE MOTORS

Where a three-phase supply is not available most small power applications can be met by using single-phase motors. In some circumstances even if there is a three-phase supply there may be economic benefit to be gained from using a single-phase machine in the way of simpler wiring and control gear.

In order to be self-starting an electric motor must have a rotating magnetic field. The phase displacement in a three-phase

345

supply produces this, but a single-phase motor requires an auxiliary (starting) winding designed to give a displacement similar to a two-phase supply before this effect is produced. It can be achieved in various ways, each of which produces a motor with a different set of characteristics. Most of the motors described below are of the induction type, only the series and repulsion motors having wound rotors.

Split-phase motor. The starting winding of a split-phase motor, Figure 14.24 uses fine wire and thus has a high resistance. It is also arranged to have a low reactance. The current in the start winding thus leads that in the main winding and a rotating magnetic field is set up similar to a poor two-phase motor. The start winding works at a high current density and it must be switched out as soon as possible when the machine reaches about 75% of full load speed. This type of motor is suitable for low inertia loads and infrequent starting. Starting current is relatively large and account should be taken of this when installing, to avoid excessive voltage drop.

Capacitor start motor. A capacitor is inserted in series with the start winding to reduce the inductive reactance to a low or even a negative value, Figure 14.25. The start winding current therefore leads the main winding current by almost 90 degrees. A large a.c. electrolytic capacitor is used and, since this is short-time rated, it must be switched out as soon as the motor runs up to about 75% of full load speed. The motor is suitable for loads of higher inertia or more frequent starting than is the split-phase motor; starting torque is improved and starting current reduced.

Capacitor start and run motor. A paper capacitor is permanently connected in series with the start winding, Figure 14.26. Starting torque is low but running performance approaches that of a two-phase machine. Generally quieter than split-phase or capacitor start motors; efficiency and power factor are improved.

Capacitor start, capacitor run motor. For this machine a large electrolytic capacitor is used for starting but this is switched out when the motor runs up to speed and a smaller paper capacitor is left in circuit while the machine continues to operate, Figure 14.27. Thus the good starting performance of the capacitor start motor is combined with the good running performance of the capacitor start and run machine.

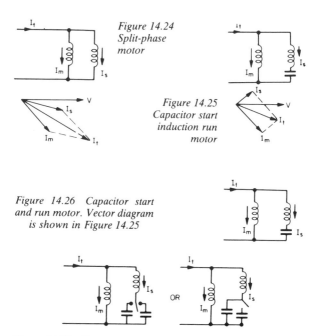

Figure 14.24 Split-phase motor

Figure 14.25 Capacitor start induction run motor

Figure 14.26 Capacitor start and run motor. Vector diagram is shown in Figure 14.25

Figure 14.27 Capacitor start, capacitor run motor. Vector diagram is shown in Figure 14.25

Shaded-pole motor. A short-circuited copper ring is placed round a portion of each pole, and this ring has currents induced in it by transformer action, these cause the flux in the shaded portion to lag the flux in the main pole and so a rotating field is set up. Starting torque is small and efficiency is poor since losses occur continuously in the shading ring (Figure 14.28).

Typical performance details. Table 14.4 gives performance details of the various motors discussed and compares them with a three-phase machine of a similar size. Small power motors only are considered.

347

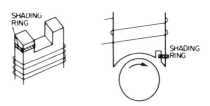

Figure 14.28   Shaded-pole motor

Open-circuiting devices. Reference has been made to the open-circuiting devices in split-phase, capacitor-start induction-run and capacitor-start, capacitor-run motors. These machines are fitted with a device to open-circuit the starting winding once the load has been accelerated to a certain speed.

In some cases the device is a relay, the operating coil of which is connected in series with the running winding or in parallel with the starting winding to contacts which open the starting winding. In most cases however a speed-sensitive arrangement is used designed to operate as near as possible to the cross-over point between start and run winding speed-torque curves, varying for motors of different polarities or supply frequencies.

Speed variation. The induction motor is a constant speed machine but, in conjunction with fan drives, capacitor start and run and shaded-pole motors, can be controlled to give a limited speed variation. See later how the static inverter enables a cage machine to operate economically at variable speed.

Under constant load conditions a motor will stall or fail to start when the voltage is lowered below a certain value. However, the load presented by a fan varies as the square of the speed, and stable running conditions continue at lower voltages despite loss of torque, see Figure 14.29.

Voltage variations may be achieved in a number of ways.

1 *Resistors*. Heat losses occur in resistors, and voltage drops occur at starting.

2 *Tapped windings*. Careful matching of fan and motor is essential. Voltage supply variations cause changes in speed which cannot be corrected. Manufacturing problems are created.

348

3 *Auto-transformer or variac.* These devices are generally too expensive and automatic control is difficult.

4 *Thyristors.* Thyristors can be costly but as their use increases the costs fall. They are economic in multiple control systems. Negligible losses are incurred. They can be easily controlled through external circuits such as thermostats to give automatic regulation. Some problems exist with harmonic distortion of the public supply. When using thyristors special attention must be paid to peak currents at about 2/3 full speed, which might result in excessive heating. Thyristor control is possible down to 10% of full load speed, but between 10% and 40% full load speed noise may be a problem. The motor is usually derated to 80–90% of full output.

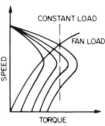

Figure 14.29   Speed control of fan load

Single-phase series commutator motor. This type of motor is similar to a d.c. series machine and it will run satisfactorily on a single-phase a.c. supply, Figure 14.30. The torque-speed characteristic is similar to its d.c. counterpart. Such motors are commonly used on domestic appliances like vacuum cleaners, power tools, etc.

Repulsion motors. There are many forms of repulsion motors, but the main principle is that a stator winding similar to a series motor is used with a wound rotor having a commutator which is short-circuited. The brushes are set at an angle (about 70°) to the main field and by means of transformer action the field and armature fluxes are such that they repel each other and the rotor produces a torque, Figure 14.31.

Repulsion motors can be started either by a variable series resistance or by auto-transformer, and a fair starting torque can be obtained. On this account the principle is used for starting in

349

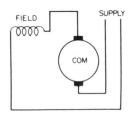

Figure 14.30   Simple series motor

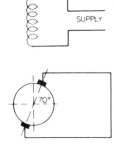

Figure 14.31   Simple repulsion motor

the *repulsion-start single-phase induction motor*. In this motor the rotor is as used for a repulsion motor, but after starting the two brushes are lifted and the commutator short-circuited all round by means of a copper ring. The motor then runs as an induction motor.

The speed of a repulsion motor driving a constant-torque load may be controlled either by movement of the brushes or by variation of voltage applied to the motor. The former method is usually adopted because of its simplicity and the avoidance of additional control gear.

The performance of a repulsion motor start induction motor is generally similar to that of a high torque capacitor start machine. These motors are now available from only a small number of specialist manufacturers.

British standards.   Most single-phase motors are in the small power range. Performance is covered by BS 5000 Part 11 99, IEC 34–1 and dimensions by BS 2048.

SPEED VARIATION OF A.C. MOTORS

Standard types of a.c. motors do not permit of any real speed variation as their speeds are fixed by the frequency of the supply on which they operate. The synchronous motor has a definitely constant speed and the induction motor can be assumed as constant speed as the maximum slip is not usually more than 5% except when supplied at variable frequency by a static inverter. A limited speed variation can be obtained by rotor resistances.

Rotor resistance control. The characteristic of the slip-ring machine with rotor resistance control approaches very closely to that of the series d.c. motor in that the speed rises as the load falls off and, therefore, it can only be satisfactorily employed for speed regulation where the load is fairly constant. Speed reduction by rotor resistance control is wasteful because of the power dissipated in the resistance, resulting in a low overall efficiency. In general, accurate control of speed cannot be obtained with any degree of satisfaction below 30–40% of full speed due to the fact that slight variation in the load causes wide fluctuations in the speed. In addition, difficulty may be experienced in maintaining constant speed due to the change of resistance with temperature of the external resistances, the speed tending to fall as the temperature of the controlling resistance increases.

With a suitable external resistance of the liquid type, stepless variation between limits may be obtained, but with the metallic resistance grids and a drum controller, the number of steps of speed is, of course, limited to the number of settings in the controller, the resistances being graded to suit these steppings.

Pole-changing motors. As the speed of an induction motor depends on the number of poles, two, three or even four different speeds can be obtained by arranging the stator winding so that the number of poles may be changed Speed ratios of 2:1 may be obtained from a single tapped winding and other ratios are available from two separate windings. The two systems may be combined, to give, on a 50 Hz supply, speeds of say 3000, 1500, 1000 and 500 rev/min. No intermediate speeds can be obtained by this methods. These motors are sometimes used for machine tools, as, for example, drilling machines, and this method gives a very convenient speed change.

P.A.M. motors. Pole amplitude modulation is a system of winding that offers various speed ratios by pole-changing without the need of separate windings. The system is patented by Prof. Rawcliffe of Bristol University and is a development of the original single-tapped two-to-one ratio Dahlander winding of 1897.

A single tapped winding is used and by reversing one half of each of the phases the flux distribution in the machine is changed, thereby producing a resultant field of different polarity. Main advantage of this system is the better utilisation of the active material in the motor, resulting in a smaller frame size for a particular speed combination than would obtain using the two-winding method. Both the efficiency and power factor are higher with this method than for the equivalent two-winding machine.

Table 14.4 Performance details of single-phase machines compared with similar sized three-phase motors

Item	Three phase	Split phase	Capacitor start	Capacitor start and run	Shaded pole
Full load efficiency %	60/80	50/60	50/60	50/60	30
Full load power factor	0.6/0.8	0.6/0.7	0.6/0.7	0.9	0.5/0.7
Start current × full load current	6	8–11	5	4	2
Start torque %	200	180	280	40	30
Run-up torque %	200	165	190	35	25
Pull-out torque %	250	200	200	150–250	120
Speed variation	Some*	None	None	Some*	Some*
Starting cycles/hour	60	10–½ sec	15–3 sec	60	60
Limitations	–	High start curr.	Short time rating of caps.	Low start torque	Low start torque and low efficiency
Typical applications	–	For general use when high start torque or freq. start not needed	High start torque; low start current. High inertia drives, pumps, compressors	Generally confined to fan drives or where low starting torque acceptable	

Some speed variation possible with fan drives by reducing the voltage, see text

Change in speed both up and down can be made without imposing undue transients on the supply system. Speed ratios of 3:1 or greater are not practical. The p.a.m. principle can be applied to slip-ring as well as cage machines.

Cascade induction motors. Induction motors can be arranged in cascade form to give intermediate speeds. In this arrangement two motors are arranged so that the rotor of one motor is connected in series with the stator of the second, Figure 14.32. Only one stator is fed from the supply.

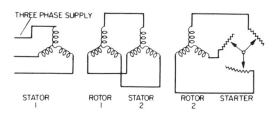

THREE PHASE SUPPLY

STATOR	ROTOR	STATOR	ROTOR	STARTER
1	1	2	2	

Figure 14.32   Three-phase cascade connections

The speed of the common shaft will be equal to that of a motor having a number of poles equal to the sum of those of the two motors. The speed of a cascade arrangement is thus a low value – usually an advantage for driving heavy machinery. For speed variations they can be arranged so that either the main motor can be used separately or in cascade. For instance a combination of a 4-pole and 6-pole motor will give either 1500 or 1000 rev/min separately, or combined the speed will be

$$\frac{50 \times 60}{2 + 3} = 600 \text{ rev/min}$$

The three-phase commutator motor (Schrage). A fully variable speed motor for use on three-phase is the commutator motor, one type of which is that due to Schrage, Figure 14.33. The primary winding is situated on the rotor and is fed by means of slip-rings. The rotor also carries a secondary winding which is connected in the usual way to the commutator and through the brushes to another secondary winding on the stator.

353

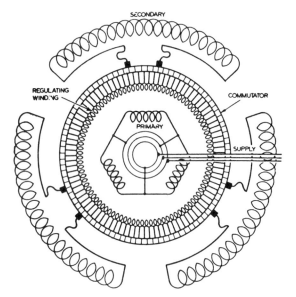

Figure 14.33   Diagram showing windings of three-phase Schrage commutator motor

Three pairs of brushes are required, each pair feeding one phase of the stator winding, as shown in the diagram. Speed variation is obtained by moving each pair of brushes relative to each other, this being done by a hand or automatic control through suitable worm-gear.

The speed range is roughly 3 to 1 for normal load – this ranging from 40% above to 60% below synchronism. The speed varies from 5–20% with the load, but this can of course be counteracted by further movement of the brushes. Motors are usually started by placing brushes in lowest speed position, giving a starting torque up to 1½ times full load torque.

These motors are expensive in first cost but have proved very satisfactory for driving machinery requiring speed control, such as printing machines, textile mills, etc.

The higher initial cost over the slip-ring motor of corresponding rating is soon recovered because of the elimination of rheostatic losses.

Motors for a speed range of 3:1 are available with ratings from 2¼ W to 187 kW. Larger motors up to about 300 kW can be built for a smaller speed range. Larger ranges of speed up to 15:1 are possible, but they involve a more costly motor, as the frame size for a given torque is governed by the speed range.

Variable-speed stator-fed commutator motor. Many applications requiring a variable-speed a.c. motor may be met by the stator-fed commutator type, Figure 14.34.

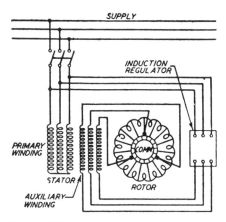

Figure 14.34   Internal and external connections of stator-fed variable-speed commutator motor

This motor is similar to a slip-ring induction motor, with the difference that the rotor winding is connected to a commutator instead of to slip-rings. Speed regulation is obtained by means of a separate induction regulator connected between the mains and the brushes on the commutator.

An auxiliary winding is often employed on the stator connected in series with brushgear and the regulator, the purpose being to

355

Table 14.5 Characteristics of a.c. motors

Type	Speed variations on load	Possible speed control	Starting torque % Full Load	Starting current % Full Load	Notes
Polyphase induction cage	Falls up to 5 to 6% from no load to full load	No variation or control possible*	120–150 33–55 20–80 % Full Load	500–600 150–250 100–400 % Full Load	Direct starting Star-delta starting Auto-transformer Standard industrial motor. Speed assumed constant for general use.
Polyphase induction high torque	As above	None	180–300 60–100 % Full Load	390–525 130–175 % Full Load	Direct starting Star-delta Several makers supply specially designed motors with high starting torque with moderate starting current.
Polyphase induction slip-ring wound-rotor	As for cage	Can be varied by rotor resistance but variation depends on load. Up to 30% approx.	100–150 % Full Load	150–250 % Full Load	For general use where motor must start up on heavy load. Otherwise performance similar to cage.

Polyphase commutator (Schrage)	Varies up to 20% for one brush position	Variation 3 to 1 by moving brushes	150–200 % Full Load	150–200 % Full Load	A.C. variable speed motor for moderate or large sizes. Expensive in first cost. Speed control is ideal.
Polyphase commutator (stator fed)	Varies about 20%	Variation 3 to 1 by induction regulator	175 % Full Load	125 % Full Load	Starting characteristics at minimum speed position. Motor ratings from 1 to 375 kW. May be designed for high-voltage operation.
Synchronous	Constant [power factor can be varied.]	None	None	–	For large continuously running drives. The auto-synchronous starts as an induction motor giving F.L. torque.
Single-phase induction split-phase	Falls slightly as load increases up to 6%	None	200–220 % Full Load small powers only	650–750 % Full Load	Used for small powers only on account of high starting current. Speed assumed constant

* Except when a variable frequency supply is available

increase the power factor of the motor throughout its range of speeds.

The regulator acts as a variable-ratio transformer, supplying a variable voltage at supply frequency to the brushes. As the regulator secondary voltage is reduced, the speed of the motor rises; the motor speed will rise above synchronous speed as the voltage is increased in the opposite direction.

In the majority of cases, operation of the regulator by a handwheel for speed adjustment is all that is necessary. As the regulator needs only a cable connection to the motor, remote control of the motor is very simple.

The speed of a stator-fed commutator motor can be varied without steps within the limits of the speed range. Normal ranges of speed regulation are 1:1.5, 1:2, 1:3, 1:4, but machines can be designed for ranges of 1:10, or greater. The machines for 1:3 speed range are usually designed so that the maximum speed is about 35% above synchronism and the minimum about 55% below.

The speed of the motor, like that of any shunt motor, is higher at no load than at full load. If very close speed regulation is desired, it may be obtained either by electrical compensation in the motor, or by the provision of automatic control of the regulator.

For most forms of drives, stator-fed motors develop sufficient starting torque when directly switched on to the line with the induction regulator in the lowest speed setting. Starting with the regulator in the other positions increases both starting torque and current.

Motors are available with ratings from ¾ kW up to about 375 kW for a speed range of 3:1. The size is a little larger than a slip-ring induction motor for the same duty. The motor is readily built with totally enclosed or totally enclosed fan-cooled enclosure because of the fixed brushgear and absence of slip-rings. Since the motor is stator-fed it may be designed for high-voltage supplies.

'Tandem' motors.   For lift and crane operation it is sometimes desirable that the power unit should be capable of giving a stable creeping speed in addition to the normal hoisting or lowering speed. To meet these conditions the 'Tandem' motor has been developed for use where the electric supply is alternating current.

The 'Tandem' motor consists of two component motors, built as one complete unit. The slow-speed motor is of the cage type and forms the main supporting part of the set. A slip-ring motor forms the high-speed unit.

Where the available power supply is 50 Hz, the two speeds usually selected are 960 rev/min for the slip-ring machine and 150 rev/min for the cage motor, but other speeds can be provided if desired. Speed changing can be effected whilst the motor is running and is carried out by transferring the supply from one stator winding to the other.

On changing down from the high speed to the low speed, the latter is reached by powerful regenerative braking during which the total reverse torque may amount to 400% of full load torque. Under normal conditions the fairly high inertia of the rotor absorbs much of this torque, thereby preventing too rapid deceleration and excessive stress in the gears and other mechanism of the drive.

Further adjustment of the braking may be effected by inserting a choke or buffer resistance in series with the stator winding of the cage low-speed motor.

Slip energy recovery schemes. If, instead of wasting the energy in the resistances associated with the speed control of slip-ring machines the power can be fed back into the supply then an improvement in the efficiency of the drive at reduced speeds can be obtained. A direct connection between the slip-rings and supply is not possible because both the slip-ring voltage and frequency vary with motor speed.

In the modified Kramer scheme a thyristor bridge is connected in the rotor circuit as shown in Figure 14.35. This converts slip-ring frequency power to d.c. which is fed to a separate d.c. motor driven asynchronous generator which returns the slip power to the supply. A completely static system is available in which the motor generator is replaced by a static inverter. In another system the d.c. output from the thyristor bridge may be fed to a d.c. motor coupled to the main slip-ring motor so that the combined torques of both machines provide a constant power output to the drive over the speed range.

Variable frequency drives. The speed of an induction motor can be altered by varying the supply frequency. This variable frequency can be provided in a number of ways.

Until the advent of the thyristor the most common method of obtaining a variable frequency was by a motor generator set as

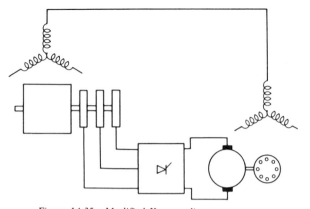

Figure 14.35   Modified Kramer slip recovery system

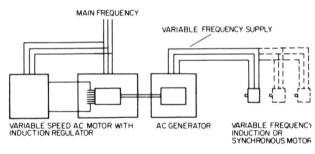

Figure 14.36   Motor generator set supplying variable frequency motors

shown in Figure 14.36. As such machines are expensive it is usual to use one set to supply a number of motors as indicated. Three full power electromechanical power conversions are involved in the use of a motor generator set and it is therefore advantageous to use systems in which the frequency conversion is accomplished electrically within the converter. Output frequency is still altered by changes in converter speed, but this is accomplished using only

360

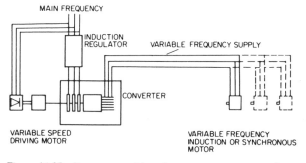

Figure 14.37   Parry motor-driven frequency converter supplying variable frequency motors

a small driving motor. The converter itself is an unwound stator within which rotates a rotor whose windings are connected to both slip-rings and a commutator, Figure 14.37.

Self-driven converters are also available which are basically stator-fed commutator motors in which the output is taken not as mechanical power at the shaft, but as slip frequency power from a set of slip-rings connected to the rotor windings, Figure 14.38. The induction regulator in such an equipment controls the converter speed and therefore the frequency, the output voltage inherently varying with the frequency to maintain a constant ratio between voltage and frequency.

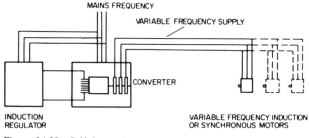

Figure 14.38   Self-driven frequency converter supplying variable frequency motors

361

The availability of thyristors makes it possible to change from a fixed frequency to a varying frequency by purely static means. Where the required frequency range is well below the mains frequency the cyclo-converter system can be employed. The required frequency can be synthesised by appropriate sequential switching of the motor supply terminals to successive supply phases. Two sets of thyristors are needed, supplying respectively the positive and negative half-cycles of the a.c. supply to the motor. Regenerative operation of the motor is possible by allowing power flow back to the supply from the kinetic energy of a high inertia drive during rapid deceleration. Figure 14.39 shows the connections for a cyclo-converter drive.

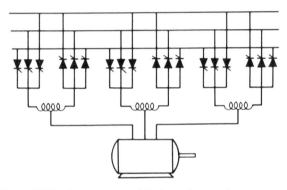

Figure 14.39   Arrangements of thyristors for a cyclo-converter to supply a variable frequency motor.

Static frequency converter drive.   Essentially a cage induction motor is a fixed speed machine, the speed being related to the supply frequency and number of poles. The advent of variable frequency static control systems enables a standard 50 Hz cage machine to be driven at any speed corresponding to between 0.5 and 100 Hz, i.e. twice rated speed. One such system is the SAMI made by ABB (Stromberg, ASEA and Brown Boveri) and widely used in industries such as steel, petrochemical, paper, water and heating and ventilating. Wherever variable speed drives are required the static frequency converter and cage motor can usually be employed. Energy saving is a feature of such installation because the optimum speed for a given output can be attained.

Essentially a SAMI frequency converter consists of a 6-pulse diode bridge rectifier connected to the a.c. supply network and a pwm inverter fed with d.c. from the rectifier, to provide the variable frequency to the motor, Figure 14.40. The d.c. link between the two static components includes an LC filter to smooth out any remaining a.c. from the rectifier and also to act as an energy store.

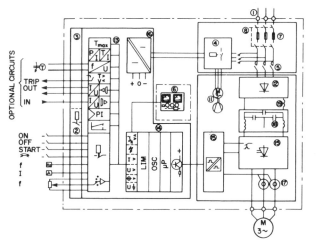

Figure 14.40   Block diagram of SAMI frequency converter

MOTOR DIMENSIONS

Dimensions and outputs of electric motors for use with a cooling medium with a temperature not exceeding 40°C are covered by BS4999 Part 141 and BS5000 Part 10. BS4999 gives general

363

information in which is included the nomenclature for identifying frame sizes and lays down standard dimensions and symbols that are mandatory and non-mandatory. BS 5000 specifies outputs in kilowatts and the associated shaft dimensions allocated to the frame sizes defined in BS 4999, output is related to speed, type of enclosures, class of rating, supply conditions and type of rotor. Part 141 of BS 4999 also indicates slide rail dimensions.

System of frame nomenclature. This is laid down in BS 4999 Part 103 and frame sizes up to and including 400 which are primarily intended for low voltage induction type motors the nomenclature is as follows. Firstly a letter indicates the type of enclosure: C is enclosed ventilated; D is totally enclosed except flameproof; and E is for all types of flameproof enclosures. Secondly a number of two or three digits indicates the height in millimetres of the shaft centre above the feet on a foot-mounted frame. Thirdly a letter, S, M or L characterises the longitudinal dimensions where more than one length is used. Finally, for other then foot-mounted types, a letter is employed to indicate the type of mounting: D for flange mounting; V for skirt mounting; C for face flange mounting; P for pad mounting; and R for rod mounting. For example a t.e.f.c. motor of frame size 160M, suitable for flange mounting is designated D160MD.

Single letter symbols for dimensions are specified in Table 14.6 Double letter symbols for dimensions are shown in Table 14.7. They are all shown on Figure 14.41. It is not possible to reproduce all the dimensions relating to the different types of induction motors covered by the standard but Table 14.8 shows the fixing dimensions of foot-mounted frames (all enclosures) for low voltage a.c. induction motors. The stator terminal box is positioned on the left hand side of the motor, looking at the non-driving end, with its centre on or about the centre line of the motor. Tables 14.6–14.8 are reproduced on pages 378–381.

Outputs and shaft numbers for single-speed, t.e.f.c.cage rotor, MCR motors with Class E or Class B insulation are shown in Table 14.9. They are suitable for a three-phase 50 Hz supply not exceeding 660 V. Other tables in the standards give similar information for enclosed ventilated, slip-ring, enclosed ventilated slip-ring and flameproof t.e.f.c. motors. A further table provides the same information relating to t.e. frame surface cage rotor MCR motors.

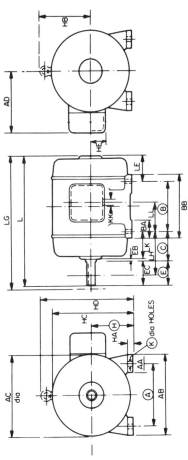

Figure 14.41 Symbols for foot-mounted frames

Notes:
The ringed symbols are defined in Table 14.6. Unringed symbols are defined in Table 14.7. The above diagram is intended to be purely illustrative of the dimensions defined in Tables 14.6 and 14.7 on pages 369–71.

365

MOTOR CONTROL GEAR

For safety, all motor starters should automatically return to the 'off' position in the event of failure of the supply and for this purpose an undervoltage or 'no-volt' release must be fitted. The undervoltage release forms an inherent part of the starting switch in all electro-magnetically-operated starters, but is additional to the starting-switch in hand-operated starters.

In faceplate d.c. starters the no-volt release takes the form of an electromagnet which holds the starter arm in the 'full on' position by magnetic attraction (see Figure 14.4, page 320).

With drum-type a.c. starters having a manually-operated starting handle, the starting switch is fitted with a spring which biases the switch to the 'off' position, but is retained in the 'on' position by a mechanical latch. Fitted in the starter is a shunt-wound electromagnet, or solenoid, excited by the supply voltage. In the event of supply failure, the plunger or armature of the solenoid or electromagnet is released, and is arranged to knock off the hold-on catch and so return the switch to the 'off' position. A push-button for stopping the motor can be connected in series with the no-volt coil (see Figure 14.42).

Contactor control gear. In contactor starters, the no-volt feature is inherent as shown in Figure 14.43. Pressing the starter button energizes the operating coil which closes the contactor. In order to keep the contactor closed when the 'Start' push-button is released, retaining contacts are required. These are closed by the moving portion of the contactor itself and thus maintain the operating-coil circuit, once it has been made and the contactor closed, irrespective of the position of the 'Start' push-button. Depression of the 'Stop' push-button, or failure of the supply to the operating coil, immediately causes the contactor to open. By energising the operating coil from the same circuit that supplies the motor, such an arrangement is equivalent to a no-volt release.

When starting and stopping is automatically provided by means of a float, pressure or thermostatic switch, as in the case of motor-driven pumps, compressors, refrigerators and the like, two-wire control is used as in Figure 14.44. In these cases, it is desirable that some form of hand reset device be incorporated with the overload release, to prevent the starter automatically reclosing after tripping on overload or fault until the reset button has been pressed after clearance of the fault.

366

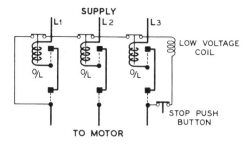

Figure 14.42   Diagram of drum type hand-operated direct-on-line a.c. starter

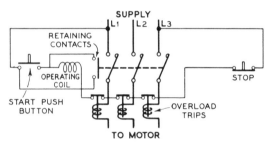

Figure 14.43   Diagram of direct-on-line contactor starter

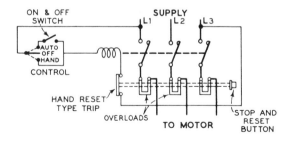

Figure 14.44   On and off switch control

Overload protection. An important function of motor control gear is to prevent damage due to excessive current. This may be due to either mechanical overload on the motor or to a defect in the motor itself. In either case it is essential that the supply be disconnected before any damage is done to the motor. An overload device thus usually operates by releasing the latching-in device by disconnecting the supply to the no-volt coil, or, in contactor starters, by opening the operating coil circuit.

Time-lag considerations. It is essential to guard against unnecessary operation from temporary overloads due to the normal operation of the machines which are being driven. The detrimental effect of an overload on a motor is a matter of time – a slight overload taking considerable time to develop sufficient heart to do any damage, whereas a heavy overload must be removed much more quickly.

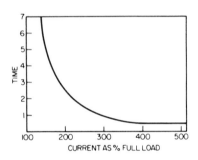

Figure 14.45 Curve showing variation in speed of operation with load for overload time lag

Overload devices therefore usually have a time-lag feature giving a curve of which Figure 14.45 is typical. Both the overload value should be adjustable and also the time feature for important motors. For d.c. or single-phase one overload device is usually sufficient, while for three-phase at least two of the lines must be protected, and it is considered advisable by many engineers to have an overload device in each phase. A three-phase motor will often run as a single-phase machine if one line gives trouble, and this is often avoided by protecting all three lines with an overload relay which incorporates a single-phase tripping device. If the current in any one phase fails the relay trips out.

Overload devices.   These are generally of two types – electro-magnetic and thermal. The electro-magnetic type consists of a coil or solenoid carrying the line current (or a proportion of it) with an armature which when attracted sufficiently operates the release circuit or latch. A time-lag feature is obtained by means of a dash-pot or similar arrangement as otherwise the action would be practically instantaneous.

Thermal overload devices have been developed to a considerable extent due to their low cost. They may be bi-metal strips or solder pot elements, and in either case, as the action is due to their heating up, a time element is always present.

The action of a bi-metal strip overload release depends on the movement resulting from the different rates of expansion of the two metals forming the combined strip when heated. The bi-metal strip may be directly heated by the current, or indirectly heated by a coil of resistance wire which carries the current. In the 'solder pot' form of release, a spindle carrying a ratchet wheel is embedded in a low-melting-point fusible alloy. This alloy is heated and melted by excess currents and, when molten, permits the ratchet wheel to rotate and so trips the starter. This type of overload device has never enjoyed much popularity in the UK but has been used in the USA successfully for a number of years. Thermal trips are usually of the hand reset type; the reset feature may be combined with the stop push-button.

Thermal overloads are usually confined to the smaller control units up to 15–22 kW, but it will be realized that, with modern industrial tendencies, small motors represent a very large proportion of the machines now being installed.

Thermistor protection.   Thermistors are semiconductors which exhibit significant changes in resistance with change in temperature, i.e. they are thermally sensitive devices. They are based on barium titanate and are formulated to produce negligible increase in resistance with change in temperature until the Curie point is reached. At this point any further increase in temperature causes the resistance to change rapidly. This characteristic is used to operate as a protective device.

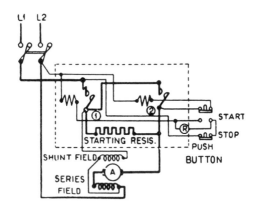

Figure 14.46 Typical d.c. contactor starter

Contactor No. 1 is the line contactor
Contactor No. 2 is the running contactor

The diagram shows d.c. starter with two-stage operation and controlled by push-buttons. The connections are for a compound-wound motor, but the starter can be used for shunt, series or compound. The pressing of the start button closes contactor No. 1, starting the motor with the resistance in circuit. The operating coil of the second contactor is connected across the motor armature and operates as soon as the motor voltage reaches a certain value. The main contactor acts as a no-volt device and overload protection can be added.

The thermistor is housed in the stator windings of the motor at that point which is considered to be the 'hot-spot', i.e. the highest temperature point for a given overload current. The associated control gear is generally housed in the starter and takes the form of an amplifier and possibly a relay. Thermistors handling large currents are available and these can operate directly a small interposing relay.

Thermistors are claimed to offer better protection than any other single system because they measure directly the temperature of the motor winding. But the disadvantage is that since they have to be embedded in the winding they introduce a weakness into the insulation system. This is one of the reasons that they do not find application in h.v. motors being restricted in the main to 415 V machines.

Multi push-button control. In the case of electrically operated contactor gear both starting and stopping can be controlled from any number of points by connecting additional push-buttons

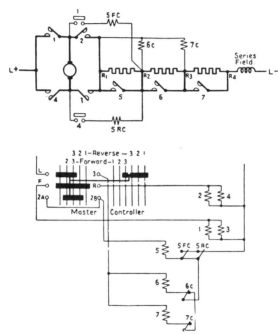

Figure 14.47 Typical d.c. contactor equipment for reversing heavy-duty motor
Relay 5RC controls contactor 5 in reverse direction
Relay 5FC controls contactor 5 in forward direction
Relay 6C controls contactor 6 in either direction
Relay 7C controls contactor 7 in either direction

371

either in series or parallel, as required. This has a definite advantage when it is inconvenient to have the control gear mounted close to the place where the operator is situated. For emergency stopping any number of 'Stop' buttons may be used to save time in cases of emergency.

The same principles are used in connection with interlocking for lifts and similar machinery where it is necessary for certain items to be in position before the machine can be started up.

Multi-stage starters. Direct-on starting is allowed for motors up to a certain size – the limit being fixed by the supply system. For 'strong' systems l.v. motors of up to 100 kW or more can be started and at 3.3 kW the rating can rise to 300 kW or greater. For 'normal' systems the limit at 415 V may be anywhere between 5 kW to 30 kW. The supply authority will advise on this matter. Above these limits it is necessary to employ some method of starting which restricts the starting current to between two and three times full-load current. For cage motors using either star-delta or auto-transformer starting, the change over from 'start' to 'run' is effected in contactor starters by employing two contactors with either a series or time relay for timing the moment of the change over.

Wound-rotor starters. Where slip-ring motors have to start up against severe load conditions liquid resistance starters are found very satisfactory. By varying the electrolyte the added resistance is under control, and liquid starters have the advantage that the resistance is reduced continuously and smoothly instead of by steps as in other systems.

For wound-rotor motors hand-controlled starters are in more general use, but contactors can be used – one main contactor for the stator and two to five stages for cutting out the rotor resistance.

The change over from one stage to another may be controlled by current or time. For current control a current or series relay is used to operate the change over from one contactor to the next when the current has fallen to a certain value by virtue of the motor speeding up. This method ensures more correct starting but requires fairly accurate adjustment.

The time relay control is by means of a dash-pot time-lag or similar device, and as soon as the first contactor is operated the relay comes into circuit. After the set time has expired the change-over takes place and the process is repeated for subsequent stages.

Air-break limits. Manually-operated air-break switchgear and starters are only satisfactory up to a certain size. For d.c., however, air-break gear is used up to large power outputs, but it is necessary to renew the contacts fairly frequently.

For controlling a.c. motors manually-operated air-break gear other than contactor gear is satisfactory up to about $1\frac{1}{2}$ kW, but for larger outputs vacuum, SF_6 or air-break circuit-breakers are generally specified.

Fuses. Overload devices in starters are designed to interrupt the circuit in the event of the current rising above a predetermined value due to the mechanical overloading of the motor, and are not usually designed to clear short-circuits. It is very desirable to include a circuit-breaker, or fuses, of sufficient breaking capacity to deal with any possible short circuit that may occur. The required ratings of high-breaking capacity cartridge fuses for motor circuits can be obtained from fuse manufacturers' lists.

Quick-stopping of motors. The application of electric braking, that is, braking by causing the motor to develop a retarding torque, to certain classes of industrial drive, such as rolling mills, electric cranes and hoists, is normal practice. It is being used on many other industrial drive systems and there are two fields of aplication:

(1) Drives where a large amount of energy is stored in the rotating parts of the driven machine.
(2) Drives requiring rapid and controlled deceleration in the event of an accident or emergency.

There are two ways in which a polyphase induction motor may be made to develop a braking torque and stop quickly:

(1) By causing it to cease to act as a motor and to operate as a generator.
(2) By reversing it, so that it is, in effect, running backwards until its load is brought to rest. The first method is exemplified by the method of d.c. injection, and the second by 'plugging.'

373

Braking by d.c. injection. In this arrangement, when the motor is to be braked, it is disconnected from the a.c. supply and direct current is fed to the stator winding. The effect of this is to build up a static magnetic field in the space in which the rotor revolves and the current thereby generated in the rotor winding produces a powerful braking torque in exactly the same way as is done in a shunt-wound d.c. motor.

In a workshop where a d.c. supply is available, braking is particularly easy, as little additional apparatus is required. In all other cases, the necessary d.c. can be provided by a rectifier built in or adjacent to the control box. No mechanical connections are required, all the braking control being included in the controller cabinet. When the stop button is pressed, the a.c. contactor (see Figure 14.48) opens, causing the d.c. contactor to close, and connect the stator of the motor to the rectifier. The motor is rapidly brought to rest. By means of a timing relay, which is arranged to give varying periods of application of the d.c. to suit the loading of the motor, the d.c. contactor is then tripped out.

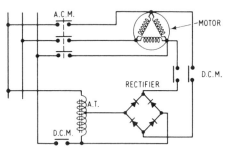

Figure 14.48   D.C. injection
A.T. Auto transformer
A.C.M. A.C. main contactor
D.C.M. D.C. main contactor

The method is applicable both to cage and slip-ring motors, although more usually applied to cage. When applied to slip-ring motors, the rotor circuit is made through the rotor resistance, a tapping on this being selected to give the best results. The d.c. may be arranged for injection on either the stator or rotor winding, or on both.

Braking by reversal (plugging). Rapid braking of a.c. (and d.c. motors) may be obtained by reconnecting the windings to the supply in the order corresponding to reversed direction of rotation. The changeover is usually made quickly, as by the operation of contactors, and the method is widely employed with automatic control systems where particularly rapid stopping is desired. When deceleration is completed and the motor stops, it will of course reverse unless the windings are disconnected.

Generally speaking, the modifications necessary to the normal starting gear required for an induction motor are not great in order to make it suitable for plugging the motor.

Plugging with direct-on starter. All that is necessary is to replace the usual starter switch by a changeover reversing switch, which may be hand-operated, or automatic, as demanded by the needs of the drive. Figure 14.49 shows a direct-on push-button starter adapted to carry out the plugging operation, the only alterations being the addition of the reversing switch, the use of a special stop button, and usually a variable limiting resistance fitted in each phase feed of the plugging circuit. The resistances

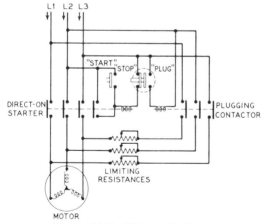

Figure 14.49 Wiring of direct-on contactor starter for plugging

375

allow of a variable time factor, and control the value of the braking torque and current so as to avoid damage to any part of the motor or driven machine.

Plugging with auto-transformer or star-delta starter. The position is rather more difficult, because means must be provided to make sure that the motor is connected to the starting position of the starter before it is plugged. This can be done by interlocking the starter and reversing switch, or by providing a changeover switch for the reversal that also reduces the voltage to the motor.

Figure 14.50 shows the curves for a cage motor with a high resistance or double cage rotor when plugging on full voltage, and on reduced voltage such as would be obtained with an auto-transformer.

Plugging slip-ring motors. The switch for the stator must be a changeover reversing switch and some form of interlock is necessary to ensure that all rotor resistance is in circuit before the motor is plugged. In certain cases an extra plugging step may be necessary, although the starter will be simplified if the starting resistance on the first step be made sufficiently great to limit the braking current to the desired value. It should be remembered that the current passed when plugging will be almost twice that passed by the same resistance at standstill. In Figure 14.51 curves are given for a slip-ring motor, with and without rotor resistance. It will be seen that if a large step of resistance is used, the resulting curves differ somewhat from those normal for the machine and that current and torque tend to increase above their standstill values. It is usual to determine the value of a star-connected resistance necessary to limit the braking current to any desired value from the formula:

$$R = 1.8 \times E/1.73 \times I$$

where E is the open circuit rotor voltage at standstill, and I the permissible braking current.

Plugging switch. The disadvantage of plugging is that the motor may carry on and start to accelerate in the reverse direction. On drives where this would be objectionable, however, it can be prevented by including a plugging switch on the motor shaft. This is a simple switch of the centrifugal type which opens when the motor reaches zero speed, and so trips the main stator

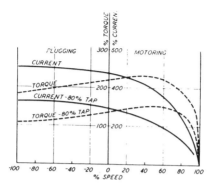

Figure 14.50 Torque/speed and current/speed curves for double cage motor on full and reduced voltage when motoring and plugging

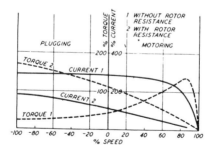

Figure 14.51 Torque/speed and current/speed curves for slip-ring motor with and without rotor resistance when motoring and plugging

377

switch of the motor by interrupting the no-volt or hold-on coil circuit. Another method of controlling the time the reverse contactor remains closed on plugging is by means of a timing relay set to drop out the reverse contactor just before the motor reaches zero speed.

With both d.c. injection and plugging the load is not held when the equipment comes to rest. If this is necessary, braking and stopping by means of an electro-mechanical brake must be used.

Table 14.6 Symbols for dimensions (BS 4999 Part 103)

Letter symbol	Dimension description
A	Distance between centre lines of fixing holes (end view)
B	Distance between centre lines of fixing holes (side view)
C	Distance from centre line of fixing holes at driving end to shaft shoulder
D	Diameter of shaft extension
E	Length of shaft extension from shoulder
F	Width of keyway
G	Distance from bottom of keyway to opposite side of shaft
H	Distance from centre line of shaft to bottom of feet
H^l	Distance from the centre line of shaft to mounting surface, e.g. bottom of feet for machines with low shaft centres
J	Radius of circle to which mounting pads (or faces for rods) are tangent
K	Diameter of holes in the feet or mounting pads (or faces)
L	Overall length
M	Pitch circle diameter of fixing holes
N	Diameter of spigot
P	Outside diameter of flange
R	Distance from surface of mounting flange to shaft shoulder
S	Diameter of fixing holes in flange
T	Depth of spigot

Table 14.7 Symbols for dimensions (BS4999 Part 103)

Letter Symbol	Dimension description
AA	Width of end of foot or pad (end view)
AB	Overall dimension across feet (end view)
AC	Overall diameter
AD	Distance from centre line to extreme outside of terminal box (end view) or other most salient object mounted on side of machine
AL	Overall length of slide rail excluding adjusting screw
AT	Thickness of slide rail foot
AU	Size of mounting holes in slide rail
AX	Height of slide rail
AY (max.)	Maximum extension of adjusting screw of slide rail
AZ	Width of slide rail at base
BA	Length of foot (slide view)
BB	Overall dimension across feet or pad (side view)
BT (min.)	Horizontal travel on slide rail
DH	Designation of tapped hole or holes (if any) in shaft extension
DJ	Centre distance of the two tapped holes in shaft extension (if appropriate)
EB	Distance from end of bearing housing to shaft shoulder
EC	Length of shaft from end of bearing housing to end of shaft
ED	Minimum length of keyway
HA	Thickness of feet
HB	Distance from centre line to top of lifting device, terminal box or other most salient object mounted on top of the machine
HC	Distance from top of machine to bottom of feet
HD	Distance from bottom of feet to top of lifting device, terminal box or other most salient object mounted on top of the machine
KA	Usable tapped depth of hole in pad (or facing) of pad-mounted or rod-mounted machine
KK	Diameter of holes in terminal box for cable entry
LA	Thickness of flange
LB	Distance from fixing face of flange to non-drive end
LD	Centre line of terminal box to fixing face of flange

Table 14.7 (*cont.*)

Letter Symbol	Dimension description
LE	Distance from extreme point of non-drive end to centre line of fixing holes in nearest foot
LF	Distance from end of shaft to fixing face of skirt-mounting flange
LG	Overall length including pulley
LH	Distance from centre line of pulley to centre line of fixing holes in nearest foot
LK	Distance from end of bearing housing (drive end) to centre line of fixing holes in nearest foot
LL	Distance from centre line of terminal box to centre line of fixing holes in foot nearest to shaft extension
XA (max.)	Centre line of bolt at adjusting screw end of slide rail to the beginning of the platform
XB	Width of slide rail at top
XC	Bolt diameter for which clearance is provided in the slot of the slide rail
XD	Height of adjusting screw centre line above the platform
XE	Distance between the centre lines of the mounting-bolt holes (side view)
XF	Distance between the centre line of the mounting-bolt hole at the adjusting screw end and the adjacent end of the slide rail

Table 14.8 Fixing dimensions of foot-mounted frames, all enclosures (BS4999 Part 141) (Primarily l.v. induction motors) *All dimensions are in millimetres.*

Frame number	H Nominal	H Negative tolerance*	A	B	C	K†	Fixing bolt or screw
56	56	0.5	90	71	36	5.8	M5
63	63	0.5	100	80	40	7	M6
71	71	0.5	112	90	45	7	M6
80	80	0.5	125	100	50	10	M8
90 S	90	0.5	140	100	56	10	M8
90 L	90	0.5	140	125	56	10	M8
100 S	100	0.5	160	112	63	12	M10
100 L	100	0.5	160	140	63	12	M10
112 S	112	0.5	190	114	70	12	M10
112 M	112	0.5	190	140	70	12	M10
132 S	132	0.5	216	140	89	12	M10
132 M	132	0.5	216	178	89	12	M10
160 M	160	0.5	254	210	108	15	M12
160 L	160	0.5	254	254	108	15	M12
180 M	180	0.5	279	241	121	15	M12
180 L	180	0.5	279	279	121	15	M12
200 M	200	0.5	318	267	133	19	M16
200 L	200	0.5	318	305	133	19	M16
225 S	225	0.5	356	286	149	19	M16
225 M	225	0.5	356	311	149	19	M16
250 S	250	0.5	406	311	168	24	M20
250 M	250	0.5	406	349	168	24	M20
280 S	280	1.0	457	368	190	24	M20
280 M	280	1.0	457	419	190	24	M20
315 S	315	1.0	508	406	216	28	M24
315 M	315	1.0	508	457	216	28	M24
315 L	315	1.0	508	508	216	28	M24
355 S	355	1.0	610	500	254‡	28	M24
355 L	355	1.0	610	630	254‡	28	M24
400 S	400	1.0	686	560	280‡	35	M30
400 L	400	1.0	686	710	280‡	35	M30

*There are no positive tolerances.
†The K dimensions are selected from the coarse series in BS4186.
‡Dimension C assumes ball and roller bearings.

Table 14.9 Single-speed TEFC cage rotor MCR motor (BS 5000 Part 10). Suitable for a 3-phase, 50 Hz supply not exceeding 415 V

Frame number	Output, kW				Shaft number	
	Synchronous speed, rev/min					
	3000	1500	1000	750	3000	1500 or less
D80	1.1	0.75	0.55	–	19	19
D90 S	1.5	1.1	0.75	0.37	24	24
D90 L	2.2	1.5	1.1	0.55	24	24
D100 L	3.0	2.2 and 3.0	1.5	0.75 and 1.1	28	28
D112 M	4.0	4.0	2.2	1.5	28	28
D132 S	5.5 and 7.5	5.5	3.0	2.2	38	38
D132 M	–	7.5	4.0 and 5.5	3.0	38	38
D160 M	11 and 15	11	7.5	4.0 and 5.5	42	42
D160 L	18.5	15	11	7.5	42	42
D180 M	22	18.5	–	–	48	48
D180 L	–	22	15	11	48	48
D200 L	30 and 37	30	18.5 and 22	15	55	55
D225 S	–	37	–	18.5	55	60
D225 M	45	45	30	22	55	60
D250 S	55	55	37	30	60	70
D250 M	75	75	45	37	60	70
D280 S	90	90	55	45	65	80
D280 M	110	110	75	55	65	80
D315 S	132	132	90	75	65	85
D315 M	150	150	110	90	65	85

15

Switchgear and protection

SWITCHGEAR

The choice of suitable switchgear depends to a larger extent on the actual duty than any other type of electrical plant. In addition to switching on or off any section of an electrical installation, the switchgear generally includes the necessary protective devices which are desirable in order that the particular section may be automatically isolated under fault conditions.

Originally all switchgear consisted of open knife switches mounted on a slate or composition panel and operated by hand. The protective device consisted of a fuse which was generally mounted close to the switch. The use of high voltage a.c. and the great increase in total power in a system necessitated the use of oil-break, air-break, vacuum, SF_6 or air blast switchgear.

For low voltages (up to 1000 V) knife-type switches are still used, and in some instances open-type boards are being installed, but generally most switchgear is to-day enclosed. Metalclad switch or combined switch and fuse units are used either singly or grouped to form a switchboard. For the smaller capacities insulated cases are obtainable in place of the metal type, these being particularly popular for domestic installations.

The knife switches are usually spring controlled, giving a quick make and break with a *free handle* action which makes the operation of the switch independent of the speed at which the handle is moved. In all cases it is impossible to open the cover with the switch in the 'on' position. The normal limit for this type of switch is from 300 to 400 A, but larger units can be made specially.

Miniature circuit-breakers are used widely as protective devices in consumer premises and for group switching and protection of fluorescent lights in commercial and industrial buildings. Moulded case circuit-breakers with ratings up to 3000 A and capable of interrupting currents up to 200 kA (for the larger ratings) are becoming popular for control of l.v. networks.

When any breaking capacity rating is required, together with automatic operation it is necessary to use special control devices to interrupt the fault current. Low voltage three-phase systems of up to 415 V are usually controlled by air circuit-breakers with or without series fuses. For higher voltage systems up to 36 kV, oil, air break, vacuum and SF_6 breakers are available. For higher voltages, SF_6 circuit-breakers take over and are used up to 420 kV. For l.v. systems the air-break circuit-breaker is usually a moulded case unit. The air-break circuit-breaker for 3.3 to 11 kV has an arc control device which is suitable for motor switching and is used mainly in power stations. Its cost makes general use in industry and distribution systems unusual.

Oil circuit-breakers which still are popular for h.v. distribution systems despite the perceived fire risk, consist of an oil enclosure in which contacts and an arc control device are mounted. The arc is struck within the control device and the resultant gas pressure sweeps the arc through cooling vents in the side of the pot. High reliability and simple maintenance are available from these devices.

Figure 15.1 shows a typical truck oil circuit-breaker with its cubicle for 12 kV. Vacuum and SF_6 circuit-breakers are also available in a similar form of truck mounting with similar cubicles.

Vacuum circuit-breakers were the first type of oil-less circuit breaker to be available and have been used in industrial situations since the later 1960s. Vacuum interrupters are sealed-for-life ceramic 'bottles' containing movable contacts in a high vacuum. The circuit breaking performance of this design is very high and a large number of short-circuit operations can be achieved before any replacement is necessary. In fact, in most cases, this will never be required.

SF_6 circuit-breakers come in a number of forms, all utilising the good dielectric and arc extinguishing properties of this gas to provide another type of oil-less circuit breaker. While the life of the contacts is not as great as those in a vacuum, the SF_6 circuit-breaker has other advantages that make it equally as acceptable for industrial and distribution use. Special designs of oil-less units are available such as shown in Figure 15.2 which illustrates a double tier unit of vacuum circuit-breakers which allows a compact arrangement for speedy assembly on site.

All these new units are designed to minimise the maintenance that is required to the interrupting unit – although regular checks

384

Figure 15.1   11 kV switchgear with oil circuit-breaker withdrawn from its housing (Reyrolle Distribution Switchgear)

Figure 15.2   Double tier 12 kV vacuum circuit-breakers permitting speedy assembly on site (Reyolle Distribution Switchgear)

on the mechanical operation and the cleanliness of exposed insulation is always advisable.

All circuit-breaker systems up to 36 kV are three-phase units, but for higher voltages up to 420 kV, three separate single-phase breakers are sometimes used to facilitate single-phase opening and closing for transient faults.

A major part of circuit-breaker cubicle cost is the protection and instrumentation systems that are associated with the particular panel protection and its interlocking with adjacent circuits. Protection from overload is obtained by means of a device which releases the mechanism and opens the breaker. For small breakers, the protection is provided by overload coils or thermal releases inside the unit itself. For large units which are protected by complex relay systems, the operation of one of the relays of the protection system releases the tripping mechanism in a similar manner.

Some essential features of all switchgear are:

(*a*) Isolation of the internal mechanism for inspection. This is important and full interlocks are always provided to prevent the opening of any part of the enclosure unless access to the higher voltage supply is prevented.

(*b*) Insulation from breaker contacts to the side of the enclosure must be adequate for the voltage and maximum load which the breaker will be called upon to deal with.

(*c*) Provision for manual operation in case the electrical control (if provided) fails to operate.

(*d*) Provision for instruments which may be required. These may be in the form of either an ammeter or voltmeter on the unit itself or the necessary current and voltage transformers for connecting to the main switchboard or a separate instrument panel.

For switchgear up to 11 kV and most circuit-breakers up to 33 kV, isolation is effected in the following ways:

(*a*) By isolating links in or near the busbar chamber.

(*b*) Draw-out type of gear in which the whole of the circuit-breaker is withdrawn vertically from the busbar chamber before it can be opened up.

(*c*) Truck-type, in which the circuit-breaker with its connections is isolated in a horizontal manner before inspection or adjustment.

It should be noted that in certain cases, double isolating devices are necessary, i.e. both on the incoming and outgoing side.

Isolation is, of course, always required on the incoming side, but it is also necessary on the outgoing side if that part of the network can be made alive through any other control gear or alternative supply. With conventional switchgear, isolation of both sides takes place automatically.

The difference between a switch and circuit-breaker is that the switch is a device for making and breaking a current not greatly in excess of its rated normal current, and the circuit-breaker is a device capable of making and breaking the circuit under both normal and fault conditions. Oil switches, however, have to be capable of making onto a short-circuit.

A breaker is usually classified according to the voltage of the circuit on which it is to be installed; the normal current which it is designed to carry continuously in order to limit the temperature rise to a safe value; the frequency of the supply; its interrupting capacity in kA; its making capacity in kA(peak), i.e. the instantaneous peak current; and the greatest r.m.s. current which it will carry without damage for a specified length of time, usually 1 or 3 secs.

Table 15.1   Number of operating cycles for the mechanical endurance test (from BS 4752)

1	2	3	4	5	6	7
			Number of operating cycles			
		All circuit-breakers	Circuit-breakers designed to be maintained‡		Circuit-breakers designed not to be maintained	
Rated thermal current in amperes	*Number of operating cycles per hour**	*With current without maintenance†*	*Without current*	*Total*	*Without current*	*Total*
		n	n'	$n + n'$	n''	$n + n''$
$I_{th} \leqslant 100$	240	4000	16000	20000	4000	8000
$100 < I_{th} \leqslant 315$	120	2000	18000	20000	6000	8000
$315 < I_{th} \leqslant 630$	60	1000	9000	10000	4000	5000
$630 < I_{th} \leqslant 1250$	30	500	4500	2500	3000	
$1250 < I_{th} \leqslant 2500$	20	100	1900	2000	900	1000
$2500 < I_{th}$	10	(Subject to agreement between manufacturer and user)				

* If the actual number of operating cycles per hour does not correspond to values in column 2, this shall be stated in the test report.
† During each operating cycle, the circuit-breaker shall remain closed for a maximum of 2 seconds.
‡ The manufacturer shall supply detailed instructions on the adjustments or maintenance required to enable the circuit-breaker to perform the number of operating cycles in column 5.

I_{th} is the rated thermal current of the circuit-breaker.

BS 4752:1977 covers circuit-breakers of rated voltage up to and including 1000 V a.c. and 1200 V d.c. It supersedes BS 862:1939 covering air-break circuit-breakers for systems up to 600 V and BS 936:1960 Oil circuit-breakers for m.v. alternating current systems. It also supersedes those part of BS 116:1952, Oil circuit-breakers for a.c. systems, and BS 3659:1963, Heavy duty air-break circuit-breakers for a.c. systems, which relate to breakers having rated voltages up to and including 1000 V. BS 4752 follows very closely IEC Publication 157–1, 2nd edition, 1973.

All circuit-breakers shall be capable of carrying out a given number of mechanical and electrical operating cycles, each one consisting of a closing operation followed by an opening operation (mechanical endurance test) or a making operation followed by a breaking operation (electrical endurance test). The numbers of cycles for the mechanical endurance test are shown in Table 15.1.

OVERLOAD AND FAULT PROTECTION

Protection against electrical faults may be broadly divided into fusegear or circuit-breakers. In some instances fuses are used with circuit-breakers to take over the interruption of higher short-circuit currents. This is so with miniature or lower rated moulded case circuit-breakers.

In h.v. distribution circuits another arrangement is that of fuses fitted in series with an oil switch; upon the blowing of a fuse link a striker pin is ejected which trips the oil switch.

Types of fuses. The simplest and cheapest form of protection against excess current is the fuse. Two types of fuse are in use:

1. The semi-enclosed type, comprising removable plastics or porcelain holder with handle through which the tinned copper fuse-wire passes.

2. The totally enclosed or cartridge type fuse in which the fuse itself is enclosed by a cylinder of hard, non-combustible material having metal capped ends, and is filled with non-flammable powder, or other special material. Regulation 533–01–04 says fuses should preferably be of the cartridge type.

Rating of fuses. Fuse ratings for given circuits should be selected with care. In low voltage general purpose circuits, with non-inductive loads such as heating and lighting, the selected fuse

389

should have a current rating which exceeds the full load current of the circuit, but which is less than the cable current rating. The cable is then fully protected against both overload and short-circuit faults.

On the other hand, in circuits which contain inductance or capacitance, the resulting inrush currents dictate the choice of fuse current rating. Thus in a motor circuit, the starting current and its duration have to be related to the fuse time/current curves, and a fuse chosen to withstand this surge will have a current rating up to three times the motor full load current. In such a circuit, the fuse provides overcurrent protection only, and overload protection is provided by other means (see section on fuse selection).

Cartridge fuses.   Cartridge fuses are available for systems up to 660 V in current ratings from 2 to 1600 A and breaking capacity ratings in excess of 50 kA. Some fuses can interrupt currents up to 200 kA. Higher kA ratings can be obtained but they are not standard. Figure 15.3 shows the cross-section of a l.v. fuse.

The breaking capacity rating at 11 kV is not less than 13.1 kA and current ratings are available up to 350 A.

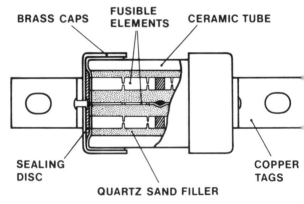

Figure 15.3   Construction of a modern cartridge fuse (GEC Installation Equipment Ltd)

390

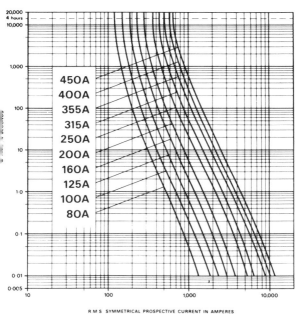

Figure 15.4   Typical 'T' range: time/current characteristics (GEC
Installation Equipment Ltd)

Other advantages of well-designed cartridge fuses are:

Discrimination.   The fuse nearest the fault will operate, thus
ensuring that only the faulty circuit is isolated and healthy circuits
are unaffected. This discriminating property is inherent in car-
tridge-type fuses, as a glance at the time/current curves for
different sizes of fuse in Figure 15.4 will show. It will be seen that
the speed of operation for any particular value of overload or fault
current increases as the fuse gets smaller.

High speed of operation on short-circuit.   This property enables fuses to be used for the back-up protection of motor starters
and low breaking capacity circuit-breakers. For such purposes

high breaking capacity in itself is not sufficient. The speed of operation of the fuses used for back-up protection must be faster than the speed of operation of the motor starter or circuit-breaker to be protected; otherwise the apparatus under protection would be damaged or destroyed before the fuses had time to act.

Selection of fuses. When selecting suitable fuses for any particular situation the following factors should be considered.

1. *Short-circuit ratings of cartridge fuses.* Cartridge fuses for l.v. industrial applications are most commonly proved capable of interrupting fault currents up to 80 kA at 415 V, and some designs will safely interrupt such fault currents at 660 V. They therefore have more than adequate breaking capacities for the overwhelming majority of applications, but they must not be used at voltages above their assigned rating.

2. *Current rating of a fuse.* This is the maximum current which the fuse can carry continuously without deterioration. As already indicated the fuse current rating selected for any circuit should have a current rating not less than the full load current of the circuit.

3. *Overload protection.* A general purpose type fuse (gG) to BS 88: Part 2 will protect an associated PVC insulated cable against overload if its current rating (I_N) is equal to, or less than, the current rating of the cable (I_Z). See Reg. 433–02–01 of the 16th Edition of the IEE Wiring Regulations.

4. *Motor circuits.* In a motor circuit, the starter overload relays protect the associated cable against overload, and the fuses in circuit provide the required degree of short-circuit protection. Manufacturers usually make recommendations regarding the fuse ratings needed to cope with motor starting surges, and also indicate minimum cable sizes needed to achieve short-circuit protection. Reg. 434–03–03 of the IEE Regulations is applied to both fuse and cable selection in order to accomplish satisfactory protection.

In a well-engineered combination, the starter itself interrupts all overloads up to the stalled rotor condition, and the fuses should only operate in the event of an electrical fault. The starter manufacturers indicate the maximum fuse rating which may be used with a given starter to ensure that satisfactory protection is achieved.

392

The automatic circuit-breaker. Circuit-breaker equipment, whilst considerably more expensive than fusegear protection, possesses important features which render it essential on all circuits where accurate and repetitive operation is required. It is usually employed on all main circuits, fusegear being reserved for sub-circuits.

Overload releases. The simplest automatic release used is the direct acting over-current coil, which is a solenoid energized by the current passing through the unit, and calibrated to operate when this reaches a predetermined value. The trip setting is adjustable, from normal full load current up to 300% for instantaneous release and 200% for time lag release. Therefore a setting as near as is desired to full load current can be obtained. On a three-phase insulated system, coils in two phases will give full protection, since any fault must involve two phases: but with an earthed system, release coils must be provided in all three-phase, unless leakage protection is also provided, when two overcurrent coils will again suffice. In small current sizes it is usual to make the overcurrent coils direct series connected: but in larger sizes the coils are often operated from current transformers.

Time lags. If full advantage is to be taken of the facility to obtain close overcurrent settings that is offered by circuit-breakers, some form of restraining device is necessary to retard the action of the releases, as this is normally instantaneous. This retarding device, or time lag, usually takes the form of a piston and dashpot. Its purpose is to delay the action of the trips so as to allow for sudden fluctuations in the load, and also to give a measure of discrimination. In practice, a characteristic closely resembling that of a fuse is obtained, and a typical characteristic for an oil dashpot time lag – perhaps the most common type in general use – is given in Figure 15.5.

Use of time-limit fuses. When the overcurrent coils are transformer operated, it is possible to use a shunt fuse in place of the mechanical retarder. These fuses are designed to short-circuit the trip coils, which, therefore, cannot operate until the fuse has blown. The fuse itself is usually of alloy-tin wire or some other non-deteriorating metal, and is accurately calibrated as regards blowing current. When time-limit fuses are used, no advantage is to be gained by making the release coils themselves adjustable:

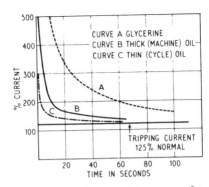

Figure 15.5 Characteristic curves for series trip coil with oil dashpot

and alternative trip settings are obtained by varying the blowing current of the fuses.

Cartridge fuses and circuit-breakers. With a circuit-breaker, a certain time must elapse between the operation of the tripping mechanism and the actual breaking of the current, and with commercial types of breaker this time is usually of the order of 0.1 sec. On the other hand, a cartridge fuse is capable of clearing a very heavy fault in less than half a cycle; and, therefore, it will operate in these conditions long before a breaker has had time even to start the opening operation.

RELAYS AND PROTECTIVE GEAR

The protection of plant, distribution and transmission lines has reached a high state of perfection and faulty sections can now be automatically isolated before a fault or oveload causes much damage to the section itself or the remainder of the system of which is it a part.

In all the different methods the essential feature is that of isolating the faulty section, and in considering the principles of the various systems of protection it is usually understood that the actual isolation is carried out by circuit-breakers which, in turn, are operated by means of currents due to the action of the protective gear. Usually the required tripping current for the breaker is controlled by relays which are in turn operated by the protective gear.

Similar principles are used for the protection of machines such as alternators, transformers, etc., as for overhead lines and cables. The essential difference is only a matter of adaption, the most important difference being the fact that on overhead lines and feeders certain protective systems may require pilot wires connecting the protective apparatus at each end of the line and these are sometimes undesirable. With machines this point is unimportant as the two sets of protection – one on each side – are not separated by any real distance.

With all systems one or both of the following two undersirable features are guarded against – namely, *overload* and *faulty insulation*. The overload conditions which make it necessary to disconnect the supply may be due to faulty apparatus or to an overload caused by connecting apparatus of too great a capacity for the line or machine. The faulty insulation or *fault* conditions may be either between the conductors or from one or all the conductors to earth.

In connection with all protective gear the following terms are generally used:

Stability ratio. This may be termed the measure of the discriminating power of the system. The stability is referred to as the maximum current which can flow without affecting the proper functioning of the protective gear. Stability ratio can also be defined as the ratio between stability and sensitivity.

Sensitivity is the current (in primary amperes), which will operate the protective gear. In the case of feeders and transmission lines this is a measure of the difference between the current entering the line and the current leaving it.

Overload relays. The simplest form of protection is that where the circuit-breaker is opened as soon as the current exceeds a predetermined value. For low voltage a.c. systems the overload trip may operate the circuit-breaker opening mechanism direct or through a protective relay usually mounted on the switchboard above the circuit breaker. For high-voltage a.c. systems, oil, air, air blast, SF_6 or vacuum circuit-breakers must be used and the trip coils of these breakers are operated by protective relays sometimes mounted on a separate panel.

The time-current characteristics of inverse definite minimum time lag relays (i.d.m.t.l.) will be seen from the curves in Figure 15.6, which show how the time lag between the overload occurring and the gear operating is inversely proportional to the amount of the overload. The setting or time lag generally used to indicate how the relay has been adjusted is that of short-circuit conditions so that the settings of the relay for which the curves are drawn would be as marked.

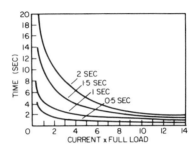

Figure 15.6   Inverse time-lag characteristics of an overcurrent relay

The principle of grading on a system is that of decreasing the setting as we proceed away from the source of supply. The settings of the relays for which the curves have been drawn would be as shown in the diagrams in Figure 15.7 and 15.8. Figure 15.7 represents a distributor fed at one end, whereas Figure 15.8 shows a ring main. A distributor fed at both ends is similar to Figure 15.8.

Electromechanic overcurrent relays are usually of the induction type. The action is similar to that of an induction wattmeter. A metal disc rotates against a spring, the angle of rotation being proportional to the current. As soon as the disc has rotated through a certain angle

Figure 15.7   Grading of overcurrent relays for radial feeder.

the trip circuit is closed. As the speed of rotation is proportional to the load the inverse time element is obtained and the time-setting is adjustable by altering the angle through which the disc has to turn before making the trip circuit.

The latest generation of overcurrent and earth fault relays use microprocessor technology which has the following benefits:

(a) Selection of time/current characteristics built into the relays.
(b) Wider range of settings.
(c) Increased accuracy allowing shorter grading intervals.
(d) In situ relay testing by simply pressing a push button.

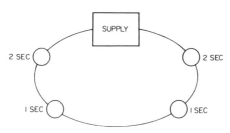

Figure 15.8   Grading for ring main.

Directional protection.   In addition to overload protection it is often desirable for immediate interruption of supply in the case of a reverse current. In this case directional relays are used and these can be combined with overcurrent relays when required as shown in Figure 15.9. The contacts on the directional element are so connected that, under fault conditions and with the current flowing in the right direction, they complete the i.d.m.t.l. element operating coil circuit. The use of directional and non-directional relays is shown in the system reproduced in Figure 15.10.

Pilot wire systems.   Protective systems requiring pilot wires for use on some overhead or feeder cable lines operate on the principle of unequal currents and are referred to as differential systems. The same methods are used for machine and transformer protection, the pilot wires being, of course, much shorter and thus not referred to as pilot wires in this case.

The essential principle is that if the current entering a conduc-

397

tor is the same as that leaving it there is no fault, but as soon as there is a difference more than is considered advisable the supply is disconnected.

Both *balanced voltage* and *balanced current* systems are in use. Current transformers are situated at each end of the conductor and either the voltage or current in the secondaries is balanced

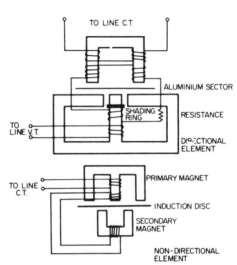

Figure 15.9   Directional overcurrent relay

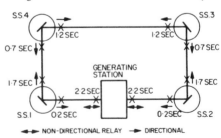

Figure 15.10   Discriminative time protection of ring main

398

one against the other. For balanced voltage (Figure 15.11) these are placed *back-to-back* and no current flows for correct conditions. For balanced current, the secondaries are placed in series and current flows in the pilot wires. In this case fault conditions are determined by relays connected across the pilots as shown in Fault 15.12.

The original balanced protection system was known as the Merz-Price system. There have been several modifications and most manufacturers have their own particular method of applying this principle and improving the actual performance compared with the original simple Merz-Price system.

A typical modification is the Translay S system. Figure 15.13 shows the basic circuit arrangement for Translay S differential feeder protection from GEC Measurements.

A summation current transformer T1 at each line end produces a single-phase current proportional to the summated three-phase currents in the protected line. The neutral section of the summation winding is tapped to provide alternative sensitivities for earth faults.

The secondary winding supplies current to the relay and the pilot circuit in parallel with a non-linear resistor (RVD). The non-linear resistor can be considered non-conducting at load current levels, and under heavy fault conditions conducts an increasing current, thereby limiting the maximum secondary voltage.

At normal current levels the secondary current flows through the operation winding T_o on transformer T2 and then divides into two separate paths, one through resistor R_o and the other through the restraint winding T_r of T2, the pilot circuit and the resistor R_o of the remote relay.

The resultant of the currents flowing in T_r and T_o is delivered by the third winding on T2 to the phase comparator and is compared with the voltage across T_t of the transformer T1. The

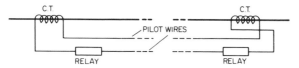

Figure 15.11   Balanced voltage system

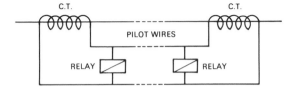

Figure 15.12   Balanced current system

e.m.f. developed across T_t is in phase with that across the
secondary winding T_s, which is in turn substantially the voltage
across R_o.

Taking into account the relative values of the winding ratios
and circuit resistance values, it can be shown that the quantities
delivered for comparison in phase are:

$$(\bar{I}_A + 2\bar{I}_B) \quad \text{and} \quad (2\bar{I}_A + \bar{I}_B)$$

where I_A and I_B are the currents fed into the line at each end. For
through faults $I_A = -I_B$. The expressions are of opposite sign for
values of I_B which are negative relative to I_A and are between $I_A/2$
and $2I_A$ in value. The system is stable with this relative polarity
and operates for all values of I_B outside the above limits.

If the pilot wires are open-circuited, current input will tend to
operate the relay and conversely if they are short-circuited the
relay will be restrained, holding the tripping contacts open.

In order to maintain the bias characteristic at the designed
value it is necessary to pad the pilot loop resistance to 1000 ohms
and a padding resistor P_r is provided in the relay for this purpose.
However, when pilot isolation transformers are used the range of
primary taps enables pilots of loop resistance up to 2500 ohms to
be matched to the relay.

Reyrolle Solkor-R high speed feeder protection.   Solkor-R high
speed feeder protective systems belong to the circulating current
class of differential protection in that the current transformer
secondaries are connected so that a current circulates around the
pilot-loop under external fault conditions. The protective relay
operating coils are connected in shunt with the pilots, across
equipotential points when the current circulates around the pilots.

400

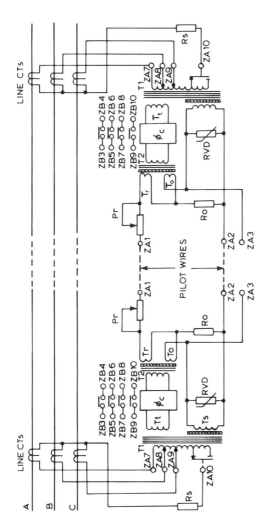

Figure 15.13 Circuit diagram for Translay S differential feeder protection (GEC Measurements)

401

In this particular scheme equipotential relaying points exist at one end during one half cycle of fault current and at the other end during the next half cycle. During the half cycles when the relays at either end are not at the electrical mid-point of the pilots, the voltage appearing across them causes them to act in a restraining mode.

With the current transformers connected as shown in Figure 15.14 a single-phase current is applied to the pilot circuit so that a comparison between currents at each end is effected. The tappings on the summation transformers have been selected to give an optimum balance between the demands of fault setting and stability.

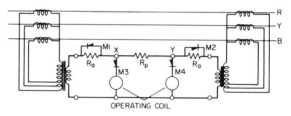

Figure 15.14   Connections of Solkor-R pilot wire feeder protection system

In the diagram the pilot wire resistance is shown as a single resistor R_p. The rest of the pilot loop is made up of two resistors R_a and two rectifiers M1 and M2. The operating coils are made unidirectional by the rectifiers M3 and M4 and are connected in shunt with the pilots at points X and Y.

During external fault conditions an alternating current circulates around the pilot loop. Thus on successive half cycles one or other of the resistors R_a at the end of the pilots is short-circuited by its associated rectifier. The total resistance in the pilot-loop at any instant is therefore substantially constant and equal to $R_a + R_p$. The effective position of R_a alternates between both ends, being dependent upon the direction of the current. This change in the effective position of R_a makes the voltage distribution between the pilot cores different for successive half cycles. The resulting potential gradient between the pilot cores when R_a is equal to R_p is shown in Figures 15.15a and 15.15b.

From these two diagrams it will be seen that the voltage across the relays at points X and Y is either zero or in the reverse direction for conduction of current through the rectifiers M3 or M4. When R_a equals R_p a reverse voltage appears across a relay coil during one half cycle and zero voltage during the next half cycle. The voltage across each relaying point is shown in Figure 15.15c.

In practice, resistors R_a are of greater resistance than R_p, and this causes the zero potential to occur within resistors R_a as shown in Figures 15.16a and 15.16b. The voltage across each relaying point, X and Y throughout is shown in Figure 15.15. From this it will be seen that instead of having zero voltage across each relay operating coil on alternate half cycles, there is throughout each cycle a reverse voltage applied. This voltage must be nullified and a positive voltage applied before relay operation can take place. This increases the stability of the system on through faults.

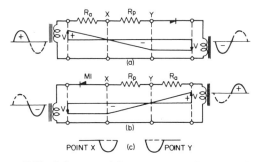

Figure 15.15   Behaviour of basic circuit under external fault conditions when $R_a = R_p$.

Diagrams (a) and (b) show the effective circuit during successive half-cycles: diagram (c) indicates the voltage across relaying points X and Y during one cycle

During internal fault conditions with the fault current fed equally from both ends, the effective circuit conditions during each half cycle are shown in Figure 15.17a and 15.17b. Pulses of operating current pass through each relay on alternate half cycles, and the relay at each end operates.

With a fault fed from one end the relay at the end remote from

403

the feed is energised in parallel with the relay at the near end. The relay at the feeding end operates at the setting current and the relay at the remote end at approximately 2½ times the setting current. Thus the protection operates at both feeder ends with an internal fault fed from one end only provided the fault current is not less than 2½ times the setting current.

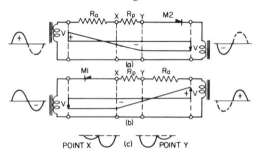

Figure 15.16 Behaviour of basic circuit under external fault conditions when $R_a > R_p$
Diagrams (a) and (b) show the effective circuits during successive half-cycles; diagram (c) indicates the voltages across relaying points X and Y during one cycle

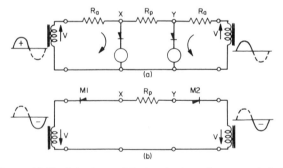

Figure 15.17 Behaviour of basic circuit under internal fault conditions, fault fed from both ends. Diagrams (a) and (b) show the effective circuits during successive half-cycles

In addition to the basic components mentioned above the complete protective system includes at each end two non-linear resistors, a tapped 'padding' resistor and a rectifier. The non-linear resistors are used to limit the voltage appearing across the pilots and the operating elements. The padding resistors bring the total pilot loop resistance up to a standard value of 1000 ohms in all cases. The protection is therefore working under constant conditions and its performance is largely independent of the resistance of the pilot cable.

The rectifier connected across the operating coil is to smooth the current passing through the coil on internal faults, thereby providing maximum power for relay operation.

The operating element is of the attracted armature type with three pairs of contacts, each pair being brought out to separate terminals. They are suitable for direct connection to a circuit-breaker trip coil, thus no repeat relay is necessary.

The above description relates to the original design of Solkor R protection based on a half-wave rectification principle shown in Figure 15.14. The latest development is Solkor Rf which is based on a full wave principle. This has the advantage of faster speed of operation, and the facility of including isolating transformers in the pilots to allow pilot voltages up to 15 kV. Without pilot isolating transformers Solkor Rf is suitable for use with 5 kV pilots as was the original Solkor R.

Current balance protection schemes can equally be applied to rotating machines and transformers. Figure 15.18 shows a typical scheme for generators where an instantaneous unbiassed relay can be used to give adequate protection. The characteristics of transformers makes them unsuitable for using in unbiassed relay, because they cannot take into account tap-changing, magnetising inrush currents and phase connection of the transformer. For these reasons the CTs are connected in phase opposition to the transformer connections and a biassed relay is used as shown in Figure 15.19. In addition a harmonic bias feature is added to the relay to prevent operation under magnetising inrush currents.

Distance protection. Pilot wire protective schemes are uneconomical for lines longer than 24 km to 32 km. For such lines it is necessary to use either distance protection or carrier current protection or a combination of both. These do not require pilot wires.

The principal feature of a distance scheme is the use of a relay

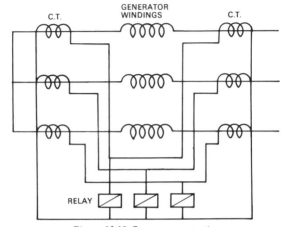

Figure 15.18 Generator protection

that measures the fault impedance, which, if it is below a certain value means that the fault is within the protected zone and the relay operates to trip out the circuit-breaker. The relay is energised by current and voltage and operates when the ratio of E/I gives an impedance below the relay setting.

Standard distance protection schemes protect the overhead line in three stages. The first stage has a zone of operation extending from the relaying point at the first substation to a point 80% of the way to the next substation. A fault occurring on this part of the line causes the protection at the first sub-station to trip immediately.

The second stage has a zone of operation extending from this 80% point past the second substation along the next feeder to the third substation. A fault in this second zone causes the circuit-breaker at the first substation to trip after Zone 2 time limit provided the protection on the second feeder has not operated.

The third stage has an operating zone extending a little way past the third substation and provides back-up protection with a longer delay.

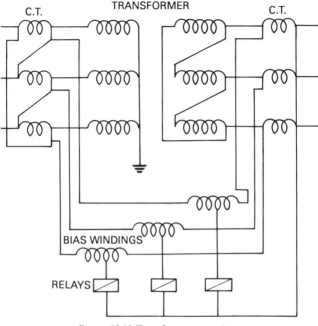

Figure 15.19 Transformer protection

The means by which these operating zones and times are obtained in the Reyolle type THR range of distance protection schemes is shown typically on a single-phase basis in Figure 15.20. In this particular arrangement each fault type and each zone has an individual measuring element so that 18 elements are required to make up the scheme covering phase and earth faults in three zones. The first two zones are directional and are usually cross-polarised mho relays, but alternative characteristics for special requirements are obtained by directly replaceable modules. The third zone has an off-set characteristic shaped against load ingression, or is plain off-set mho and where switched schemes are employed acts as the starting relay. In a full scheme the Zone 3

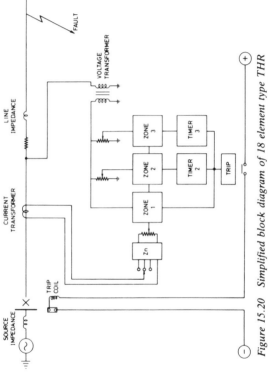

Figure 15.20 Simplified block diagram of 18 element type THR three-stage distance protection from NEI Reyrolle Protection

relay is not required for starting but is included for power swing blocking and line check facilities.

The measuring elements of the three zones have definite settings relative to the line impedance which are referred to the primary side of the voltage and current transformers. Zone 1 is set to 80% of the line impedance by coarse and fine settings on the replica impedance ZN. Voltage dividers are used to set the Zone 2 to typically 120% and Zone 3 to between 200 and 300% of the line impedance.

If the line fault occurs within the first zone the polarised mho relay operates to trip the circuit-breaker at the first substation instantaneously. If its distance is in the zone of the second stage the Zone 2 times out, and the polarised relay trips the circuit-breaker. If the fault lies within the third zone of protection the circuit-breaker at the first substation is tripped when the stage 3 timer has completed its operation.

16

Heating and refrigeration

WATER HEATING

Hot water systems for domestic, commercial and industrial premises are available in a variety of forms, the choice depending on
the number and distribution of draw-off points, whether it is to be
combined with space heating, and in some cases the nature of the
water supply. Electricity is particularly suitable as a sole energy
source for water heating, specially so when advantage can be
taken of off-peak tariffs. It is cheap and convenient, requires no
fuel storage or flues, involves small capital cost, and operates at a
high level of efficiency. When tanks or storage vessels are
properly lagged electric water heating is competitive with other
forms of fuel, especially on off-peak tariffs.

Various types of electric water heating appliances are available
and the most suitable type for a given set of circumstances is best
decided in consultation with the local electricity board. The
heating element is always completely immersed in the water to be
heated so that heat transference is always 100% even if the
element is encrusted with scale. For a given consumption of hot
water the efficiency of a system depends only on the thermal
losses from the storage vessel and the associated pipework.

Tariffs. In many cases it is possible to arrange for the
electricity to be provided during the cheap off-peak period,
usually during the night. This is best suited to those systems where
water is stored in large quantities in suitably lagged vessels.

A popular tariff known as Economy 7 is one whereby the user
has electricity at two rates, a day rate and a much cheaper night
rate. Electricity used for 8 hours between midnight and 8 a.m. is
charged at the low rate. The time for the cheap night rate varies
from area board to area board. A special Economy 7 boiler has
been designed for this tariff and is described on page 416.

Types of water heaters. Basically there are two types of water heaters, non-pressure and pressure. The non-pressure water heater is one in which the flow of water is controlled by a valve or tap in the inlet pipe, so arranged that the expanded water can overflow through the outlet pipe when the valve or tap is opened. This type of water heater can be connected direct to the main water supply.

The pressure type of water heater is supplied from a cistern in the normal way with provision for the expanded water to return to the feed cistern. A heater of this type must not be connected direct to a water main. Hot water tap connectors are made in the normal way to draw-off piping.

Non-pressure water heaters. Examples of non-pressure heaters are those that are installed over a sink and fitted with swivel outlets, Figure 16.1. Capacities range from 7–14 litres. They are available in cyclindrical or rectangular shape, in various colours and in some cases are fitted with a mirror.

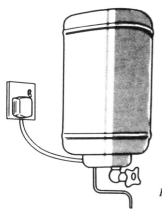

Figure 16.1   Oversink storage water heater

A small under-the-sink storage water heater can be of the non-pressure type. It is plumbed with a choice of special taps to the wash basin or sink which have to be designed for use with it. Normal capacity is between 7–14 litres. This type is also available in either rectangular or cylindrical shape. Undersink or basin types are produced, Figure 16.2.

An instantaneous water heater of the flow type, fitted with a swivel outlet is a further unit falling within the class of non-pressure. Typically a unit can deliver approximately up to 0.23 litres per minute at a hand wash temperature of about 43°C, see Figure 16.3.

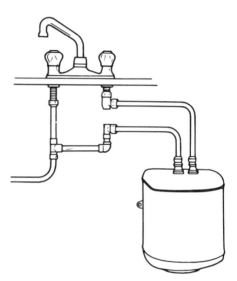

Figure 16.2 Undersink storage water heater with choice of special taps

If a larger vessel with higher rated heating elements is employed than for the small instantaneous water heater it is possible to provide a self-contained instantaneous shower unit. It is fitted with an adjustable shower rose, water flow restrictor, pressure relief device and a thermal cut-out. Figure 16.4 shows typical units which will deliver water at a temperature about 43°C at a rate of 0.47–0.57 litres per minute. Loading for a shower unit at 240V would be about 6kW and the weight of a unit about 3kg.

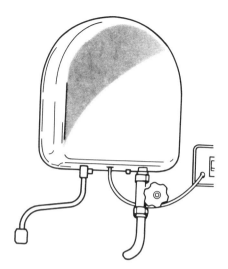

Figure 16.3
Instantaneous 'flow' type water heater

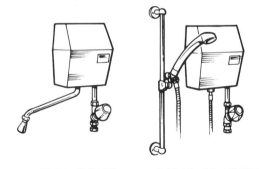

FIXED SHOWER
SPOUT

PILLAR ATTACHMENT

Figure 16.4   Instantaneous shower units: (left) fixed shower spout
and (right) pillar attachment (Heatrae – Sadia Heating Ltd)

413

Pressure cistern types. Cistern type water heaters must be mounted above the highest tap otherwise the hot water will not flow. There must also be 200–300 mm clearance above the cistern to allow access to the ball valve for servicing. Because the hot water storage is of the pressure type, taps or valves can be fitted in an outlet pipe draw-off system. The ball valved cistern is connected to the main water supply.

One of the most popular of the pressure types for use in homes is shown in Figure 16.5. The main characteristic of this unit is that it is fitted with two heating elements, one mounted horizontally near the base and the other horizontally but near the top of the vessel. The top heater keeps about 37 litres of water at the top of the tank hot for general daily use while the other is switched on when larger quantities of water are required. Total storage

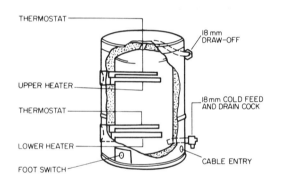

OUTER CASING 825 mm HIGH 508 mm DIAMETER

Figure 16.5   'Two-in-One' 180 litre heater

capacity is typically 108 litres. Both elements are controlled thermostatically. The heater can be stored under a draining board, alongside a working surface or in a corner. The associated plumbing must be vented back to a high level cistern, and a number of hot water taps can be served.

Larger capacity storage water heaters are available ranging from 182 litres to 455 litres as standard equipments, and even larger ones can be made to order fitted with two immersion heaters as in Figure 16.5; these are connected to off-peak supplies. Economy 7 for example is described on page 416. A large water heater can be fitted with a number of elements, the associated thermostats being set if necessary, to operate in a pre-selected manner to provide an ordered sequence of heating. When very large quantities of stored water are required it is preferable to install a number of smaller heaters than a few large ones. This permits the adoption of smaller pipework and component parts and also allows a heater to be taken out for servicing without affecting the overall performance too much.

Advantage can be taken of the benefits of the Economy 7 tariff (see page 410) by modifying an existing installation or by installing a new high performance Economy 7 cylinder. The three ways are shown diagrammatically in Figure 16.6. The three main ways of converting to Economy 7 showing progressively improving running costs are as follows:

(*i*) Installing an Economy 7 water heating controller or Economy 7 boost controller to control the existing 685 mm immersion heater and adding an 80 mm insulating jacket to the existing storage vessel (Figure 16.6a).

(*ii*) Installing an Economy 7 controller and a dual immersion heater and adding an 80 mm insulating jacket to the existing storage vessel (Figure 16.6b).

(*iii*) Installing a new high performance Economy 7 cylinder (Figure 16.6c).

Economy 7 water heating controllers are produced by three manufacturers but all models operate in much the same way. The controller Figure 16.7 incorporates a fixed cam timing arrangement and automatic changeover switch. In this way the controller automatically switches on the immersion heater during the night period while the customer can operate the immersion heater during the day period by means of the run back or electronic boost timer, should the hot water become depleted.

With the dual immersion heater, the Economy 7 controller operates the long element overnight and the short element during the day using the boost timer. This allows a smaller quantity of water to be heated at the more expensive day rate than with the existing single immersion heater.

Two of the models of Economy 7 controller manufactured have a quartz timer with a rechargeable battery reserve of at least 100 hours. These controllers can be expected to keep reliable time without frequent inspection.

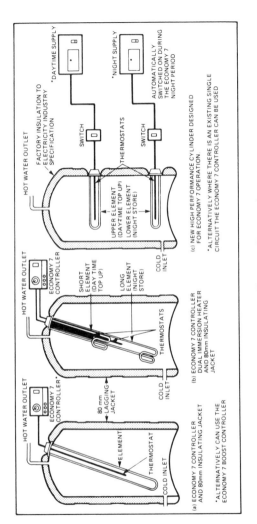

Figure 16.6 Economy 7 conversions

Where a specifier wants to ensure that the night-time electricity supply to the immersion heater is timed by the supply authority's timeswitch, the Economy 7 boost control is an alternative. In this case the controller is normally sited at or near the electrical distribution board, e.g. in the kitchen or hall.

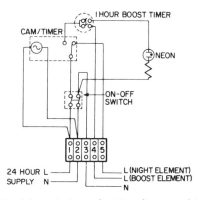

Figure 16.7   Schematic internal wiring diagram of Economy 7 water heating controller

Although this control is wired with both an off-peak circuit (via the REC's timeswitch) and a 24 hour supply, a single immersion heater is controlled, normally fed from the former. If extra water is required during the day the customer turns a run back timer and for the period required the immersion heater is boosted at day rates.

Where the hot water storage vessel is being replaced or a system is being installed in a new home a high performance hot water storage unit can be utilised as shown in Figure 16.6c. It may take the form of a cylindrical or rectangular combination unit or most commonly a replacement or new cylinder that will be plumbed into an existing cistern. Three sizes are available that will meet most needs the BS type reference 7 (120 litre), 8 (144 litre) and 9E (210 litre), and for convenience they are called small, medium and standard. Type 9E is the preferred size for off-peak operation.

Typical arrangements of the pipework, cistern, heater and outlets for houses and flats are shown in Figures 16.8, 16.9 and 16.10. As can be seen the hot-water draw-off pipe is continued upwards to turn over the side of the ball-valve cistern. The arrangement of pipework is very important if single-pipe circulation is to be avoided. Single pipe circulation takes place in any

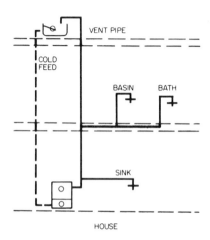

Figure 16.8   The use of dual element pressure-type electric water heaters (house)

pipe rising vertically from a hot water storage vessel due to the water in contact with the slightly cooler wall of the pipe descending being replaced by an internal column of hot water rising from the storage vessel. It is easily prevented by running the draw-off pipe horizontally immediately it leaves the top of the storage vessel for about 450 mm before any pipe, such as the vent pipe, rises vertically from it. Figure 16.11 shows how this can be done.

Figure 16.10 is a self-contained cistern type electric water heater which is made with an integral cold feed cistern controlled by a ball valve above the hot water tank. It is thus a complete water heating system in itself, and only needs connection to the cold feed, hot draw-off pipe, overflow and electrical supply.

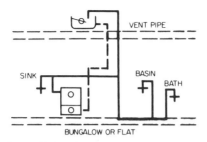

*Figure 16.9   The use of dual element press-
ure-type electric water heaters (bungalow or
flat)*

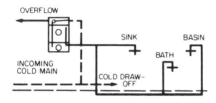

*Figure 16.10   The use of self-contained cistern
type electric water heaters*

A packaged plumbing unit is made suitable for providing both
hot and cold water for bath, basin, kitchen sink and w.c. It
comprises a cold water cistern, hot water container and pipework
built into and within a rigid framework, Figure 16.12. The method
of installation when serving two floors is shown in Figure 16.13.
This also shows how it is used in conjunction with a back boiler.
Many of the cistern type systems discussed can be used in
conjunction with fossil fired boilers.

Immersion heater.   An immersion heater cannot heat water
below itself. Water is heated by convection and not conduction
unless there is fortuitous mixing. The installation of an immersion
heater into an existing hot water tank is the cheapest method of

419

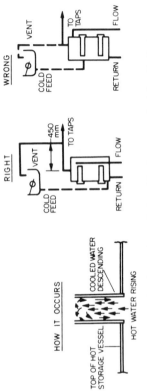

Figure 16.11 'Single pipe' circulation. The layout on the left will eliminate 'single pipe' circulation, whilst this type of circulation will result from the layout on the right

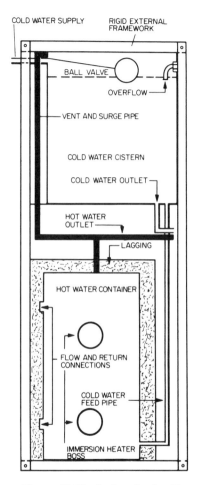

COLD WATER SUPPLY

RIGID EXTERNAL FRAMEWORK

BALL VALVE

OVERFLOW

VENT AND SURGE PIPE

COLD WATER CISTERN

COLD WATER OUTLET

HOT WATER OUTLET

LAGGING

HOT WATER CONTAINER

FLOW AND RETURN CONNECTIONS

COLD WATER FEED PIPE

IMMERSION HEATER BOSS

Figure 16.12 Packaged plumbing unit. Approx. dimensions: 1.98 m high × 0.7 m wide × 0.61 m front to back

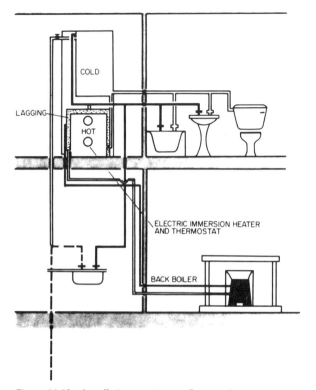

Figure 16.13   Installation serving two floors and showing alternative connection with back boiler

converting a non-electric water heating system to an electric method. It thus can augment an existing solid fuel system so that the latter can be shut down during the summer months, allowing the electrical element to provide the hot water required.

An immersion heater mounted horizontally at low level heats all the water above it, to a uniform temperature. A small amount of water cannot be produced quickly by this method, so when this is required a second heater is positioned near the top of the storage vessel.

An immersion heater mounted vertically cannot produce a tank full of hot water at uniform temperature. A temperature gradient is produced ranging from hot at the top to cold at the bottom. When a vertically-mounted immersion heater is used it should be long enough to reach within 50–75 mm of the cold water inlet.

A compromise is the 'dual' or twin heat arrangement comprising a short and a full length element on one boss each with its own thermostat and interlocked to provide an alternative choice.

There are basically two types of immersion heater, a withdrawable and a non-withdrawable type. The withdrawable type is constructed so that the heating element can be withdrawn from the enclosing sheath without disturbing or breaking a water joint. This type of element is used mostly in industry. For the non-withdrawable type the heating element cannot be withdrawn without breaking the water joint.

Calculating energy required for water heating. It is a comparatively simple matter to calculate the amount of energy required to raise the temperature of water through a given number of degrees Centigrade.

$$\text{Energy (kWh)} = \frac{k \times \text{kg} \times \text{temperature rise in } °C}{\text{efficiency}}$$

It requires 4180 J to raise the temperature of 1 kg of water through 1°C and 4180 J equals 1 kcal. 1 kWh = 3.6×10^6 J.

An example will illustrate how the energy is calculated:

An electric heater raises the temperature of 10 kg of water from 15°C to boiling point. Only 75% of the energy intake is used to heat the water, the rest is lost by radiation. Calculate the energy intake of the heater.

Heat energy of water content = $10 \times (100 - 15)$ = 850 kcal

Energy intake = 850/0.75 = 1133 kcal

Energy in joules = 1133×4180 = 4.75×10^6 J

Energy in kWh = $\dfrac{4.75 \times 10^6}{3.6 \times 10^6}$ = 1.32 kWh

From the above formula it is easy to derive all other related quantities: it should be noted that k in the above formula is $4180/3.6 \times 10^6$.

$$\text{Efficiency} = \frac{4180 \times \text{kg} \times \text{temperature rise in }°C}{\text{kWh} \times 3.6 \times 10^6}$$

If it is required to know the time in hours it would take to raise a given quantity of water through a given temperature the formula would be:

$$\text{Time in hours} = \frac{4180 \times \text{kg} \times \text{temperature rise in }°C}{\text{loading in kW} \times 3.6 \times 10^6 \times \text{efficiency}}$$

SPACE HEATING

Space heating can be classified into direct acting or storage systems. Storage systems are operated on lower cost off-peak supplies and can be designed to take full advantage of the thermal storage properties of the building and its contents. Direct acting appliances, which use electricity at the standard rate, have the advantage of immediate response to temperature changes and intermittent programming. The two systems can be used independently or may be complementary to each other. Storage heating would account for about 90% of the electricity consumed over the winter system in a combined installation.

Storage heater. Storage heaters operate on the principle of retaining the heat that is taken in during the off-peak period (usually overnight), and releasing it at a later time, i.e. during the day, when the supply is disconnected from them. Storage heaters are filled with a material that will absorb heat, and are so constructed that emission of heat is both gradual and continuous throughout the 24 hours. The basic types of off-peak systems are: storage fan heaters; Electricaire; Economy 7 boiler; hot-water thermal storage vessel with radiators; and floor warming.

Storage radiators. The electric storage radiator consists of heating elements embedded in a storage core, surrounded by a layer of insulating material, the whole being in a metal casing. The heating element is automatically switched on during off-peak hours, and conserves its heat in the surrounding storage material and discharges it through the insulating layer for the full twenty-four hours. Control may be effected by a combination of charge controllers, external controllers and programme timers. Under ideal conditions the outer casing remains warm throughout the charging and discharging periods.

The rating of a storage radiator is the input in kilowatts taken by the heating element. It is of the order of 2–2.625 or 3.375 kW. The charge acceptance is the quantity of heat which the radiator can absorb and emit and is measured in kWh. Acceptance ratio is the ratio of actual to maximum charge acceptance. Most radiators on the market have an acceptance ratio of about 83%. The nearer this figure is to unity the more economical is the radiator to use.

Net storage capacity is the quantity of heat retained within the storage radiator when fully charged over an eight-hour period. Retention factor is the ratio of the net storage capacity to the eight-hour charge acceptance. For the average storage radiator this figure is about 66%.

There are two ways of obtaining a boosted output from a storage radiator. The first is by incorporating a damper controlled by a bi-metal strip which allows heat to be extracted direct from the radiator core by convection during the late afternoon. Radiators incorporating this are known as 'damper' models. The second way is to employ a fan to operate a closed air circuit between the core and the inside of the front panel and a separate electricity supply is required for this.

Storage fan-heaters. Storage fan-heaters are basically similar to storage radiators except that they have more thermal insulation and incorporate a fan to discharge warm air into the room as required. A small amount of heat is given off from the heater all the time and this is sufficient to take the chill off a room. These heaters lend themselves to time and temperature control so that rooms can be quickly brought up to the desired heat and constant temperatures maintained.

Electricaire. Basically an Electricaire system consists of a central heater unit which supplies air to the rooms through a ducting installation. These systems are completely automatic, the electricity board's time switch determining the period during which energy is supplied to the core and a core thermostat controlling the quantity. Air is discharged from the unit to the insulated ducts by a fan which gives the warm air a velocity of up to 7 m³/min. The fan is fed from an unrestricted supply source and is controlled by a wall-mounted thermostat. The location of this thermostat is important for successful operation of the system.

Storage capacities are available up to about 3.7×10^7 J per day and most models have a boost device to permit higher output over short periods of 20 minutes of so.

The core of an Electricaire unit is formed from a number of spirally wound elements of about 1 kW rating fitted to refractory formers. These are used to heat the storage blocks. These blocks are made of cast iron or refractory material and insulated to prevent loss of heat through the cabinet. A high temperature thermostat with manual adjustment controls the charge to the core. The outlet temperature of the unit is maintained automatically by a regulator which mixes the right amount of cool air with heated air from the store from which the distribution ducts originate. Fully automatic models are now available requiring no manual attention from the user.

There are three basic layouts for the ductwork of an Electricaire system, stub-duct, radial-duct and extended plenum. With a stub-duct the Electricaire unit is usually installed in a central position and heat is conveyed to the rooms through short ducts. In some cases these ducts can be extended to upper rooms but quite often this is not necessary due to the natural convection of the air.

In a radial duct system warm air from the plenum chamber feeds through a number of radial ducts to the outlet registers. A riser for this type of system comes direct from the plenum chamber which allows air to be taken into a number of rooms in a larger or more complex building layout. A riser may be taken from any point on the plenum to serve upper floors.

Input loadings of Electricaire units range from 6–15 kW. The heater is considered to be exhausted when the temperature of the air leaving the heater falls to about 43°C. At this point there is still a useful quantity of heat left in the core and by arranging for a direct-acting element to supplement the output of the depleted core the active life is extended. The direct acting element is positioned across the air flow before it enters the plenum and can be switched on automatically by monitoring the air outlet temperature or the core temperature.

Economy 7 boiler. Hot water radiators are a familiar and long-established means of heating homes and an economic and competitive way to raise the temperature of the water is by an electric boiler specially designed for use with the Economy 7 cheap night rate tariff.

Most of the hot water needed to warm the radiators can be heated at the low cost night rate and then stored in the well-insulated electric boiler (see page 416 Figure 16.6 for the design). When the hot water is needed, a room thermostat controls the central heating pump which drives it around the radiators. Heaters in the top of the boiler can give an extra boost during the day if required or during spells of extreme or prolonged cold weather.

The boiler can be sited almost anywhere provided it is protected from the weather. A garage, outhouse, conservatory or a kitchen are all suitable. The boiler does not need a flue and requires little annual maintenance.

Size of boiler depends on the requirements of the home and its insulation level. A standard 680 litre boiler measuring $710 \times 813 \times 2134$ mm high is suitable for a $90 \, m^2$ well-insulated detached house. A much larger house could use two standard units, or a special tailor-made unit to fit an available space. Trials conducted by the Yorkshire Electricity Board proved that an electric boiler central heating system is more economic than both oil and solid fuel.

Floor warming.   Floor warming systems use elements directly embedded in the floor concrete. Most installations are of the embedded type. Capital outlay of a floor warming system is low compared with other forms of central heating and it can be operated on the Economy 7 tariff. In principle the floor becomes a low-temperature radiating device. Concrete floors provide the heat storage medium and they may be of the following types of construction; hollow pot, hollow concrete beam, prestressed plank and in-situ reinforced concrete. Insulation of the edge and underface of the concrete floor is essential.

Floor warming elements have conductors made of chromium, iron, aluminium, copper, silicon or manganese alloys. Insulation can be asbestos, mineral, PVC butyl, silicon rubber and nylon. Installations in the UK are designed for an average floor surface temperature of 24°C.

When installing, the elements should be arranged as far as possible to cover the whole of the floor area and in general the loading falls between $100 \, W/m^2$ and $150 \, W/m^2$. Elements should be chosen so that the spacing between adjacent elements does not exceed 100 mm or less than screed depth, so as to avoid temperature variations on the floor surface. The 'density' of the heating elements should vary according to location, with more heating elements being used near external walls, Alternatively the floor area may be zoned, each zone being thermostatically controlled.

Some specialised outdoor areas may also be provided with this form of heating. They include car park ramps, airport runways and pedestrian ramps.

Direct acting systems.   A wide range of direct-acting heaters is available such as radiant fires, convectors, radiant/convectors, panel heaters, domestic fan heaters, industrial unit heaters and combined light and heat units. In addition ceiling heating, an extended panel system deserves to be mentioned separately. All these systems are described in the following paragraphs.

Ceiling heating.   Ceiling heating may be provided by prefabricated panels or installed as a complete entity. In the latter case the installation of the elements is carried out in a similar way to that of floor heating except that the element wiring is fixed in position on the underside of the ceiling before the ceiling board is erected.

The low thermal capacity of a ceiling heating installation enables it to respond more quickly to thermostatic and time controls than a thermal storage system. Where each room installation is individually controlled response to changes in internal temperature, i.e. solar gains or heat gains from lighting or occupants, is rapid. Because of the economy possible in this way the system can be competitive though it uses on-peak electricity.

Ceiling surface temperatures are between 18°C and 21°C and heat emission approximately 3.6–4.3 W/m^2 deg C.

Radiant heaters.   There is a wide variety of domestic radiant heaters ranging from 740 W bowl fires up to 3 kW with multiple heating bars which are switched separately. Often these fires incorporate a convector facility and are decoratively designed to provide an attractive fireplace (see page 429).

They are portable and instantaneous in response.

Infra-red heaters.   The infra-red heater consists of an iconel-sheathed element or a nickel-chrome spiral element in a silica glass tube mounted in front of a polished reflector. Loadings vary from 1 kW to 4 kW and comfort can be achieved with a mounting height of between 6 m and 7½ m. There are intended for local heating in large areas.

A smaller rated version, 500 W or 750 W is made suitable for bathrooms; such a unit may be combined with a lamp to form a combined heating and lighting unit.

Oil-filled radiators. Oil-filled radiators are very similar in appearance to conventional pressed-steel hot-water radiators. They are equipped with heating element to give loadings between 500 W and 2 kW. They can be free-standing or wall mounted and are generally fitted with thermostatic control.

These radiators heat up more slowly than radiant or convector panels and retain their heat for longer periods after switch-off. Being on thermostatic control they avoid rapid temperature swings. Surface temperature is around 70°C, comparable to a hot-water radiator.

Tubular heaters. These are comparatively low temperature heaters used to supplement the main heating in a building.

A heater is made from mild steel or aluminium solid drawn tubing with an outside diameter of around 50 mm and varying in length from 305 mm to 4575 mm. The tube is closed at one end and has a terminal chamber at the other. Elements are rated at 200 W to 260 W per metre run and the tube surface temperature is of the order of 88°C, slightly higher than the conventional low-pressure hot-water radiator.

Skirting heaters. The skirting heater is a refinement of the tubular heater but shaped like conventional skirting. Loadings vary widely from 80 W to 500 W per metre run.

Convectors and convector/radiant heaters. Convectors can be free-standing, wall-mounted or built in. They consist basically of a low-temperature non-luminous wire heating element inside a metal cabinet. The element is insulated thermally and electrically from the case so that all the radiant heat is converted to warm air. Cool air enters at the bottom of the unit and warm air is expelled at the top at a temperature of between 82°C and 93°C and at a velocity of about 55 m/min. Integral thermostatic control is usually provided.

Combined convector and radiant heaters incorporate in addition to the convecting element a radiant element usually of the silica tube type. Some models have three 1 kW silica tube elements together with a range of heats from the convector. Control allows a combination of convected and radiant heat to be obtained to suit a wide variety of requirements. Integral thermostatic control is also usually provided for these units.

For economical operation, particularly where convector or radiant convector panels comprise a control heating scheme, thermistor integral thermostats or wall-mounted remote air thermostats must be used to control them.

Fan heaters. Fan heaters operate on the same principle as convectors but employ a fan to circulate the warm air. Domestic types are available in loadings up to 3 kW. The fan is often mounted direct on the shaft of the small motor which may have two speeds. Fan blades are tangential in order to keep noise to an acceptable level. For domestic use these heaters are floor standing models and can be as small 380 mm × 150 mm × 100 mm.

There is a design of fan heater which has both water/air and direct electric heating elements. They are suitable for use with all heating systems using water as the heating medium. The direct acting element can be used to provide supplementary heating to the central source or even independently if so desired.

THERMOSTATIC TEMPERATURE CONTROL

Thermostatic switches are widely employed for automatic temperature control of water and space heating appliances. They generally operate on the bi-metal principle in which the unequal expansion between two metals causes distortion which actuates a switch.

Room thermostat. The TLM room thermostat from Satchwell Sunvic is designed for heavy current circuits suitable for the direct switching of heating loads up to 20 A. They are therefore suitable for the control of space temperature. Heaters, valves, drive motors and other heating and cooling equipment can be controlled with equal simplicity. The thermostat is an on-off control unit, the circuit closing with a fall in temperature.

Other models in the range can be fitted with changeover contacts for controlling motorised valves or for reverse acting applications. By installing the thermostat in a weatherproof case it is suitable for outside use.

The switch mechanism of the TLM thermostat ensures a high contact pressure up to the instant of opening so that even with loads up to 20 A any self-heat is reduced to a minimum. An accelerator heater is incorporated in the design so that rise in temperature is anticipated by the instrument when 'calling' for heat. Noticeable overshoot is thus eliminated ensuring precise maintenance of room temperature.

Immersion thermostat. The Satchwell Sunvic W immersion thermostat is a stem type model, Figure 16.14. It is designed for

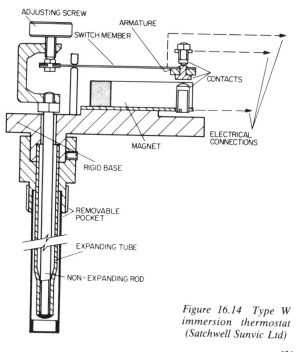

Figure 16.14 Type W immersion thermostat (Satchwell Sunvic Ltd)

the control of the temperature of hot water installations, industrial processes and oil heaters. The temperature sensitive stem is made from aluminium brass and nickel-iron, and the resulting differential expansion is used to operate a micro-gap switch. Different stem lengths are provided according to the temperature range and sensitivity required. Large switch contacts are fitted to provide a long trouble-free life and the switch mechanism is enclosed in a dust-excluding moulded phenolic cover. Contacts 1 and 3 (Figure 16.15) break circuit with rise of temperature, contacts 2 and 3 make with rise of temperature.

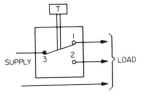

Figure 16.15   Connection diagram for thermostat shown in Figure 16.14

Various modifications are available to the standard design to give a variety of operating techniques. A hand-resetting device can be fitted which ensures that once the control circuit is broken by the thermostat (contacts 1 and 3), it cannot be remade until the thermostat is manually reset by means of a pushbutton adjacent to the control knob. This facility allows the thermostat to be used as a high limit control device with complete safety. Contacts 2 and 3 can be used to operate an alarm warning device if required.

Installation of thermostats.   Room thermostats should be located in a position representative of the space to be controlled where it will readily be affected by changes in temperature. When used to control input to storage heating systems the accelerator heater should be left disconnected.

Knockouts are provided in the base of the TLM room thermostat for surface wiring allowing either top or bottom entry.

When ordering thermostats give the manufacturer as much information as possible about the installation in which it is to be used, particularly the control temperature and the differential required, i.e. the difference between the cut-in and cut-out points.

ELECTRIC COOKERS

Modern electric cookers are usually fitted with an oven, a grill and, usually four boiling rings. Apart from the conventional design, there are double-width and split-level cookers. Split-level cookers are those that have the grill mounted at eye-level. The rings are generally of the radiant pattern and consist of a flat spiral tube with the heating element embedded in it. Dual rings are available in which the centre portion of the ring can be separately controlled from the outer section; alternatively the whole of the ring can be switched as one entity. A disc ring type of plate is available consisting of a flat circular metal plate, usually with a slightly recessed centre, enclosing the heating element. The whole unit is then sealed into the cooker hob.

A development of the split-level design is where the oven and grill units are separate from the hob and are built into the wall or into housing kitchen furniture at any position convenient to the user. Another development is the ceramic hob with four integral cooking areas.

Boiling rings are normally 180 mm in diameter but they are also available in 150 mm, 200 mm and 215 mm diameters. Electrical loadings from 1.2 kW to 2.5 kW are common. Rings are controllable from zero to full heat by means of rheostats operated from control knobs on the escutcheon plate of the cooker. Some models have thermostatic control of one or more of the rings, to keep the pan temperature at the desired setting.

Grills can either be fitted at waist-level below the hob, at eye-level position above the rear panel or separate from the hob as mentioned above. Some eye-level grills are fitted with a motor-driven spit. Occasionally a second, separately controlled element is fitted below the spit for use when cooking casseroles and similar dishes. Heating elements for grills range from 2 to 3.5 kW. Control is in the same way as for boiling rings, i.e. with energy regulators. Some models have facilities for switching only one half of the grill on.

The latest designs in ovens include either stay-clean facilities or are of the self-clean type. In many cases the oven is arranged for complete removal to ease cleaning. Ovens are almost totally enclosed with only a small outlet for letting steam escape. The heating elements are usually fitted behind or on the side walls and sometimes beneath the bottom surface. Oven loadings are of the order of 2.5 kW. Oven controls allow temperature to be preset between 90°C and 260°C and maintained by thermostatic switching which is described later. An indicator light is often fitted to show when the oven is switched on. Modern cookers are generally fitted with a clock that can be used to automatically switch the cooker on and then off to provide delayed control.

Microwave cookers are now on the market. The electromagnetic radiation enables rapid cooking or reheating of food to be carried out. The waves, which range from 897 to 2450 MHz, penetrate the food to a depth of about 65 mm all round and cause the molecules to move around at high speed. The heat for cooking is produced by molecular friction inside the food.

Installation. It is important to install an electric cooker with due regard to electrical safety. Usual practice is to have a completely separate circuit for the cooker from all other circuits in the house. It is fused at either 45 A and even 60 A although some older installations may be fitted with 30 A fuses. The supply from the consumer unit is terminated in a cooker control unit mounted just above the cooker itself. A cable is generally internally

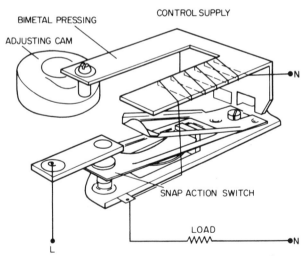

Figure 16.16 The Simmerstat (Satchwell Sunvic Ltd)

connected to the cooker control unit and brought direct into the cooker itself with no intervening break. It is best to let a qualified electrician or the local electricity board wire up a cooker.

Cooker control units are generally fitted with a neon indicator to show whether the main switch on the control unit is on, but if this neon is glowing it does not necessarily mean that the cooker is switched on. Many cooker control units incorporate a 13 A socket outlet although latest practice is to dispense with the socket-outlet and just have a main control switch. The view is held that the presence of a socket outlet can encourage carelessness with trailing leads across the cooker hob with possible dangerous consequences. Never install a cooker near to a sink or tap; a distance of at least 1.8 m should be aimed at.

Earthing. It is important that both the cooker and main switch are efficiently earthed, either through the conduit and flexible metallic tubing where this is used or through separate earth leads where lead-covered cable or plastics conduit is used. The use of the lead sheath as an earth connection is not to be recommended except as a last resource.

Cooker control. The most common method of heat control is the 'Simmerstat' controller.

The 'Simmerstat' controller is shown in Figure 16.16 and the associated wiring diagram in Figure 16.17. Here the heat control is obtained (given a certain amount of heat storage as in a boiling plate) by periodically interrupting the circuit. If a boiling plate is switched on for, say, ten seconds and then switched off for ten

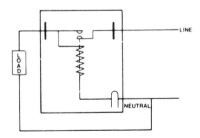

Figure 16.17   Wiring diagram of the Simmerstat

seconds repeatedly, over a period the amount of energy consumed is, of course, half that consumed if switched on full all the time giving half or 'medium' heat. If the ratio of on-and-off periods is altered to give five seconds on and fifteen seconds off the result is 'low' or quarter heat. Two seconds on and eighteen seconds off would be one-tenth heat, and so on.

By turning the control knob up or down, any desired variation in energy fed to the boiling plate is achieved. The periodical operation of the switch contacts is effected by a bi-metallic strip which is made to move by a minute amount of heat applied to it by a very small heating element. When the strip has reached the end of its travel it opens a pair of contacts, breaking the circuit to the heating elements. The strip cools and resumes its former position when the contacts close again, and the cycle of movement is repeated indefinitely.

If a boiling plate is connected in parallel with the controlling heating element it, too, will be subject to the periodic application of energy and by varying the distance of travel of the bi-metallic strip, by turning the control knob, any desired ratio of on-and-off periods can be obtained. The control knob is numbered from 0 to full with an 'off' position below the 0. Full heat is obtained when the control is turned to full on position, the switch contacts remaining together continuously. As soon as the control is turned to about 4 the contacts begin to interrupt the circuit periodically, and heat is reduced to about one-half as variation is not required above this amount. An even and smooth reduction of heat is obtained by further turning the control towards zero, at which point the heat input is reduced to approximately 8%. This is equivalent to about 150 W, with an 1800 W boiling plate, reducing heat input to a saucepan well below simmering point.

HIGH-FREQUENCY HEATING

Induction heating is used extensively today by industry in four main areas covering process heating applications: induction heat treatment, billet and bar heating, metal melting, and scientific equipment applications. A typical furnace is shown in Figure 16.18.

Where mass-produced parts require selective hardening to close tolerances, for example automobile engine and transmission components, induction heating is an ideal tool. The same process

436

Figure 16.18   A Radyne 900 kW 1.5 tonne melting furnace powered by a 1 kHz solid-state generator in use at English China Clays, South West England

can be used for brazing, soldering, annealing, tube welding and many other applications.

The forging industry utilises induction heating for heating billet and bar stock both continuously and selectively along the lengths of the bars concerned up to a temperature of 1250°C prior to forging by mechanical presses or horizontal forging machinery.

Ferrous and non-ferrous materials are melted in induction furnaces with capacities from a few kilograms up to more than 70 tonnes. These furnaces can be small tilt, push-out, rollover, platform tilt and vacuum furnaces, depending on the particular requirement.

Scientific applications include epitaxial deposition on silicon for micro circuits, crystal growing, zone refining, plasma analysis and etching, and carbon and sulphur determination.

Induction heating applications are normally categorised by frequency: r.f. (radio frequency) 250 kHz to 10 MHz; m.f. (medium frequency) 1–10 kHz; and l.f. (low frequency) 50–750 Hz. The frequency of the power source to be selected depends

437

entirely upon the application; the higher the frequency the more shallow the heating effect achieved. For through-heating applications low frequencies are used.

The static inverter has greatly increased the scope for medium to low frequency induction heating applications because of its reliability, compactness and high efficiency, in some cases up to 95%. Efficiency of the valve generator for r.f. applications rarely exceeds 60%.

There are purpose built coating machines used in the paper and board industry that incorporate high velocity hot air dryers, sometimes with steam-heated drying cylinders. One of the most promising developments for non-contact drying is the use of radio frequency energy (r.f.). The selective feature of r.f. energy, where most of the energy is absorbed in the web or coating, as well as the through-drying effect, gives a moisture profile levelling effect. This prevents moisture being trapped inside the material. Equipments having about 30 kW of r.f. output have been installed in the UK.

Both dielectric and microwave heating are being used in food processing. Both rely on the same phenomena, namely dipole agitation, caused by the alternating electric field. The main difference between the two techniques is that the former uses frequencies in the range 10–150 MHz while microwave heating employs frequencies above this range. The two bands currently used are 897–2450 MHz.

In dielectric heating, the material to be treated is placed between electrodes whereas in microwave heating waveguides are used for the same purpose. Greater power densities with less electrical stress on the food product can be achieved at microwave frequencies. This can be of particular advantage when used on relatively dry food products such as potato crisps.

One of the best established applications of dielectric heating is in the post-baking of biscuits and some bakery products. The process reduces the moisture content to a pre-determined level and completes the inside cooking. There is usually no advantage in applying high frequency heating in the early stages of drying and baking where traditional methods are clearly effective. The best area of application is when the moisture level has dropped to around 15% for at this level the conventional ovens cannot remove the moisture at anything like the rate required. Rapid drying down to very low moisture content levels is possible with dielectric or microwave heating.

438

Dielectric heating is widely used for setting synthetic resin adhesives, for 'welding' thermoplastics materials and for preheating rubber and moulding powders. Since the heat is developed evenly throughout the thickness of the material the temperature rise is uniform throughout so that the centre of the material is at the same temperature as the rest.

The welding of thermoplastics materials such as PVC is the largest active area for dielectric heating covering such applications as stationery items, medical products, inflatables, protective clothing, the shoe industry, leisure items, etc. A major user is the car industry for seating, door panels, sun visors and headlinings.

ELECTRIC STEAM BOILERS

There are many industrial plants where steam is required for process work and an electric boiler is worthy of consideration as an alternative to a fuel-fired boiler.

Comparison of boiler running costs should not be made on a fuel calorific value basis only. There are many other considerations to take into account, i.e. flue, fuel storage, is automatic boiler-plant operation wanted, summertime process loads, nighttime and weekend loads. These are a few instances where electrically raised steam has advantages.

The principle of the electrode boiler is simple. An alternating current is passed from one electrode to another through the resistance of the water which surround the electrodes. A star point is formed in the water which is connected through the shell of the boiler to the supply system neutral, which, in accordance with the IEE recommendation, should be earthed. A circuit-breaker provided with the boiler gives overcurrent protection.

Generally electric boilers are connected to low voltage (415 V three-phase supply) but for very large loads high voltages may be used. Figure 16.19 is a diagram of the GWB Jet Steam high-voltage electrode steam boiler, but smaller steam process loads are provided from low-voltage supplies. Loads up to 60 kW may be provided from immersion-heated steam boilers. Between 40 kW and 2300 kW electrode boilers are used, and Figure 16.20 shows the GWB Autolec Junior range rated up to 170 kW.

For all electric steam boilers the specification includes the boiler complete with its mountings, a feed pump and circuit-breaker, which in the case of the immersion heated boiler is a

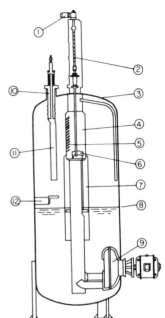

1. Control motor.
2. Control sleeve drive shaft.
3. Boiler shell.
4. Jet column.
5. Jets.
6. Control linkage.
7. Control sleeve.
8. Water level.
9. Circulating pump.
10. Insulator.
11. Electrode.
12. Counter electrode.

Figure 16.19 GWB Jet Steam h.v. electrode steam boiler

contactor or thyristor directly controlled from water level and steam pressure.

All GWB electrode steam boilers are provided with a patented feed reservoir, which improves the performance of the boiler. All steam boilers require a feed tank. This is included in the Autolec Junior specification.

The principle of operation of low-voltage electrode steam boilers is to control the steam output by varying the submerged depth of electrode in the water. This is achieved by releasing water from the boiler shell back to the feed tank through a solenoid valve. The valve is controlled as a function of the current flowing in the main electrode circuit or as a function of the boiler pressure. By these means a steady pressure is maintained under

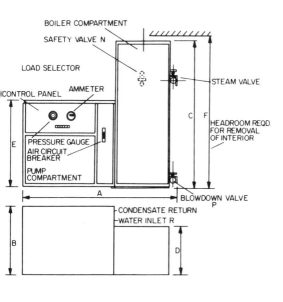

Dimensions	SE.40	SE.65	SE.105	SE.170
A	1730 mm	1730 mm	1784 mm	2108 mm
B	724 mm	724 mm	851 mm	930 mm
C	1444 mm	1530 mm	1880 mm	1111 mm
D	508 mm	508 mm	508 mm	686 mm
E	914 mm	914 mm	1054 mm	914 mm
F*	2591 mm	2794 mm	3531 mm	2032 mm
Weight	293 kg	486 kg	573 kg	795 kg

*Add dimensions of lifting
gear

Figure 16.20 Autolec Junior packaged electrode steam boiler
(GWB Ltd)

441

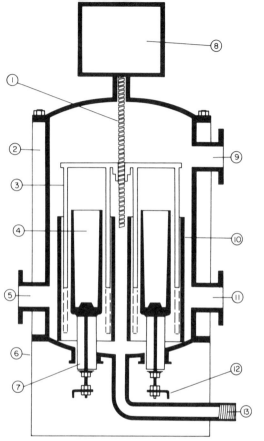

Figure 16.21   Electrode hot water boiler

1. Load control shaft.
2. Lagging.
3. Control shield.
4. Cast iron electrode.
5. Return or s/v correction
6. Terminal guard.
7. Insulator.
8. Geared motor.
9. Flow connection point.
10. Shields.
11. Return or s/v connection.
12. Terminals.
13. Drain valve connection.

fluctuating steam demand conditions and the boiler is modulated to a virtual zero output when no steam at all is being drawn.

Water feed can form scale, or the electrical conductivty of the water can be increased by concentration of salts. These, in the case of the electrode boiler, may be removed from the shell through solenoid valves, and where the situation merits it a more sophisticated equipment automatic control may be fitted.

In addition to steam boilers GWB make electric steam super-heaters ranging from 6kW to 200kW in a single shell. Operating pressures and temperature of superheaters are 20.4 bar and 350°C maximum.

The control of the high-voltage electrode steam boiler is by varying the number of water jets which issue from the jet column (item 4, Figure 16.19) on to the electrodes (item 11). The water not generated into steam falls back in the water in the bottom of the boiler shell. The control sleeve (item 7) is modulated as a function of current flowing in the electrode circuit and the boiler pressure.

ELECTRIC HOT WATER BOILERS

Electric water boilers mainly provide hot water for central heating systems and the supply of domestic hot water. Where the hot water is to be used through an open outlet, i.e. taps or swimming pool heating, it is necessary to use a storage or non-storage heat exchanger (calorifier).

Boilers may be connected to an 'on peak' or 'off peak' electrical system. In the latter case the energy is normally stored in thermal storage vessels in which the water is stored at a high temperature and mixed to the required secondary system conditions. This principle is shown in Figure 16.22.

The actual heating load may vary from straightforward central heating and domestic hot water services to swimming pool heating, top-up heating for energy recovery systems and process solution heating.

The electric boiler operates at a very high thermal efficiency and provides a rapid response to varying demands which, in the case of a thermal storage plant, is by retaining the unused heat during the discharge cycle within the vessel, making the heat input less during the charging cycle.

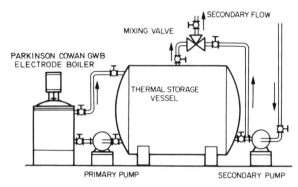

Figure 16.22   Thermal storage heating. The static head available is not sufficient for the required storage temperature. A false pressure head is created by pressurisation plant. The water drawn from the vessel is mixed and circulated through the secondary system

Advantages of an electrical boiler plant are no need for space for fuel storage; no chimney requirement; no ash handling; no smoke or grit emission; minimum attention required for long plant life.

The principle of the electrode boiler is that from each phase of the three-phase supply connected to an electrode system a current flows to the neutral shield. This flow is controlled automatically by a porcelain control sleeve which is raised or lowered over the electrode to increase or decrease the current flow. The power input control is a function of the current flowing and of the temperature of the outlet water from the boiler. The electrical connection is star, and in accordance with the IEE recommendations the star point is connected to the system neutral and earthed at that point (Reg. 554-03-05). A circuit-breaker is provided with the boiler which gives overcurrent and no voltage protection.

Figure 16.21 shows a cross-section of an electrode boiler. The gear motor unit (item 8) raises or lowers the load control shields (item 3) by rotating the load control shaft (item 1). When the control shields are in their minimum position, i.e. fully down over the electrodes, the boiler input is at a minimum and (dependent upon the water temperature) will be in the order of 4% of the boiler rating.

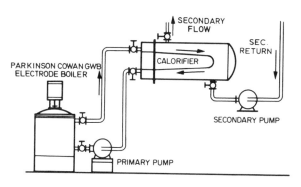

Figure 16.23   Swimming pool heating and hot water services. The calorifier is connected to the primary side of the electrode boiler. The water to be heated flows through the secondary side of the calorifier

Electric boilers are designed for three-phase low-voltage (415 V) networks; but in certain circumstances, where the loading and tariff situations warrant, high-voltage boilers can be made. Figure 16.24 is a diagram of a high-voltage electrode water boiler (6.6 to 18kV). The kW rating of low-voltage boilers ranges from 144 kW to 1600 kW and hot water boilers from 1 MW to 18 MW. For small heating requirements, ranging from 12 to 72 kW, resistance wire or immersion heated boilers are used. The elements are directly switched through a contactor.

As in all cases of boiler plant, the water used should be as free of scale-forming properties as possible and the conductivity of the water within the design of the boiler and system. For low-voltage boilers this will probably be below 1500 Dionic units and for high-voltage boilers 200 Dionic units.

LAMP OVENS FOR INDUSTRY

Many items of furniture and domestic appliances are finished with cellulose lacquer. This finishing process usually consists of several distinct operations including coating and recoating with lacquer with drying periods between each coat. Rapid drying of the lacquer can be achieved by using radiant heating ovens.

445

The main source of heat in an oven of this nature is tungsten lamps which emit a considerable proportion of the energy consumed in the form of short-wave infra-red radiation. Short wave radiant heat passes through air and most gases in a manner similar to light, without any appreciable loss of energy. It is only when

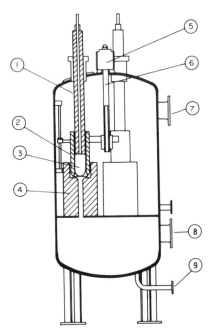

Figure 16.24   Diagrammatic cross-section of h.v. electrode hot water boiler

1. Bushing insulator.
2. Load control shield.
3. Electrode.
4. Jet tube.
5. Load control mechanism.

6. Load control shaft.
7. Flow connection.
8. Return connection.
9. Lower drain.

446

the radiant heat falls upon an object in its path that energy is absorbed and the temperature of the object rises.

The efficiency of a radiant heat oven is very high and with quartz glass heat lamps, temperatures in excess of 1300°C can be achieved.

Advantages associated with this form of heating are speed; compactness; rapid response; ease of control; and cleanliness.

Heat lamps divide into two basic types. The first is substantially a parabolic shaped bulb having an integral reflector inside the envelope. In a typical range from Phillips Electrical there are three ratings suitable for industrial applications, 150 W, 300 W and 375 W. These are used widely in industry for paint baking and low temperature process heating up to about 300°C.

The second type of heat lamp, suitable for temperatures up to about 1000°C, is a tubular quartz glass type. Elements range typically from 500 W to 3 kW and are used with external reflectors. Lamps with reflectorised coatings are also available. Other lamps with ratings up to 20 kW having an halogen additive are suitable for temperatures above 1000°C.

In general plane or flat reflectors are the best when treating flat or sheet materials. When solid objects are to be processed or where radiant energy is to be concentrated on to a small area parabolic or elliptical reflectors are better.

For drying and paint baking applications it is necessary to have an extraction plant to remove the saturated air. Since there is no contamination of the air by combustion products it is possible to recover any expensive solvents being carried away in the air stream.

Radiant heat ovens can be built into existing production plants with the minimum of modifications.

REFRIGERATION AND AIR CONDITIONING

While the electrical engineer's interest in the field of refrigeration may generally remain confined to the design, installation, maintenance or repair of electrical equipment already familiar to him from other applications, knowledge of the 'production of cold' will be of assistance to him.

The basis of all refrigeration is that heat is extracted either directly or indirectly from the item to be cooled (usually food) and used to evaporate a liquid known as the refrigerant. All liquids require heat to cause evaporation, for example water evaporates at 100°C and ammonia at −33°C at atmospheric pressure. This heat is known as the latent heat of the substance and is made use of in refrigeration. A low boiling refrigerant is contained in a component called an evaporator and heat is taken from the item to be refrigerated to boil the refrigerant.

In many cases an intervening medium is used to act as the refrigerating agent. It may be blown over the coils of an evaporator in an air cooler, or it may be employed in a domestic refrigerator or display cabinet. In other cases the medium may be water in an ice flake machine or fish in a contact freezer. This extraction of heat from the ambience is called the 'production of cold'.

Apart from ammonia which is now only used as a refrigerant in industrial installations, there are many other suitable refrigerants. They include a whole family of halocarbons, classified as R12, R22 and so on. They are used in both domestic and commercial refrigerators and in air conditioning installations.

To perform its duty economically in a closed cycle, the vapour formed by evaporation of the refrigerant and loaded with the heat picked up at a low temperature, must be allowed to discharge the heat and thereby re-liquefy or condense. For this purpose, the refrigerant must be brought into heat exchanging contact with a heat sink such as atmospheric air or cooling water at ambient temperature. This means that the refrigerant leaving the evaporator must be elevated to a temperature higher than that of the heat sink.

In the vapour compression cycle this duty is assigned to a compressor, which for small and medium installations may be a reciprocating piston type type or a rotary type. For large installations a centrifugal compressor is generally employed. The useful part of the energy spent results in a pressure and temperature rise of the vapour. During subsequent dissipation of heat the refrigerant is liquefied but still remains at high pressure. For example, ammonia condensing at 25°C the saturation pressure is 10 atmospheres. To return the liquid to the evaporator and thus complete the cycle, the high pressure liquid must be expanded to low pressure whence cooling takes place. A throttling device is used and it may take the form of a simple capillary tube or a throttle valve the action of which may be electrically controlled in response to either pressure or temperature difference. A further method may be a metering device which opens or closes a small bore orifice in response to a liquid level.

Compressors may be of the open or closed type. For the former the shaft, sealed against loss of refrigerant and lubricating oil by a gland, protrudes through the casing. The driving motor is either directly coupled to the compressor shaft or fitted with a belt drive.

In a closed type compressor, the rotor is directly coupled to, or mounted on, the compressor shaft and thus the motor windings are exposed to the refrigerant which is mixed with oil vapours. The assembly forms either an hermetically sealed or a semi-hermetic unit. The latter is a unit which can be opened in a factory by undoing a flange connection while the former can only be opened by cutting.

Though based on the same principle of evaporation of a refrigerant at low temperature, a different form of refrigerating cycle is obtained; when replacing the mechanical energy (of the compressor) by heat energy and removing the vapour in the evaporator by absorbing it in an appropriate liquid to which it has chemical affinity. That liquid is contained in the absorber. Ammonia may be used with water as the absorbent, or water vapour may be used with lithium-bromide as the absorbent. As the solution becomes stronger and saturated, it is pumped to the higher pressure of a boiler from which the refrigerant (ammonia or water) is boiled off. It is subsequently condensed, drained and expanded through a throttling device into the evaporator.

Absorption systems are extensively found in large size air conditioning installations where the steam generated in winter for heating purposes can be used in summer for heating the strong solution and expelling the vapour of the refrigerating system. Likewise this system is found in the chemical industry, where the steam from back-pressure turbines is used, and in freeze-drying installations for food processing or preparation of instant coffee or tea. It does, however, require a large area for installation of the plant as compared with the space occupied by a vapour compression cycle. For this reason absorption plants are not in common use.

While the electrical equipment, such as motor starters, controls, relays and circuitry may vary according to the manufacturer or country of origin, the principles of construction are mainly those described in the appropriate sections of this book.

Microprocessors are playing an increasing part in the control of refrigerating plants, particularly in the starting and stopping of compressors in a multi-compressor plant. Such control is designed to reduce the amount of manual attention required.

A fairly recent newcomer to the refrigeration field is the screw compressor. This consists of two large screws or helical gears meshing together which pump the refrigerant and oil into a separator and thence into a condenser. Having no reciprocating parts it is smooth running and can be operated at speeds of 2500–3000 rev/min. It is compact for its output and has a wide range of duties from −29°C to −1°C. This type of compressor is particularly useful for dealing with large duties in air conditioning and with low temperature, and its output can be controlled down to as low as 10% of full load.

Another type of compressor, quite recently introduced into this field uses only one screw and two star rotors. It has the advantage of eliminating the need for an oil pump by generating a pressure difference between the suction and delivery lines to circulate the oil and the refrigerant.

Suffice it to say that an important role is assigned to low-voltage electrical heating matting in the floor of low temperature freezing chambers and storage rooms built at ground floor level. Even the best type of thermal insulating material applied under the floor of such an installation will, within the lifetime of the cold store, allow its low temperature to spread gradually into the soil and, with its high moisture content, particularly in the presence of a high water table, will gradually form a solid block of ice. The expanding mass of ice will exert great pressure on the subfloor with its thermal insulation and on the concrete and granolithic flooring.

This is known as 'frost heave' and must be prevented otherwise it would cause irreparable damage to the floor. One method consists in electrically heating the subfloor to a controlled temperature above freezing point, say +1 to +3°C. Depending on the temperature of the chamber and on the thermal conductivity and thickness of the insulating material, the specific heating energy required is of the order of $12.5 \, \text{W/m}^2$ of floor area.

AIR CONDITIONING AND VENTILATION

Air conditioning in the UK is not applied to buildings on the same scale as in the USA, but its use is becoming more common in the large buildings now being erected. The amount of fresh air

450

required in a given building will depend on the activity therein, and the IHVE guide sets out the recommended amounts for given occupations.

Air conditioning takes various forms depending upon the requirements, but essentially it is a system of delivering cooled air to a given space and maintaining the area at a given temperature, and sometimes humidity. This can be achieved by a large plant cooling water that is circulated to cooling coils within the rooms of the building. Air is circulated over the coils, and the temperature is controlled by a thermostat in the room which cuts off the water supply as required. Humidity is more difficult to control within fine limits, but this can be done by injecting water spray into the air stream.

Another method is to have a large plant cooling air that is conveyed by trunking to the various spaces. Temperature is regulated by controlling the amount of air delivered to the space. This system is known as variable air volume. Such plants are particularly suitable for incorporating centrifugal or screw type compressors.

Where a comparatively small room is to be air conditioned – or three or four offices of average size – individual air conditioning units may be used in each room, each independently controlled by a room temperature thermostat. These incorporate their own refrigerating plant and are designed to look like standard electric convector heaters, although they are slightly bulkier.

17

Building automation systems

Over the years centralised monitoring and control systems have developed almost beyond recognition from the uniselector and diode matrix systems first introduced more than twenty years ago. New technology and falling hardware prices have resulted in systems which offer many times the capability at a fraction of the cost of those earlier systems.

Today the majority of Building Automation Systems (BASs) are based on the relatively inexpensive personal computer (PC) and it is only with the extremely large systems, in terms of monitored points, that the mini-computer is used.

Intelligence and stand alone capability have been pushed further out into the field equipment to the point where the latest systems offer individual controllers, no larger than, and costing little more than, conventional analogue controllers. In addition to control these devices offer full monitoring capability. They communicate with the system in the same way that the larger and more costly Field Processing Units (FPU) do.

The provision of good environmental conditions in building requires the operation of all the individual systems to be co-ordinated in the most efficient and economic way. Also performance has to be monitored so that alternative or remedial action can be taken automatically should there be a significant deviation from the normal situation.

A BAS then, must satisfy both the need to monitor and control although it is true that in some circumstances monitoring without control is a requirement. The reverse is seldom true. Monitoring without control advises the operator of conditions from which he takes the appropriate action manually. Depending on the nature of the action required it may or may not be implemented through the BAS.

More and more the BAS or Energy Management System (EMS) will take action based on predefined routines which have proved over the years to be reliable and effective. Thus, for example, the failure of the duty pump would result in the BAS starting the standby, or electrical services may be switched automatically to a standing power supply, or loads may be shed to avoid demand charges and so on.

In addition, a BAS as opposed to an EMS may also monitor and control other services such as fire detection, fire safety and building security. They are capable of monitoring, optimising, co-ordinating

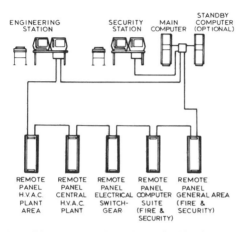

Figure 17.1   Block diagram of typical centralised building automation system

and controlling in real time as the dynamics of the situation demand, see Figure 17.1.

The use of computers to achieve this has two main advantages over other approaches. The first is that equipment can be standard-ised so that the differences between individual projects becomes largely a matter of quantities of compartments and how they are grouped.

The second is that computerised systems now offer a single, very cost-effective means of controlling heating, ventilating and air-conditioning plant, electrical services and lighting compared with each of these services on an individual basis. Incorporating fire safety and security does not necessarily reduce the overall capital cost but it may be the best way to achieve the overall reliability and integrity required when the possibility of adverse interaction between services exists. For example, there is obviously conflict between the needs of the control system, the energy management system and the fire safety system when it comes to the control of fans and dampers.

A large centralised BAS will typically have a single computer and possibly a standby as shown in Figure 17.1 if the application is of a critical nature. The CPU shown in Figure 17.2 is based on PC. It is more compact and costs less than the mini-computer which would have been used even for smaller projects up to a few years ago. (Figure 17.3).

The communication links between the Central Processing Unit (CPU) and the FPUs usually take the form of a shielded twisted pair. Coaxial cables are sometimes used for very high transmission speeds. Different configurations are used by different manufacturers. These include multidrop which is probably the most common, star and loop configurations.

Figure 17.2   A personal computer (PC) based BAS from Johnson Controls JC85/40 range

Figure 17.3   A JC/85/40 system consisting of a colour VDU with full colour graphics, keyboard, printer and mini-computer (Johnson Control System Ltd)

Between FPUs, sensors and actuators, discrete wiring is normally used. This also is usually screened. Some systems, such as the Johnson DSC1000 have the facility to use addressable actuators which means that they also can be multi-dropped off a single two-wire bus.

In the main sensors are either binary, that is they have opening or closing contacts, or they are analogue which means they produce a signal which varies in magnitude in some relationship to the variable they are sensing. With the exception of temperature most analogue sensors produce a standard signal which may be 0–10 V d.c., 4–20 mA, 2–10 V d.c. etc. Temperature sensors may be to an accepted standard such as PT100 or PT1000. In commercial applications less costly resistance element sensors are used as are thermistors which are negative temperature coefficient (NTC) sensors. All of these are invariably non-standard which means that the BAS has to be capable of accepting and interpreting the non-standard signals which this group of sensors produce.

455

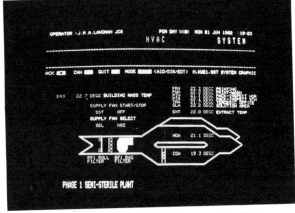

Figure 17.4 One of the earlier dynamic graphic displays

Outputs from FPUs are used to switch plant, open and close motorised valves etc. These are digital signals. FPUs also produce analogue output signals to provide precise positioning of control devices. The most common of these is 0–10 V d.c.

In recent years much emphasis has been placed on the ease of use and quality of presentation of information of the operator input/output (I/O) device. This has resulted in the extensive use of graphics to display information and input devices such as mouse to minimise the use of a keyboard device.

Figure 17.2 shows the present day approach to graphics while Figure 17.4 shows a graphic display which would have been considered quite advanced five or six years ago. Both are used as part of the Johnson JC85/40 system, which is a good example of how a well-designed BAS can be improved as component parts improve. This is not the case with all systems.

(a) Facilities which save energy cost by reducing 'on time' of equipment such as HVAC plant and lighting

Centralised manual start/stop. The ease with which plant can be started and stopped encourages the operators to save fuel by running plant only when it is needed.

Programmed start/stop. The facility to set individual start and stop on a 7-day programme encourages the achievement of reduced run times and the close timing of plant operation to the occupants, requirements.

Optimal start. This extremely flexible facility available on computerised systems again reduces plant run-time and fuel costs.

Demand monitoring prediction and control. This is an automatic programme which reduces the electrical load on a building on an 'Ideal rate' or prediction basis to reduce electrical demand charges.

Chiller plant optimisation. A programme which runs multiple chiller plant at optimum efficiency thus reducing energy costs.

The saving is multiple, in that reduction of run hours cuts not only fuel costs but also maintenance, where this is performed on an hours-run basis, and will also lengthen life of equipment.

(b) Facilities which allow the reduced plant 'On-time' to lower maintenance costs

Run-time totalisation with limits. Permits reduced plant run-time to be reflected in lowered maintenance costs by monitoring the hours run of all monitored plant and advising the operator when set limits have been reached.

Calculated point. Allows limits to be assigned to calculated parameters and thus the operator to be advised when, for example, plant efficiency starts to fall off.

Alarm message. Permits a detailed message to be presented to the operator of actions to be taken. Allows therefore lower-skill operators to be used.

Maintenance management. This detailed family of programmes, available with larger computerised systems, allows complete automation of the whole maintenance management. It issues complete maintenance task sheets at the appropriate times, sets targets for materials and labour in the tasks keeping full records. It will also allow prediction of future maintenance and labour needs dependant on experience.

The most cost effective way of doing this is to run the programme on a separate personal computer such as the IBM PC. This allows the manager separate access to all the maintenance programmes without interrupting the other building management functions which are being collected and monitored from the BAS system itself. Time and condition based data is transferred from the BAS to the personal computer.

(c) Automatic checking facilities

The many automatic checking facilities available can significantly reduce the labour involved in routine checking of plant and the analysis of problems. This can have two possible costs effects:

- The reduction in labour involved if this has to be performed manually.
- A reduction in breakdowns where manual checking has not been carried out.

(d) Applications programmes

There are many application programmes now available with sophisticated computerised systems which can reduce energy costs.

Supply air reset. A programme which automatically adjusts central air conditioning supply temperatures to the optimum point to save energy.

Outdoor enthalpy control. A programme which adjusts to use of fresh and recirculated air on air conditioning plant to optimise their use.

Calculated point. General programmes which can be used in many ways to optimise energy.

Special time reset programme. Provides automatic reset of controllers to reduce energy use at certain periods of time.

(e) Energy management software

Sophisticated software programmes are adding another dimension to energy management. For example, the Johnson Controls JC/85/40 building automation system provides three basic sets of energy management software programs to control energy. These interact with one another to determine the best management decision for each situation.

Load manager. Has the responsibility for minimising the use of electricity. It lowers electrical consumption by turning pumps and fans off for short periods throughout the day. It can reduce low priority electrical loads to cut back on the use of excessive power at any given time.

HVAC manager. Considers weather conditions and delays morning start-up of mechanical equipment. As a result of this, the comfort level is reached only when occupancy of the building

458

begins. It also eliminates overheating or cooling of air delivered to occupied areas, and determines whether outdoor or recirculated air will require the least amount of cooling. These measures serve to reduce the use of cooling energy, thereby saving on energy costs.

Chiller manager.   Aimed at one of the primary energy consumers in air conditioned buildings, the chiller plant. It keeps chilled water temperature as high as possible while remaining within established comfort standards, keeps the chiller plant operating at the most efficient point, and chooses the chiller or sequency of chillers that will handle the load most efficiently.

While each 'energy manager' has a defined area of responsibility, priority conditions will occur. To accommodate these situations, all programmes reference each other constantly to resolve control priorities and determine the most effective action.

From these programmes useful, concise reports on consumption and demand can be requested by the building manager in order to evaluate system performance.

(f) Capital cost set-offs

Potential capital cost savings on other equipment which can be set against the cost of the computerised automation system are listed below. By using power failure restart, maximum demand control and software interlocking savings are made in essential and non-essential switchgear; busbar distribution; mechanical and electrical services control and fire operation; and metalwork.

- Elimination of wiring between panels where replaced by software interlocking and centralised manual and programmed start and stop.
- Elimination of duplicate wiring for essential/non-essential services.
- Elimination of supply and installation of detectors, alarm sounders, control panels, power supplies, and associated communication/display equipment plus their installation where incorporated as part of the building automation system for fire detection systems.
- Elimination of supply and installation of specialist duplicate displays/control points for firemans's use where these are incorporated as part of the building automation system.
- Elimination of supply and installation of detectors, alarm sounders, control panels, power supplies, and associated communication/display equipment for security systems.

- Elimination of standard and special instrumentation and monitoring recording equipment where this is replaced by the building automation system.

(g) User programming language (UPL)

No standard software can provide every possible control and optimisation strategy. Every building is unique. Not only in size, but in the equipment installed and the necessary operating characteristics of that equipment. Hence the development of UPL.

The UPL programme is written in an easy-to-learn structured computer language, much like Pascal. Having determined that a particular problem may be solved by UPL, the system user designs a solution to that problem and then writes a UPL programme to implement the solution. An example of this process is a simple programme that starts an exhaust fan when a gas sensor indicates an abnormally high concentration in the area:

```
              LISTING FOR () ,GAS,EXH)
1     PROCESS @ (O,GAS,EXH):HVAC.ALARM;
2     CONSTANTS
3         GAS.DETECTOPR IS (H,BLD3,RM6,GAS);
4         EXHAUST,FAN IS (H,BLD3,RM6,EXH);
5     TRIGGER EXECUTION
6         WHEN GAS DETECTOR CHANGES TO ALARM,
          NORMAL;
7     BEGIN
8         IF STATUS (GAS DETECTOR) IS ALARM THEN
9         BEGIN
10            START (EXHAUST.FAN);
11            ADVISORY ("BUILDING3, EQUIPMENT
              ROOM 6 GAS
12                    SENSOR ALARM – EXHAUST
                      FAN STARTED")
13        END
14    ELSE
15        STOP (EXHAUST.FAN)
16  END!
```

In this example, we have named the process "O,GAS,EXH", given it an "HVAC.ALARM" priority and indicated the two points in the JC/85/40 data base to be acted upon (the GAS. DETECTOR and the EXHAUST.FAN). The process will trigger only when the detector changes to an alarm condition or returns to a normal condition. With the introduction of User Programming Language, the savings to be achieved are limited only by the user's imagination.

(h) Intangible factors

Then there are the intangible factors. The intangible 'savings' that can be realised by the system are considerable but cannot readily be costed due to the difficulty in assessing item true value. They include: reduction in frequency of breakdowns; minimising occupant dissatisfaction and disturbances; increased efficiency of all personnel due to improved environmental conditions; improved morale and job satisfaction of operators.

(i) Direct digital control

Finally, attention must be brought to the increasing intelligence to be found out in the field. A good example is the Johnson Controls DSC8500 Digital System Controller; providing either standalone capability or as part of a fully integrated building automation system.

The DSC8500 can be programmed with a unique series of advanced control routines superior to the standard Proportion (P), Proportional + Integral (PI), and floating routines typical of most digital controllers. Johnson Controls' unique approach of adding derivative and an advanced incremental control algorithm provides those specialised control modes necessary to properly meet the variety of complex control problems faced by today's commercial buildings.

Direct digital control may be simply applied to the control of temperature, pressure, etc, in much the same way as conventional analogue controllers. Alternatively with the addition of a simple communication interface, it can be fully integrated into the BAS as is the case with the DSC8500 and JC/85/40. Thus it not only controls temperatures, etc, but through its programming language will carry out all motor control interlocks, time programming and energy management as an integral part of the BAS being applied to the building.

Since it does all this at a local level the central computer is free to carry out global control strategies and provide additional management information and generate special reports.

As a standalone digital system controller, the DSC8500 will

461

perform complete control and optimisation functions for the system it serves.

As mentioned in the opening paragraphs, intelligence and stand alone capability has been pushed further and further out from the CPU. With the Johnson Controls System 91 this has resulted in small, low-cost direct digital controllers. These fall into two broad classifications – the digital terminal unit controller used to control variable air volume boxes and fan coil units in air conditioned buildings and the digital plant controller which in many ways may be considered as a smaller version of the digital system controller such as the DSC8500 described above.

With such devices out in the field the number of monitored and controlled 'points' theoretically available to the CPU may run to many thousands. Each digital terminal unit controller typically may have six analogue inputs, six binary inputs and six outputs of varying types. It is not unusual for a large building to have, say, a thousand terminal unit controllers bringing the number of theoretical physical points available to the CPU to something in the order of 20,000. In practice it is not necessary to access these points from the CPU except very occasionally, especially since the digital terminal unit controller has intelligence and stand alone capability.

The building automation section was prepared by Johnson Control Systems Ltd.

18

Instruments and meters

Most instruments and meters are based on either the magnetic principle or that of electro-magnetic induction. Some of these meters like phase angle, wattmeter, frequency meter and VAr meter use an electronic converter and a moving coil indicator rather than a direct induction effect. This should be remembered when reading the appropriate sections. In addition there are a number of other instruments which make use of electrostatic, heating and chemical effects, each of these having a special application or applications.

AMMETERS AND VOLTMETERS

The normal instruments for measuring current and voltage are esentially the same in principle, as in most cases the deflection is proportional to the current passing through the instrument. These meters are, therefore, all ammeters, but in the case of a voltmeter the addition of a series resistance can make the reading proportional to the voltage across the terminals. Types are:

1. Moving iron (suitable for a.c. and some suitable for d.c.)
2. Permanent magnet moving coil (suitable for d.c. only).
3a. Air cored dynamometers – moving coil (suitable for a.c. and d.c.) particularly used for high-precision measurements.
3b. Iron cored dynamometers – moving coil (generally only suitable for a.c.) particularly used for high-precision measurements.
4. Electrostatic (for a.c. and d.c. voltmeters).
5. Induction (suitable for a.c. only).
6. Hot wire (suitable for a.c. and d.c.).

All these instruments measure rms values except the moving coil which measures average values. Therefore when measuring rectified and/or thyristor-controlled supplies it is important to

463

know whether the load current for example as registered is the rms or average value. Heating or power loads generally require rms indication while battery charging and electrochemical processes depend on average values.

Although some of the above square law instruments will read d.c. it should be appreciated that it is still the rms value that is indicated. This will only be the same as the average value indicated by a moving coil instrument if the supply is d.c. without any ripple. The difference between the moving iron and moving coil instrument connected in the same unsmoothed single-phase rectified supply could be as much as 11%.

Note that the permanent-magnet moving-coil instrument can be used for a.c., provided a rectifier is incorporated, and this is described in another section.

Accuracy. The accuracy of instruments used for measuring current and voltage will naturally vary with the type and quality of manufacture, and various grades of meters have been scheduled by the BSI. These are set out in BS89.

There are nine accuracy classes in BS89 designated in 'Class Indices' which enable an engineer to purchase instruments with an accuracy sufficient for the purpose for which the instrument is intended. For normal use instruments of Accuracy class 1.5 and switchboard and panel instruments Accuracy class 2.5 are usually sufficiently accurate.

When specifying an instrument for a particular purpose it is important to realise that the accuracy relating to the Class Index may only be obtained under closely specified conditions. Variations of temperature, frequency, magnetic field etc. may cause the error to exceed the Class Index. The additional errors permitted are related to the Class Index and should be considered in the light of a particular application.

Many manufacturers are now producing indicating instruments, such as ammeters, voltmeters and wattmeters, with a platform scale. With this type of instrument the pointer traverses an arc in the same plane as the actual scale markings, thus eliminating side shadow and parallax error.

Moving-iron instruments. This type of instrument is in general use in industry owing to its cheap first cost and its reliability. There are three types of moving-iron instruments – the attraction type, the repulsion type and some using a combination of both

464

principles. Short scales, i.e. 90 degrees deflection are generally associated with repulsion meters, and long scales i.e. 180–240 degree with combined repulsion and attraction types (Figure 18.1).

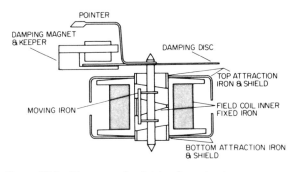

Figure 18.1 Diagrammatic sketch of moving-iron movement. This operates on the combined repulsion-attraction principle. The field coil is enclosed in nickel-iron screens, the inner edges of which project into the bore of the moulded plastics field coil bobbin forming attraction elements. The magnetic field provided by the coil polarises these elements and also the fixed and moving irons

The principle of operation is that a coil of wire carrying the current to be measured attracts or repels an armature of 'soft' iron, which operates the indicating needle or pointer. With the attraction type the iron is drawn into the coil by means of the current; and in the repulsion type there are two pieces of iron inside the coil, one of these being fixed and the other movable. Both are magnetised by the current and the repulsion between the two causes the movable unit to operate the pointer.

It will be seen, therefore, that the direction of current in the coil does not matter, making the instrument suitable for measuring any form of current, either d.c. or a.c, including rectified a.c. Accuracy is not affected by the distorted waveforms generally encountered on commercial supply systems, but peaky waveforms can cause saturation of the iron in the magnetic circuit resulting in incorrect readings.

Some causes of error.

(*a*) *Stray magnetic fields.* Owing to the fact that the deflection is proportional to the magnetic field inside the operating coil, magnetic fields due to any outside source will affect the deflection. Errors due to this are reduced by suitable magnetic screening of the mechanism.

(*b*) *Hysteresis.* The use of nickel iron in the magnetic circuit of modern instruments has reduced the error due to hysteresis such that an instrument of Class Index 0.5 can be used on both a.c. and d.c. within that class. Due to other factors the use of nickel iron is not usually possible on long-scale instruments. These are not therefore suitable for use on d.c. due to the large hysteresis error.

(*c*) *Frequency.* Ammeters are not affected by quite large variations in frequency, an error less than 1% between 50 Hz and 400 Hz being quite usual. Errors do occur on voltmeters with change of frequency due to alteration in the reactance of the coil which forms an appreciable part of the total impedance. Where necessary the error can be substantially reduced by connecting a suitably sized capacitor across the series resistance as compensation.

Permanent-magnet moving-coil instruments. This is essentially a d.c. instrument and it will not operate on a.c. circuits unless a rectifier is incorporated to permit only uni-directional current to pass through the moving coil.

The operation of a moving-coil instrument will be seen from Figure 18.2. By means of a suitably designed permanent-magnet and a soft iron cylinder between the poles a circular air-gap is formed through which a pivoted coil can move. This pivoted coil carries the current to be measured (or some proportion of it), and as the field is uniform in the air-gap the torque will be proportional to the current. The deflection is controlled by means of a torsion spring with suitable adjustment device and the current is taken to the moving coil by means of these or other coil springs.

Permanent-magnet moving-coil instruments are only suitable for small currents for actuating purposes, and thus both shunts and series resistances are used to a fair extent. Stock instruments usually have a full scale deflection of 1 mA for voltmeters (1000 ohm/V) and 75 mV for ammeters operated from shunts. These shunts may be contained inside the instruments for ranges up to around 60 A and external to the instrument for higher current ranges.

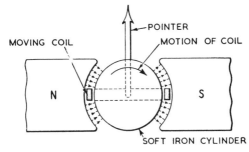

Figure 18.2   Permanent-magnet moving-coil instrument (90 degree scale).

Figure 18.3   A microamp panel meter incorporating taut-band suspension (Crompton Parkinson)

The actual calculations for values of shunts and series resistances is dealt with on a subsequent page, and an example is included of an instrument of this type.

Damping is obtained by the eddy currents induced in the metal former which is used for the moving coil. With the introduction of new magnetic materials flux densities have increased and stability greatly improved. Silicone fluid damping provides a smoother effect, free from overshoot and permits reduction in the weight of the moving element. It is now possible, with taut band suspension (see below) to produce instruments having full scale deflection with current values as low as $10\mu A$.

Although used extensively for monitoring a.c. and d.c. power supplies many taut band instruments are being employed for the read out on electronic equipment, as transducers for electrical and physical quantities, electrical tachometers, and remote position indicators for example. Figure 18.3 shows a typical scale of a taut-band silicone fluid damped instrument. The scale is substantially uniform and can be made to extend over arcs up to 250 degrees.

Taut-band suspension. This form of suspension used in instruments was introduced by Crompton Parkinson in 1962 and it eliminates the need for pivots, bearings, and control springs. In this system the moving element is suspended between two metal ribbons, one at each end of the movement. These spring-tensioned ribbons, by their twisting during deflection, provide the controlling and restoring forces. Bearing friction is completely eliminated and there nothing to wear out or go wrong. The upper portion of a taut band movement is shown in Figure 18.4.

The spigots on the end of the hub at each end of a taut band moving element work in clearance holes in a bush forming part of the fixed frame assembly. By introducing oil of suitable viscosity into the gap between the spigots and the clearance holes it is possible to control resonance in the moving assembly. The method adopted by Crompton Parkinson to contain the oil in these two areas is shown in Figure 18.5. The spigot is increased in length and an annular pad or washer of expanded polythene is fixed to the outer hub. A similar washer is fixed to the facing end of the mounting bush. The damping oil is inserted between the pads and the spigot. The oil serves a number of other functions. It provides all the damping action needed to bring the moving system to rest in its deflected position quickly. No other form of

468

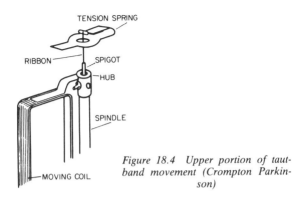

Figure 18.4 Upper portion of taut-band movement (Crompton Parkinson)

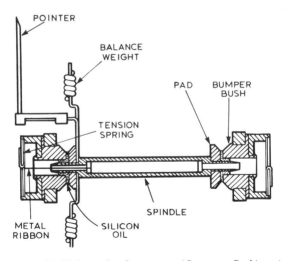

Figure 18.5 Hi-Q taut-band movement (Crompton Parkinson)

469

damping is required resulting in a very light assembly. The torque/weight ratio is therefore improved and so instruments of a higher sensitivity can be produced. Response time is also considerably reduced compared with pivoted instruments.

An important factor is that the oil-retaining pads act as built-in shock absorbers. In the latest form of Hi-Q movement, Figure 18.5, the joints between the ribbons at each end and the spindle have been eliminated. A single ribbon runs effectively from one tension spring to the other forming with a metal locating and stiffening plate the central core of an injection moulded nylon spindle. Aluminium end caps fitted to the spindle carry the shock absorbing damping pads and, at one end, the pointer and balance weight arms. In moving coil instruments they also carry the coil mounting brackets, while in the moving iron system, the injection moulding spindle is radially extended to provide a fixing for the moving iron.

Dynamometer moving-coil instruments. Instead of the permanent-magnet of the previous type of moving-coil instrument, the necessary magnetic field may be set up by passing the current through fixed coils as shown in Figure 18.6. There are thus two sets of coils – one fixed and one pivoted – the torque being provided proportional to the square of the current, making the instrument suitable for either d.c. or a.c.

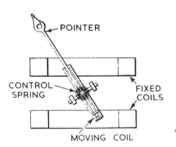

Figure 18.6   Dynamometer moving-coil instrument

In the case of an ammeter, the two systems are usually connected in parallel with suitable resistances, and in the case of a voltmeter they are in series. The leads to the moving coil are in the form of coil springs, which also act as the torque control, and in this case damping has to be provided by some type of damping device such as an aluminium disc with a braking magnet.

Owing to the necessity of keeping the current in the moving coil low the torque obtainable may be small, and these instruments are not used to any extent for general industrial purposes. They are particularly useful in the laboratory for d.c./a.c. transfer instruments. The calibration can be checked on d.c. using a potentiometer and then, if required, in conjunction with known accuracy transformers can be used as a standard for checking other a.c. instruments i.e. moving iron where the d.c./a.c. error may be large or unknown. When used care must be taken to see that they are not affected by stray magnetic fields.

The dynamometer instrument is, however, much used for wattmeters, and its use for this purpose is described later.

Electrostatic voltmeters. Electrostatic instruments are essentially voltmeters as they operate by means of the attraction or repulsion of two charged bodies. They have certain applications in the laboratory as they are unaffected by conditions and variations which give rise to errors in many other instruments. These include hysteresis and eddy current errors, errors due to variations in frequency and wave form which do not cause incorrect readings with electrostatic instruments.

For ordinary voltages the torque is small and multi-disc instruments have to be used. On this account, however, they are ideal for measuring high voltages and form practically the only method of indication where the voltage exceeds 100 kV.

For lower voltages from 400 V to a few thousand volts the *quadrant* type is used, the principle being as in Figure 18.7. The

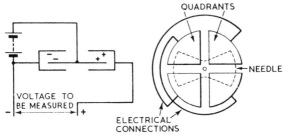

Figure 18.7   Quadrant voltmeter

moving vane is pivoted and is either repelled or attracted or both by the charges on the vane and the quadrants – these charges being proportional to the potential due to their connection to the supply.

In commercial models several sets of vanes are used in parallel to obtain sufficient torque.

Voltages of 100kV and over are measured by means of two shaped discs with an air-space between them, one disc being fixed and the other movable axially. By means of a balance the force between the two is measured and the scale of the balance marked in kilovolts. These voltmeters are not, of course, accurate to a degree that makes them suitable for indicating line voltages or for switchboard purposes. They are, however, ideal for testing purposes where cables and other apparatus is subject to high-voltage and tests to destruction. In this case they form a visual indication of the applied voltage and are a check on the value indicated by the instruments on the voltage *stepping up* apparatus.

Electrostatic instruments for laboratory use are termed *electrometers* and are usually of the suspension type with mirror-operated scales.

As these instruments consume negligible power on a.c. they have the advantage that they do not affect the state of any circuit to which they are connected. The exception is of course in the case of radio-frequency measurements.

Hot-wire instruments. The current to be measured, or a known fraction of it, is passed through a fine wire and, due to the heating effect of the current, the wire expands. If the resistance and coefficient of expansion of the wire is constant, then the heating and consequent expansion of the wire are both directly proportional to the square of the current. If the expansion is sufficient by suitable connections as shown in Figure 18.8 it can be made to move the pointer. This movement is also proportional to the square of the current. By suitable scaling the current can thus be determined.

Since such instruments obey the square law they are suitable for both d.c. and a.c. systems. Furthermore, since they only operate on the heating effect, the rms value of an alternating current is measured irrespective of frequency or waveform. They are also unaffected by stray magnetic fields.

Induction instruments. These instruments, which will function only on a.c., may be used for ammeters and voltmeters, but their

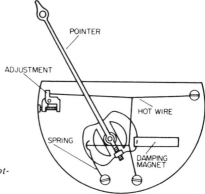

POINTER

ADJUSTMENT

HOT WIRE

SPRING

DAMPING
MAGNET

*Figure 18.8 Hot-
wire instrument*

use for the measurement of these two quantities is much less than for wattmeters and energy meters – these applications being described fully below.

In all induction instruments the torque of the moving system is due to the reaction of a flux produced by the current to be measured on the flux produced by eddy currents flowing in a metal disc or cylinder. This eddy current flux also being due to the current is arranged to be out of phase with the flux of the current being measured.

There are two methods by which these fluxes are obtained. With a cylindrical rotor two sets of coils can be used at right angles, or with a disc rotor the shaded pole principle is used (Figure 18.9) an alternative being the use of two magnetic fields acting on the disc.

Although extremely simple, with no connections to the rotor, these instruments have many disadvantages which prevent their use for general purposes as ammeters and voltmeters. The points in their favour are a long scale, good damping and freedom from stray field effects. Their disadvantages include fairly serious errors due to variation in frequency and temperature, high power consumption, and high cost. The former may be reduced by suitable compensation, but in ordinary commercial instruments the variation is still important, although Class 1 instruments can be supplied if required. See the reference to electronic converters at the start of the chapter.

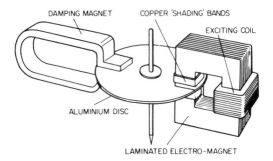

Figure 18.9   Principle of shaded-pole induction meter

WATTMETERS

Dynamometer wattmeters.   If of the air-cored type the wattmeter will give correct readings both on d.c. and a.c. It consists of a stationary circuit carrying the current and a moving circuit representing the voltage of the circuit. For laboratory instruments the meter may be either of the suspended coil or pivoted coil type. The former is used as a standard wattmeter, but the pivoted coil has a much wider scope as it is suitable for direct indicating (Figure 18.10). Long-scale wattmeters are of the iron-cored type and are not generally suitable for use on d.c. due to the large hysteresis errors.

For accurate measurements on low range wattmeters it is necessary to compensate for the flow of current in the other circuit by adding a compensating winding to the current circuit.

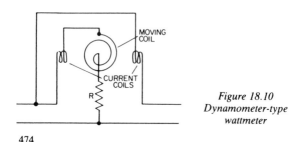

Figure 18.10
Dynamometer-type
wattmeter

This will be seen from Figure 18.11, where the connections (1) and (2) will introduce errors due to the extra current in the current coil in (1) and the volt-drop in the current coil in (2). It will be seen that the arrangement in (3) overcomes these errors.

Dynamometer wattmeters are affected by stray fields. This is true for both d.c. and a.c. measurements, but on a.c. only alternating current stray fields affect readings. This effect is avoided by using a static construction for laboratory instruments and by shielding for portable types. The use of a nickel-iron of high permeability has enabled modern wattmeters to be practically unaffected by stray fields in general use.

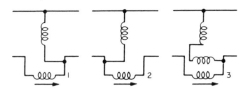

Figure 18.11   Uncompensated (1, 2) and compensated (3) wattmeters

When purchasing wattmeters of this type it is desirable to see that a high factor of safety is used as regards the rating of the coils of the instrument. This is desirable from the point of possible overload on any of the ranges of a multi-range instrument, but it also enables the indication to be made towards full-scale even at low power-factors.

For three-phase wattmeters, adequate shielding between the two sections is necessary, but even then they do not give the same accuracy as single-phase models.

Induction wattmeters.   These can only be used on a.c. and are similar to energy meters of the induction type, the rotating disc in this case operating against a torsion spring. The two circuits are similar to the induction energy meter and the connections are as shown in Figure 18.12.

Although affected by variations in temperature and frequency, the former is compensated for by the variation in resistance of the rotating disc (which is opposite in effect to the effect on the windings), and the latter does not vary considerably with the

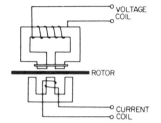

Figure 18.12 Induction-type wattmeter. The construction and principle of operation are similar to the energy meter

variations in frequency which usually obtain. Induction watt-meters must not, of course, be used on any other frequency than that for which they are designed unless specially arranged with suitable tappings.

The general construction of induction wattmeters renders them reliable and robust. They have a definite advantage for switch-board use in that the scale may be over an arc of 300 degrees of so. As with dynamometer types, the three-phase 2-element watt-meter is not as accurate as the single-phase type.

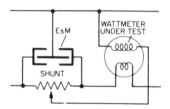

Figure 18.13 Electrostatic wattmeter used for calibration purposes

Three-phase wattmeters. These operate on the two wattmeter principle, the two rotors being mechanically coupled to give the sum of the torques of the two elements. As already stated, it is important to avoid any interference between the two sections, and there are several methods of preventing this. One is to use a compensating resistance in the connections to the voltage coils, and another, due to Drysdale, is to mount the two moving coils (of the dynamometer type) at right angles.

Measurement of three-phase power. The connections of temporarily-connected instruments for measuring three-phase power are given in Section 1. Most portable wattmeters are designed for a maximum current of 5 A, and are used with current transformers. If two such wattmeters with transformers are connected as in Figure 18.14, the algebraic sum of the wattmeter readings multiplied by the transformer ratio will give the total power in a three-phase circuit.

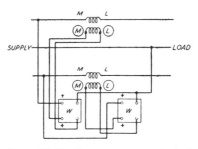

Figure 18.14   Two wattmeter method of measuring three-phase power

In most tests of the power consumption of three-phase motors, one wattmeter will give sufficient accuracy if it is properly connected, because motor loads are very approximately balanced. In this case, the total three-phase power is three times the wattmeter reading multiplied by the transformer ratio. If the supply is 3-wire without a neutral and unbalanced, two current transformers are required. Three-phase supplies usually include a neutral for lighting circuits. If the power installation is supplied from a local substation in the works, the earthed point of the neutral will be, electrically, very near the measuring position. In this condition three-phase power may be measured in balanced circuits by connecting the wattmeter voltage circuit between the line supplying the current coil of the instrument and earth. The earth connection is electrically equivalent to a connection to the supply neutral. When the power installation is some distance from the substation, there may be considerable differences of potential between the neutral of the supply and earth. In this case, the

477

following method is very convenient: Connect one voltage terminal of the wattmeter to the line in the usual way; to the other terminal join a single lead connected to one of the terminals of a lampholder adaptor. By inserting the adaptor into a switched-off lampholder, the voltage circuit can be joined to the supply neutral in the holder.

Power transducer/indicator systems. The advent of the electrical power converting transducer used with a local or remote moving-coil instrument can indicate such parameters as frequency, watts, vars, phase angle, a.c. current and voltage, position, weight, temperature, etc.

Load-independent transducers provide a remote signal facility for indicators, data loggers, computers, control systems, etc.

Neon plasma analogue instruments. The latest in CMOS circuitry has been coupled to a neon plasma display by Crompton Instruments to give an analogue indication of the quantity being measured. It has been applied to a wide range of meters including ammeters, voltmeters, kilowatt meters and angular displays.

Readability is excellent, even in poor ambient lighting conditions and is achieved by arranging a series of neon plasma bars in a strip and illuminating them consecutively so that the resultant column is proportional to the input signal. The column appears as a continuous bar and is clearly visible over a wide viewing angle.

An indicator-controller version is available with high and low adjustable set points. When the input signal moves outside the chosen limits this is clearly shown in the display, and alarm and control equipment can be operated by the output relays.

These instruments are well-suited to most metering applications where accuracy and readability are important, and are ideal where vibration or dust can be a problem or where ambient lighting may be low. Accuracy is Class 1 to BS 89, DIN 43780 and IEC 51.

VALVE VOLTMETERS

These essentially consist of a thermionic valve which has a milliammeter connected in its anode circuit. The voltage which is to be measured is normally applied to its control grid circuit, which imposes very little load on the circuit, even at a high frequency. Although the basic arrangement of a valve voltmeter has a limited range, this can be extended by the use of a potential divider.

As already noted above, the valve voltmeter takes practically no power at all from the source under test, and this factor is an important one in the measurement of voltages in a radio circuit. An ordinary moving coil meter, no matter how sensitive it may be, always draws some power from the circuit under test. In circuits where there is plenty of power available this is not serious, but when dealing with a circuit in which even a load of a few microamperes would seriously affect the accuracy of the reading, a valve voltmeter should be used.

When comparing the load imposed upon an a.c. circuit by a moving-coil volmeter, it is also necessary to consider the frequency since a moving-coil meter is very frequency conscious. By this it is meant that the instrument in the first place is calibrated at a particular frequency and will only measure alternating current accurately at this frequency. An important advantage of the valve voltmeter is that it may be designed to cover practically any frequency; commercially manufactured instruments are suitable for frequencies up to 50 MHz and above.

Secondly, the valve voltmeter will often be used on circuits where the voltage is not very high so that its input impedance can only be compared with a moving-coil meter which has been adjusted to the appropriate range. This means that it is usually impossible to determine accurately a low voltage on the high voltage range of a moving-coil meter. On the other hand, the valve voltmeter maintains a high impedance over all its ranges and is, in fact, the only type of instrument that can be used for low-voltage r.f. measurements.

A further application of the thermionic valve to voltage measurement is to be found in the diode peak voltmeter. When an instrument of this type is conected to an a.c. source, rectification takes place in the diode each half-wave surge charging a capacitor in the output circuit to the peak value of the wave. Provided the voltmeter in the circuit is of sufficiently high resistance and is large enough to prevent a loss of charge through the meter when the rectifier is not passing current, the reading on the meter will indicate the peak voltage, irrespective of the waveform.

It is an important point to note that either a.c. only, d.c. only or a.c. and d.c. can be measured on the various types of valve voltmeters in use today. An a.c. valve voltmeter which has a blocking capacitor incorporated in its input circuit cannot be used for measuring d.c. However, it is not a difficult matter to short-circuit the capacitor when it is necessary for a d.c. measurement to be taken.

Because of the steady potential on the grid in the case of d.c. the calibration will be affected. In this case a resistance is sometimes inserted between the d.c. input terminal and the range potential divider to compensate for the rise in sensitivity so that the calibration of the meter is similar for both a.c. and d.c. measurements.

SHUNTS AND SERIES RESISTANCES

The control or reduction of the actual currents flowing in the various circuits of an instrument or meter may be obtained either by means of resistances (for both d.c. and a.c.) or by the use of instrument transformers (a.c. only).

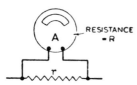

Figure 18.15   Ammeter shunt

Shunts.   Non-inductive resistances for increasing the range of ammeters are termed shunts and are connected as shown in Figure 18.15. The relative values to give any required result can be calculated as follows:

Let R = resistance of ammeter
      r = resistance of shunt
      I = total current in circuit
      i_a = current in ammeter
      i_s = current in shunt

We have $I = i_a + i_s$, and as the volt-drop across the meter and the shunt are the same we get

$$i_a R = i_s r$$

$$\therefore i_a = i_s \frac{r}{R} = I \frac{r}{R + r}$$

or

$$I = i_a \frac{R + r}{r}$$

$$= i_a \frac{R}{r} + 1$$

The expression $\frac{R}{r} + 1$ is termed the *multiplying power* of the shunt.

As an example, take the case of a meter reading up to 5 A having a resistance of 0.02 ohm. Find a suitable shunt for use on circuits up to 100 A.

As the total current has to be $\frac{100}{5} = 20$ times that through the meter, the shunt must carry 19 times. Thus its resistance must be $\frac{1}{19}$ that of the meter, giving a shunt whose resistance is $\frac{0.02}{19}$ or 0.00105 ohm.

Series resistances. In the case of voltmeters it is often impracticable to allow the whole voltage to be taken to the coil or coils of the instrument. For instance, in the case of a moving-coil voltmeter the current in the moving coil will be in the nature of 0.01 A and its resistance will probably be only 100 ohms. If this instrument is to be used for, say, 100 V, a series resistance will be essential – connected as shown in Figure 18.16.

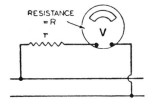

Figure 18.16 Series resistance for voltmeter

The relation between total voltage and that of the meter is simple as the two voltages are in series. Thus in the above case the full-scale current will be 0.01 A, so that the total resistance for 100 V will have to be 100/0.01 or 10000 ohms. If the resistance of

481

the meter circuit and its connecting leads is 100 ohms, then the added or series resistance must be $10000 - 100 = 9900$ ohms.

The multiplying power is the ratio between the total voltage across the instrument and that across the coil only. If

$$R = \text{resistance of meter}$$
$$r = \text{value of series resistance}$$

then multiplying power $= \dfrac{R + r}{R}$

Construction of shunts and series resistances. Shunts usually consist of manganin strips soldered to two terminal blocks at each end and arranged so that air will circulate between the strips for cooling.

As with an ammeter shunt, the impedance of the series resistance used in a voltmeter must remain constant for differing frequencies, i.e. the inductance must not vary. For this reason the resistance-coils (of manganin) are often wound upon flat mica strips to reduce the area enclosed by the wire, and hence reduce the enclosed flux for a given current.

Voltage dividers. Voltage dividers or volt-boxes can only be used with accuracy with testing equipment or measuring instruments which do not take any current, or with electrostatic instruments. The former state of affairs refers to tests using the 'null' or zero deflection for balancing or taking a reading while the electrostatic meter actually does not take any current.

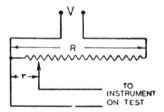

Figure 18.17   Voltage divider

The principle of the voltage divider is seen in Figure 18.17. The voltage to be measured is connected across a resistance R and the connections are taken off some fraction of this resistance as shown. The ratio is given by R/r.

A similar arrangement is possible with capacitors for use with electrostatic instruments, but the method is somewhat complicated owing to the variation in capacity of the instrument as the vane or vanes move.

ENERGY METERS

Use of d.c. meters in the UK is very limited and demands in overseas markets are diminishing. They are therefore not described.

Induction meters. The popular single-phase induction meter consists fundamentally of two electromagnets, an aluminium disc, a revolution counter, and a brake magnet, connected as shown in Figure 18.18.

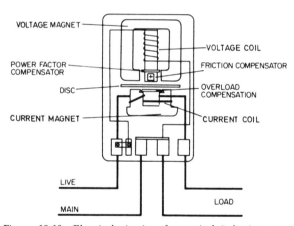

Figure 18.18   Electrical circuits of a typical induction meter connection to MMLL to BS 5685 (Ferranti Ltd)

In the Ferranti type F2Q–100 meter the voltage element is positioned above the rotor disc and comprises a three limb core with the coil mounted on the centre limb which also supports the friction compensating device. The power factor compensator

483

consisting of a copper loop encircles the centre limb and is attached to a vertically moving rack carried on the main frame.

The current element located beneath the disc has a U-shaped core with a coil supported on an insulated former. A magnetic shunt between the core and limbs compensates for the effect of current flux braking at high loads.

.The braking system consists of two anisotropic magnets located above and below the disc respectively, and enclosed in a die-cast housing. Fine adjustment is provided by a steel screw which shunts part of the magnet flux. Coarse adjustment is achieved by sliding the whole magnet system sideways. The rotor disc is supported by a magnetic syspension system.

Voltage and current magnets create two alternating magnetic fluxes which are proportional to the supply voltage and currents being measured respectively. Each of these fluxes set up eddy currents in the disc; due to the interaction of the voltage flux with the eddy currents created by the current flux, and to the interaction of the current flux with the eddy currents created by the voltage flux, a torque is exerted on the rotor disc causing it to rotate. The permanent magnet system acts as an eddy current brake on the motion of the rotor disc and makes the speed proportional to the power being measured.

When installed in a consumer's premises the voltage coil of the meter is always on circuit. To prevent the disc from rotating when the consumer's load is switched off, the disc is provided with an 'anti-creep' device. In the meter described protection against creep is provided by cutting two small radial slots in the periphery of the disc diametrically opposite to each other. This is only one of a number of ways in which compensating devices can be arranged to provide similar results. Although the construction of a single-phase induction meter is not complex it can be calibrated to a high level of accuracy which it will sustain for many years.

In metering two- or three-phase systems, two or three elements are arranged to drive a single rotor system.

TESTING OF METERS

It is not practicable to test meters under working conditions and it is therefore customary to use 'phantom loading' for test purposes.

Single-phase meters are tested on transformer circuits such as that shown in Figure 18.19. The current and voltage are supplied from separate sources, enabling the current to be used at a low voltage with consequent saving in energy. It also allows the test equipment design to be such as to permit variation in the phase angle between current and voltage so as to create conditions of inductive loading required in the calibration of meters. Polyphase test set circuits are more complex as they must provide facilities for testing of two or three elements both individually and collectively.

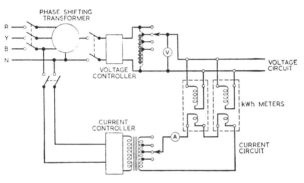

Figure 18.19   Simplified diagram for single-phase meter testing

Testing requirements for meters are given in detail in BS 5685, but in the UK the provisions of the Electricity Supply (Meters) Act 1936 govern the testing of all types of meters installed in all domestic premises and the majority of industrial premises. Appendix B of this Act sets out the methods to be used in testing meters and the actual tests to be employed by the meter examiner prior to certification of a meter.

A summary of the three methods is given below:

Method A

Long-period dial tests using sub-standard rotating meters. The load on any sub-standard rotating meter to be not less than one-quarter of or more than one and one-quarter times its full load. The duration of any such test to correspond to not less than five complete revolutions of the pointer to the last dial of the meter under test.

Method B

(*a*) *Tests (other than long-period dial tests) using sub-standard rotating meters.* The load on any sub-standard rotating meter to be not less than one-quarter of or more than one and one-quarter times its full load. The duration of any such test to correspond to the number of complete revolutions of the disc of the meter under test asertained as follows, namely, by multiplying 40 by the percentage of the marked current of the meter at which the test is being made and dividing by 100; provided that the duration of any test shall correspond to not less than five complete revolutions and need not be greater than a period corresponding to 25 complete revolutions.

Table 18.1   Tests on single-phase two wire whole current meters

	Load ranges for the four meter categories in terms of marked current			
Testing load	Basic maximum	Maximum continuous rating	Long range	Short range
High	50% max to 100% max	50% to 100% ($\frac{1}{2}$–1/1)	100%–200% (1/1–2/1)	100%–125% 1/1–1$\frac{1}{4}$)
Intermediate	One intermediate load to be taken at any part of the curve between the high and low loads actually taken			
Low	5% basic to 10% basic	1.67%–3.33% ($\frac{1}{60}$th – $\frac{1}{30}$th)	5%–10% ($\frac{1}{20}$th – $\frac{1}{10}$th)	5%–10% ($\frac{1}{20}$th – $\frac{1}{10}$th)

Provided such tests are carried out in accordance with Method B or Method C the prescribed long period dial test shall be at one of the loads specified in Table 18.1 and shall be in addition to such tests.

(*b*) *One long-period dial test.* To be made in accordance with and of duration not less than that prescribed under Method A.

Method C

(*a*) *Tests by sub-standard indicating instruments and stop watch.* The load on any sub-standard indicating instrument to be such as to give a reading of not less than 40% of its full scale reading. The duration of any such test to correspond to not less than three complete revolutions of the disc of the meter under test, or not less than 100 seconds, whichever be the longer period.

(*b*) *One long-period dial test.* To be made in accordance with and of a duration not less than that prescribed under Method A. (Method C alone to be used for testing d.c. motor meters).

Actual Tests. 1. Every motor meter other than a single-phase two-wire alternating whole current pattern to be tested:

(*a*) At 5% of its marked current, or in the case of a d.c. meter having a marked current of less than 100 A at 10% of its marked current;

(*b*) at one intermediate load; and

(*c*) at 100% of its marked current, or in the case of an a.c. meter, at 100 or 125% of its marked current.

Provided that in cases where the tests are carried out in accordance with Method B or Method C, the prescribed long-period dial test shall be at one of the loads specified for the three foregoing tests and shall be additional to such test.

Every single-phase two-wire alternating whole current pattern meter to be tested at high, intermediate and low load at unity power factor; each of the loads to be within the ranges specified in Table 18.1.

Provided such tests are carried out in accordance with Method B or Method C the prescribed long period dial test shall be at one of the loads specified in Table 18.1 and shall be in addition to such tests.

2. Every alternating current meter also to be tested at its marked current and marked voltage at 0.5 power factor (lagging), subject to a tolerance of plus or minus 10% in the power factor.

3. Every watthour meter also to be tested for 'creep' with its main circuit open and with a voltage of 10% in excess of its marked voltage applied to its voltage circuit.

4. (*a*) A three-wire d.c. or single-phase a.c. meter may be tested as a two-wire meter with the current circuits of the two elements in series and the voltage circuits in parallel.

(*b*) Every three-wire d.c. meter to be tested for balance of the elements (in the case of watthour meters with the voltage circuits energised at the marked voltage):

(i) with the marked current flowing in either of the two current circuits of the meter and no current flowing in the other current circuit; and

(ii) with half the marked current flowing in both current circuits.

Tests on polyphase meters. (*a*) Every polyphase meter to be tested as such on a circuit of the type for which the meter was designed.

487

(*b*) Every polyphase meter to be tested for balance of the elements with the voltage circuits of all the elements energised at the marked voltage with the marked current flowing in the current circuit of one element and no current flowing in the other circuit or circuits; such test to be carried out on each of the elements at unity power factor and also at 0.5 power factor (lagging) subject in the latter case to a tolerance of plus or minus 10% in power factor.

Meters operated in conjunction with transformer or shunts. Every meter intended for use with a transformer or shunt to be tested with such transformer or shunt and with the connecting leads between the meter and such transformer or shunt.

5. *Multi-rate meters.* All multi-rate meters to be tested in accordance with the foregoing paragraphs, on the rate indicated when the relay is energised. In addition with the register operating in the de-energised position, and where applicable on the high rate, a further test to be taken at low load with a Method A test at high load. For polyphase meters these additional tests shall apply only to polyphase balanced load conditions.

19

Electric welding

There are three main types of welding, namely arc welding, resistance welding and radiation welding. These can again be divided into groups as follows.

(a) Flux-shielded arc welding: This method comprises four forms:
   (1) Manual metal arc;
   (2) Automatic submerged arc;
   (3) Electro-slag;
   (4) Electro gas.
(b) Gas-shielded arc welding. There are three principal types:
   (1) Inert gas tungsten arc (TIG);
   (2) Plasma arc;
   (3) Gas metal arc.
(c) Unshielded and short-time arc processes: Two main types are:
   (1) Percussion;
   (2) Arc stud welding.
(d) Resistance welding. The four main types are:
   (1) Spot;
   (2) Seam;
   (3) Projection;
   (4) Flash.
(e) Radiation welding. Three methods fall into this category:
   (1) Arc image;
   (2) Laser beam;
   (3) Electron beam.

FLUX SHIELDED ARC WELDING

In this important series of welding processes the electric arc supplies the heat for fusion while a flux is responsible for the shielding and cleaning functions, and often also for metallurgical control. The most widely used form is a manual process known as manual metal arc welding. In this method electrodes, in the form of short lengths of flux-covered rods are held by hand. The flux on the electrode is of considerable thickness and is applied by extrusion.

The covering has other functions to perform besides that of a flux. It stabilises the arc, provides a gas and a flux layer to protect the arc and metal from atmospheric contamination, controls weld-metal reactions and permits alloying elements to be added to the weld metal. Finally, the slag left behind on the surface of the weld should assist the formation of a weld bead of the proper shape.

Metal arc welding electrodes are made with core wire diameter from 1.2 mm up to 9.5 mm. For all but exceptional circumstances the useful range is 2.5 mm up 6 mm. The length of an electrode depends on the diameter; for smaller diameters, where the manipulation of the electrode calls for the greatest control, the length may be only about 300 mm. They are usually about 450 mm long and are consumed at a burn-off rate of 200–250 mm/min. The core wire is made of mild steel.

The power source which supplies the current for the arc may be either a.c. or d.c. Regardless of the source means must be provided for controlling the current to the arc. This may be by means of a regulator across the field of a d.c. motor generator or a choke in the output side of a mains fed transformer.

Automatic submerged arc. In manual metal arc welding with covered electrodes, the current must pass through the core wire, and there is a limit to the length of electrode which can be used because of the resistance heating effect of the current itself. In the submerged arc process a bare wire is employed and the flux added in the form of powder which covers completely the weld pool and the end of the electrode wire. Extremely high welding currents can be used, giving deep penetration.

Submerged arc welding can either use a single electrode or multiple electrodes. In parallel electrode welding two electrodes are usually spaced between 6 mm and 12 mm apart and are connected to the same power source. If the electrodes are used in tandem, one following the other, an increase in welding speed of up to 50% is possible. When used side by side wider grooves can be filled. For multiple electrode welding an a.c. power source is preferred because multiple d.c. arcs tend to pull together.

Another solution to the problem of introducing current to a continuous fluxed wire is to put the flux into tubular steel electrodes drawn from strip or enclosed in a folded metal strip. A method which overcomes the disadvantage of the bulk of the cored electrode wire and the lack of visibility of the submerged arc process uses a magnetisable flux. This is in a stream of carbon dioxide which emerges through a nozzle surroundings the bare electrode wire. Because of the magnetic field surrounding the electrode wire when current is passing the finely powdered flux is attracted to and adheres to the wire, forming a coating.

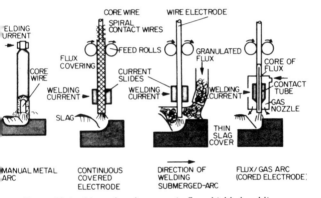

Figure 19.1   Manual and automatic flux-shielded welding

Electro-slag welding.   As the thickness of the metal welded increases, multi-run techniques become uneconomical. The use of automatic welding, however, with high current large passes in the flat position can give a weld pool so large that it runs ahead of the electrode out of control resulting in inadequate fusion. This is overcome by turning the plates into the vertical position and arranging a gap between them so that the welding process becomes rather like continuous casting. Weld pools of almost any size can be handled in this way provided that there is sufficient energy input and some form of water-cooled dam to close the gap between the plates and prevent the weld pool and the slag running away. This is the essential feature of electro-slag welding. Figure 19.2 shows the basic arrangement for the most common, wire electrode type.

For thick plates, i.e. over 100 mm, it is necessary to traverse the electrode so that uniform fusion of the joint faces through the thickness of the point is obtained. For thick materials, more than one electrode wire may be employed.

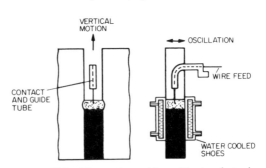

Figure 19.2   Electro-slag welding with wire electrode

There are a number of variations of electro-slag welding mainly to do with the shape of the electrode. One method, which uses a consumable guide, is much simpler than the wire electrode in that the welding head and wire feed mechanism do not need to be moved up the joint as the weld is made. Another type is that where plate or bar electrodes are suspended in the joint gap and lowered slowly as they are melted away by the slag bath. A three-phase power supply can be used with three plate electrodes connected in star fashion.

Electro-gas welding.   In this form of welding heat is generated by an electric arc which is struck from a flux-cored electrode to the molten weld pool. The flux from a flux-cored electrode forms a thin protective layer but does not give a deep slag bath as in electro-slag welding.

Additional shielding of the weld pool is provided by carbon dioxide or argon-rich gas which is fed over the area through the top of each copper shoe. Mechanically the equipment is similar to that for the wire electrode type of electro-slag welding.

GAS-SHIELDED ARC WELDING

Essentially gas shielded arc welding employs a gaseous shielding medium to protect both the electric arc and the weld metal from contamination by the atmosphere.

Inert gas tungsten arc welding.   Tungsten electrodes are used for this form of welding as being the least eroded in service. Other refractory metals have been tried without success. The rate of erosion of tungsten electrodes is so small as to enable them to be considered non-combustible. The gas which surrounds the arc and weldpool also protects the electrode. Argon or helium gas is used in preference to hydrogen because it raises the arc voltage and requires a high open-circuit voltage. It can also be absorbed by some metals affecting their characteristics.

It is usual to make the electrode negative thus raising the permissible current to eight times that with a positive electrode system. The chief advantage of having the d.c. electrode positive is the cleaning action exerted by the arc on the work.

Where this cleaning action is required and the current is over 100 A, an a.c. power supply must be used. With a.c. however the reversals of voltage and current introduce the problem of reignition twice in every cycle. When the electrode is negative there is no problem but when the electrode becomes positive the restrike is not automatic. Three ways are available to assist reignition of the arc under these conditions.

The first is with a well-designed transformer with low electrical inertia that will provide the high voltage required to cause the arc to restrike. The second is by means of a high-frequency spark unit. This may be switched off by a relay once welding begins, or alternatively it may be operated from the open-circuit voltage so that it ceases to operate when the voltage drops to that of the arc. Both methods suffer from the disadvantage that the open circuit voltage tends to be high and the power factor has to be low because this high voltage must be available at current zero.

In the third method of arc maintenance a voltage surge is injected into the power circuit to supply the reignition peak. This is done by discharging a capacitor through a switch which is tripped automatically by the power circuit, see Figure 19.3. When the arc is extinguished at the end of the negative half cycle the reignition peak begins to develop and itself fires a gas discharge valve which discharges the capacitor. Welding can be carried out at less than 50 V rms as opposed to 100 V by the first two methods.

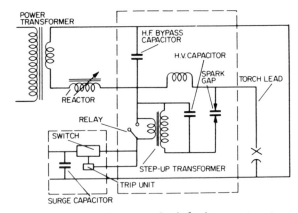

Figure 19.3 Arc maintenance circuit for inert gas tungsten arc welding. A voltage surge is injected into the power circuit to supply the reignition peak

There are several variations of the tungsten arc method for special purposes. The pulsed tungsten arc is suitable for welding thin sheet material using a d.c. power supply and a modulated current wave. The twin tungsten arc provides a wide weld bead and is used, for example, in the manufacture of cable sheathing from strip. Tungsten arc spot welding employs the same equipment as for conventional manual tungsten arc welding except that the control system includes a timing device and the torch is modified. The process is used to join overlapping sheets of equal or unequal thicknesses.

Plasma arc welding. A plasma arc heat source may be considered as a development of the inert tungsten arc. There are two types, the non-transferred and the transferred or constricted arc. If the inert gas tungsten arc torch is provided with a separately insulated water-cooled nozzle that forms a chamber round the electrode and the arc is struck from the electrode to this chamber, the arc plasma is expelled from the nozzle in the form of a flame. This is the non-transferred arc, which with a powder feed into the plasma, is used for metal spraying.

When the arc is struck from the electrode to the workpiece the arc is contracted as it passes through the orifice in the nozzle. This is the transferred plasma arc and it is used for cutting because of the high energy density and velocity of the plasma. To start an arc of this kind requires a pilot arc from the electrode to the nozzle. The transferred arc can also be used for welding: at low currents for welding sheet metal less than 1.5 mm thick and at currents up to 400 A for welding thick metal. When using the plasma arc for welding additional inert-gas shielding is required.

Gas metal arc welding. The electrode used in this form of welding is of the consumable type and a d.c. supply is employed. The electrode end melts and molten particles are detached and transported across the arc to the work by magneto-dynamic forces and gaseous streams. Size and frequency of the droplet transfer is related to the wire size and current; voltage has only a limited effect, except in one case. The best performance is obtained with small diameter wires and high current densities.

UNSHIELDED AND SHORT-TIME PROCESSES

Use of a flux or gas shield in welding is to exclude air from contact with the arc and weldpool and so prevent contamination of the weld. In unshielded welding methods, correction of the consequences of atmospheric contamination is made by adding deoxidisers such as silicon, manganese, aluminium or cerium. The welding method is the same as for gas metal arc except that no gas is employed.

Two short time and discontinuous unshielded arc processes are percussion welding and stud welding. Shielding is not required because either the whole process is carried out so rapidly that contamination is negligible or the molten metal is squeezed out of the joint. A common feature of the two processes is that they permit the joining in a single operation of parts with small cross-sectional areas to other similar parts or more massive pieces.

Percussion welding appears in several forms but in all a short-time high-intensity arc is formed by the sudden release of energy stored generally, but not always, in capacitors. Subsequent very rapid or 'percussive' impacting of the workpieces forms the weld. Three ways in which the arc is initiated are:

1. Low voltage with drawn arc;
2. High-voltage breakdown;
3. Ionisation by a fusing tip.

With each method the energy source may be a bank of capacitors, which is charged by a variable voltage transformer/rectifier unit.

Stud welding. One of the most useful developments in electric welding is stud welding. This method of fixing studs into metal components offers several advantages over the older method which entailed drilling and tapping operations to permit the studs to be screwed into position. By obviating drilling and tapping, stud welding simplifies the design of products, structures and fittings, giving a better appearance, improved construction and economy of materials.

There are two methods of stud welding, drawn-arc and capacitor discharge – both may be described as modified forms of arc welding. In both cases a special welding gun is required and also an automatic head or controller for timing the duration of the welding current.

The principle employed in the drawn-arc process is that the lower end of the stud is first placed in contact with the base metal

Figure 19.4   A Cromp-Arc TC12 electronic welding system being used to install a new anti-skid surface at a ferry terminal (Crompton Parkinson)

on which it is to be welded. A small current is then passed through the stud to burn away any scale and establish an ionised path before the full welding current is allowed to flow through the gun; a solenoid in the gun raises the stud slightly to form an arc. After a small interval the current is automatically cut off and a spring in the gun drives the stud into the pool of molten metal which has been formed by the arc.

With the capacitor-discharge method the stud has a small pip or spigot at its lower end. In the welding operation this spigot is placed in contact with the base. When the current is switched on the spigot is fused to create the arc and the stud is pressed into the molten metal.

To provide the necessary welding current for the drawn-arc process a d.c. supply from an arc-welding generator or from a static rectifier is required. Equipments of the transformer type using alternating current in the weld have also been developed. An automatic controller is included in stud welding equipment for timing the weld, the operation cycle being pre-set by the operator. Figure 19.4. shows a Cromp-Arc stud welding equipment being used to install a new anti-skid surface at a ferry terminal.

RESISTANCE WELDING

All welding methods described in the preceding pages employ an arc to produce heat in the metal to be welded. Another way of producing heat in the metal, also using current, is to pass the current through the workpiece. This method has the advantage that heat is produced throughout the entire section of the metal. The current may be introduced into the material either by electrodes making contact with the metal or by induction using a fluctuating magnetic field surrounding the metal. All resistance welding methods require physical contact between the current-carrying electrodes and the parts to be joined. Pressure is also required to place the parts in contact and consolidate the joint.

Resistance spot welding. In this process, overlapping sheets are jointed by local fusion caused by the concentration of current between cyclindrical electrodes. Figure 19.5 shows the method diagrammatically. The work is clamped between the electrodes by pressure applied through levers or by pneumatically operated pistons. On small welding machines springs may be used. Current is generally applied through a step-down transformer. The electrodes and arms of the machine usually form part of the secondary circuit of the transformer. A spot-welded joint consists of one or more discreet fused areas or spots between the workpieces.

Welding current can be controlled either by changing the turns ratio of the transformer or by phase shift means. For the former method with a single-phase supply the welding current value can be altered by changing the tapping ratio of the primary side of the transformer.

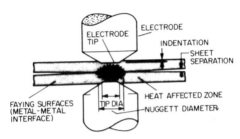

Figure 19.5 Features of the resistance spot weld process

In phase shift control the transformer primary is supplied through ignitrons which act simply as high-speed switches. At the beginning of each half cycle of primary current the ignitron is made non-conducting, but later in the half cycle it is fired. Current flows until the next current zero. Thus only a portion of the available power in the half cycle is passed through the workpiece. The ignitrons are connected in inverse-parallel or back-to-back configuration, and can be used for both single-phase and three-phase working.

The time the current is flowing is important and can be controlled either by mechanical or electronic means. Electronic timers work without regard to the phase of the welding current and because an exact number of cycles cannot be guaranteed they are not used for times less than 10 cycles.

Resistance seam welding. Two methods of continuous seam welding are available. Using spot welding equipment a series of overlapping spots can be made, a process known as stitch welding.

498

Alternatively the electrodes may be replaced by wheels or rollers so that work may be moved through the welder continuously without the necessity for raising and lowering the head between welds. Rollers are power driven and may or may not be stopped while individual welds are made. Current is generally passed intermittently while the electrodes are stationary, although continuous current is also used to a limited extent. A common technique is known as step by step seam welding, because while each weld is being made rotation is stopped and the current is then switched off while the rollers move to the next position. The amount of overlap between spots depends on peripheral roller speed and the ratio of current on-and-off time. A normal overlap would be 25–50%.

Adjustment of timing can be made to produce not a continuous seam but a series of individual welds. When this is done the process is called roller-spot welding.

Generally seam welds are made on overlapped sheets to give a lap joint but sometimes a butt joint is required. There are several methods of obtaining this and space does not allow a full description of them. Methods include mash-seam welding, foil butt-seam welding, resistance butt-seam welding, high frequency resistance welding and high frequency induction welding.

Projection welding. In this process current concentration is achieved by shaping the workpiece so that when the two halves are bought together in the welding 'machine, current flows through limited points of contact. With lap joints in sheet a projection is raised in one sheet through which the current flows to cause local heating and collapse of the projection. Both the projection and the metal on the other side of the joint with which it makes contact are fused so that a localised weld is formed. The shaped electrodes used in spot welding can be replaced by flat-surfaced platens which give support so that there is no deflection except at the projection.

The method is not limited to sheet-sheet joints and any two mild steel surfaces that can be brought together to give line or point contact can be projection welded. This process is used extensively for making attachments to sheet and pressings and for joining small solid components to forging or machined parts.

Flash welding. Flash welding is a development from resistance butt welding and it uses similar equipment. This comprises one fixed and one movable clamp, so that workpieces may be gripped and forced together; a heavy duty single-phase transformer having a single turn secondary; and equipment for controlling the current. The parts to be joined are gripped by the clamps and brought into contact to complete the secondary circuit, Figure 19.6. When the welding voltage of up to about 10 V is applied at the clamps current flows through the initial points of contact causing them to melt. These molten bridges are then ruptured and small short-lived arcs are formed. The platen on which the movable clamp is mounted moves forward while this takes place and fresh contacts are then made elsewhere so that the cycle of events is then repeated. This intermittent process continues until the surfaces to be joined are uniformly heated or molten.

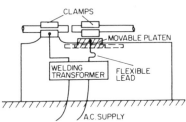

Figure 19.6   Basic arrangement for flash welding

By this time the moving platen will have advanced, at an increasing rate, to close the gap as metal is expelled. The total distance up to the point of upset is known as the 'flashing allowance'. At this point the rate of movement of the platen is rapidly increased and a high force applied to forge the parts together and expel the molten metal on the surfaces. The expelled metal forms a ragged fin or flash round the joint. Ideally all the contaminated metal produced during flashing should be removed in this way to produce a high quality joint having many features of a solid-phase weld.

RADIATION WELDING

A small group of processes employ energy for welding in the form of radiation. Two processes use energy as electro-magnetic radiation while another uses an electron beam. Radiation welding is

500

unique in that the energy for welding can be focussed on the workpiece, heat energy being generated only where the beam strikes.

Unlike arc or flame sources the work is not brought into contact with any heated media, gas or metal vapour, and the process may be carried out in a vacuum or low pressure system for cleanliness. In contrast to arc welding the melted pool is subjected to only negligible pressure.

Arc image welding. Fusion is accomplished by focussing the image of a high temperature source on the workpiece. Mirrors are used for this purpose. High pressure plasma arc sources have been developed as the heat source and outputs above 10 kW have been used for welding and brazing. Optical systems with top surface mirrors of high accuracy are necessary for focussing, but even so losses occur by dispersion because the source is finite and emits in all directions. Although the method has no future for terrestrial welding it may eventually be employed in space using the sun as the heat source.

Laser welding. A laser is a device which, when irradiated by light from an intense source is capable of amplifying radiation in certain wavebands and emitting this as a coherent parallel beam in which all waves are in phase. This beam can be focussed through lenses to produce spots in which the energy density is equalled only by the electron beam. The word laser stands for light amplification by stimulated emission of radiation.

The laser has been used for micro-spot welding and this appears to be its most promising application, particularly where minimum spread of heat is required.

Electron beam welding. Electron beam welding is a process involving melting in which the energy is supplied by the impact of a focussed beam of electrons. The essential components of a welding equipment are the electron gun, the focussing and beam control system and the working chamber which operates at low pressure.

There is a variety of types of electron guns but they all work on the same principle. Electrons are produced by a heated filament or cathode and are given direction and acceleration by a high potential between cathode and an anode placed some distance away. In a limited number of electron beam welding plants the anode is the workpiece. In other cases a perforated plate acts as the anode and the beam passes through this to the workpiece behind. The former system is known as 'work-accelerated' and the latter 'self-accelerated'. The self-accelerated type is the one used for welding.

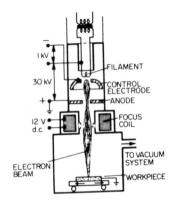

Figure 19.7 Outline of triode electron beam gun with control electrode or bias cup

Magnetic focussing is usually employed this taking place after the beam has passed through the anode, Figure 19.7. In this type the beam current is controlled by a negative bias on an electrode placed round the cathode which is called the control electrode or bias cup. This electrode behaves as a grid in that it is able to control the flow of electrons from the cathode.

Tungsten inert gas (TIG) welding. The Argonarc welding process is a method of joining non-ferrous metals and stainless and high alloy steels without using flux. An electric arc struck between an electrode and the workpiece produces the heat to fuse the material and the electrode and the weld area are shielded from the atmosphere by means of an inert gas (argon). Standard d.c. or a.c. welding equipment may be used provided the open circuit voltage is around 100 V for a.c. and 70 V for d.c.

For welding aluminium, magnesium and their alloys, stainless and high alloy steels, nickel alloys and copper alloys, up to 3 mm thick, a.c. is suitable. D.C. may be used for all other common metals, and it is essential for the welding of copper and stainless steels and alloys over 3 mm thick.

Method of welding.   The argon is supplied to the torch through PVC tubing, where it is directed round the tungsten electrode and over the weld area by either a ceramic or water-cooled copper shield.

The torch is held (or clamped) so that the electrode is at approximately 80° to the weld and travels in a leftward direction. Filler wire may be added to reinforce the weld, the angle being approximately 20° to the seam. It is important that the end of the filler wire should be pressed on to the workpiece and be fed into the front edge of the weld pool.

When comparatively long straight seams are to be welded, e.g. joining sheets, strip, or fabricating cylindrical vessels, argonarc machine welding may be used, Much of this type of work is done without using filler wire, i.e. the butted edges are simply fused together. Where maximum strength or a flush surface is require or when material over 3 mm thick is being welded, filler wire may be added.

Argonarc spot welding.   This is a supplementary process to resistance spot welding, and is used to make spot welds in positions inaccessible to resistance spot welders.

In principle, a local fusion takes place between the sheets to be joined. The torch, which is hand-controlled, contains a tungsten electrode recessed within a water-cooled shield. Argon gas is used to shield the electrode and weld area.

20

Battery electric vehicles

The expansion of battery powered vehicles into the industrial sector started soon after the second world war with the development of fork-lift and platform trucks. Diversification followed fast and today there is a wide variety of special-purpose electric vehicles available. It is not the purpose of this chapter to describe fully every type of battery-electric vehicle in use today. The list includes milk floats, delivery vans, overhead maintenance vehicles, tractors and trucks both driverless and rider-controlled, fork-lift trucks, mini-buses, mobile shops, articulated goods trucks, ambulances and interwork transporters. To this list of industrial vehicles should be added battery-electric motor cycles and more recently electric commuter cars and electrically-assisted bicycles.

Electric cars. These have advantages over their petrol- or diesel-driven counterparts. No noxious fumes are produced, and this is in keeping with the current aim to reduce atmospheric pollution. Less maintenance is required, and less noise is generated during operation. Low running costs are possible because the batteries can be charged overnight using off-peak tariff rates.

The biggest hindrance to the development of the electric commuter car is the non-availability of a battery with a better power/weight ratio than that of the lead-acid battery. With existing lead-acid battery technology the electric car range varies between 50 and 80 km, which is not considered sufficient. The editor of this book disagrees with the experts since statistics prove that such a range is ample for city use. The British Electricity Council bought sixty Enfield 8000 electric commuter cars from Enfield Automotive in 1973 for trials to be carried out by the area electricity boards, thus underlining their faith in this form of transport. Maximum speed of these vehicles is 64 km/h and their range at 96 km is more than that indicated above. The UK supply industry has recently taken delivery of large numbers of 1 tonne vans for fleet use, supplied by Bedford Commercial Vehicles and Freight Rover. The drive trains have been developed by Lucus Chloride Electric Vehicle Systems.

Delivery vehicles. The most popular vehicle in this class is used for the delivery of milk. Three or four-wheeled vehicles are available with capacities up to 2 tonnes. Battery capacity varies from 235 Ah to 441 Ah at the 5 Ah rate. Three-wheeler vehicles are for light duty and are extremely manoeuvrable. Three types of controller are available, resistance, parallel-series system of battery connections and solid state pulse type. The Pulsomatic thyristor controller provides smooth control of speed either forward or reverse. Power is drawn from the battery in short bursts, depending on the firing of the thyristors, and the absence of resistors eliminates the losses associated with resistance control methods.

Figure 20.1   CAV motor in wheel

Figure 20.2   Montgomerie Reid Series 4 forklift truck

The Commando dairy vehicle available from Electricars is
typical of the modern type of this transport. A heavy duty
enclosed, ventilated and protected series wound traction motor is
employed. Control is either by a simple resistance which increases
the voltage applied to the motors in four steps, forward or reverse
or by thyristor. Control contactors have magnetic blowout facilities and they are operated by a footswitch incorporating electroni-
cally controlled time delay. The 300 A plug and locking sockets
completely isolate the traction and auxiliary circuits during charging. A layrub propeller shaft with universal joints provides
transmission. A two-stage drive axle consists of a first stage helical

spur and the second a spiral bevel mounted on taper roller bearings.

The Electricars F85/48TS electric two speed dairy vehicle has been developed for situations where greater range and speed are required. For travel to and from an urban area the high-speed mode can be used, while for urban delivery the speed is halved.

Platform trucks. Fixed or fork-lift type platform trucks are available for pedestrian control or for carrying a driver either standing or seated. Pallet and stillage trucks are designed for the horizontal movement of unit loads on pallets, stillages and skids. They are self-loading and Lansing Linde has models capable of handling loads exceeding 3000 kg. They are particularly useful as auxiliaries to fork-lift or reach trucks.

Two basic types of pallet transporters made by Lansing Linde are:

1. The pedestrian controlled truck employed where duties are intermittent or distances are not excessive.
2. Stand-on transporters allowing the operator to ride on the truck and thereby work over much longer distances.

Stillage and skid handling trucks are also available in both forms.

The Electricars company has a range of driven-type platform trucks with pay loads up to 2000 kg and fixed or elevating platforms. On an elevating model a fully-loaded platform can be raised in 20 secs. Controller can be resistance or thyristor system depending on the model.

Forklift trucks. These are available in 3-wheel or 4-wheel rider-operated counterbalanced reach trucks. Montgomerie Reid's Series 3 range is capable of handling payloads from 1000–1500 kg, while the Series 4, four-wheel model will lift loads from 1700–2750 kg (Figure 20.2). The largest is the MR4.27, which has a maximum lifting height of 7.77 m. Forward and backward tilts of the forks of 3 and 10 degrees respectively are provided to assist picking up and dropping loads. A 72 V battery has a capacity of 431 Ah at the 5 h rate.

The Montgomerie Reid 25P counterbalanced 3-wheel pedestrian-controlled forklift truck has a capacity of 1200 kg at 3.3 m lifting height. High transmission efficiency is obtained by using a motor-in-wheel unit, and a diode contactor control system makes the best use of the battery power. Capacity of the 24 V battery at the 5 h rate is 324 Ah. Gradeability on a continuous rating is 2% laden and 5% unladen, rising to 8% and 12% respectively on a 5-minute rating.

Pedestrian-controlled, stand-on operated or seated-rider types of forklift trucks are designed for working in narrow aisles and are capable of high lifts. The FAER5.1 Turret truck made by Lansing Linde is a narrow aisle model fitted with a rotating turret and reach mechanism which can turn through 180 degrees, enabling it to pick up and deposit loads on either side of the operating aisle. It is available in three capacities, 1000 kg, 1500 kg and 2000 kg, with lifting heights up to 9 metres. A solid-state electronic control system enables the truck to be accelerated from standstill to full speed in a smooth manner. The twin accelerator pedals, one for each direction, also control the truck's electrical braking system.

To speed operation times and increase operating safety a rail-guided steering system within the racking is recommended. The truck is steered by side rollers working in rails fixed to either side of the racks, enabling the operator to release the steering wheel and concentrate on the lifting and stacking operations. This type of truck is ideal for large warehouses and stores where space utilisation and fast access to goods are of major importance.

Lansing Linde was among the first to supply a forklift truck with an integral motor/wheel arrangement. Another, provided by CAV, is available in two sizes and is designed for mounting directly onto the vehicle frame, eliminating the need for a transmission system. The d.c. motor is built inside the hub of the wheel unit and for the 330 mm diameter wheel is driven by a 24 V battery, and for the 381 mm diameter wheel a 36 V battery. The wheel can also be used in reach trucks, tow tractors and other battery electric vehicles, see Figure 20.1.

Tow tractors. Tractors or tugs do not themselves carry a load unless working as an articulated unit. Their purpose is to pull trailers or trolleys, either as single units or coupled together to form trains. The tractive resistance of the trailers can differ considerably depending in the type of trailer, its wheels and the road surface. Most manufacturers provide information on the speed attainable for given drawbar pulls. For the Electricars Model B2000, the speed with a trailed load of 2000 kg is 8.8 km/h.

Aisle width depends on the number of trailers making up a train. Typically for one trailer the aisle width should be about 2.2 m and for four trailers 3.1 m.

Some three-wheel models are obtainable having a single front wheel drive to give, it is claimed, increased manoeuvrability. Generally four wheel tractors are favoured. Ranges vary according to battery capacity.

Motor operating voltages range from 24 V to 72 V and ratings extend up to 12 kW. Series motors are generally employed and control may be one of the three types mentioned for other battery electric vehicles. Drive may be a direct coupled propeller shaft and universal joint to a double reduction heavy duty automotive type rear axle, or by a duplex primary chain with a final gear reduction to the drive wheel.

The UK industry's trade association is the Electric Vehicle Association based at 13, Golden Square, Piccadilly, London, W1R 3AG.

21

Battery systems

Lead-acid batteries

Operating principles. The active components of lead-acid cells are the lead dioxide, PbO_2, of the positive plates (chocolate brown), the spongy lead, Pb, of the negative plates (grey) and the sulphuric acid, H_2SO_4, of the electrolyte. During discharge there is a partial reduction of the positive-plate material with oxidation of the negative-plate material and combination of the product in each case with sulphuric acid. The result is a transformation of part of the material of both plate into lead sulphate, $PbSO_4$, accompanied by a lowering in concentration of the electrolyte. During recharge this process is reversed and restores the original system.

The fundamental chemical action is represented by:

Charged   *Discharged*

$$PbO_2 + 2H_2SO_4 + Pb \Longrightarrow PbSO_4 + 2H_2O + PbSO_4$$

Positive   Electro-   Negative   Positive   Diluted   Negative
          lyte

Large stationary batteries employ Planté positive plates which are of the lamelle type. They are made of chemical lead not less than 99.97% pure, which complies with the impurity levels of BS 334, Type A except that the maximum impurity levels of bismuth, iron, silver and tellurium are, in percentages respectively, 0.030, 0.001, 0.005 and 0.0005. No cadmium, sulphur and zinc are allowed. The parameters of the plates themselves are outlined in BS 6290 Part 2.

Negative plates are of the open grid type, pasted to sustain the positive plate performance as laid down in the BS quoted. Intermediate negative plates are required to have a greater capacity than their associated positive plates and their dimensions shall not be less.

Modern Planté battery containers are made of a transparent material like styrene acrylonitrile. Positive plates are hung from ledges moulded in the container and the negative plates are supported from the bottom of the container. The mass of both plate groups are transmitted vertically on to a flat base. Terminal pillars do not carry any part of the mass of either plate group. Separators between the plates are made from a microporous material designed to cover the facing surfaces of all the plates and to overlap the edges of the positive plates by an amount laid down in BS 6290 Part 2.

Rated capacity of all stationary cells and batteries is the 3 h rate (C_3) which is the ampere-hour capacity calculated from the current which when drawn from a fully-charged battery, discharges it in 3 h, i.e. $C_3 = 3I$ where I is the discharge current.

RE technology. At the end of the charging cycle of a battery oxygen and hydrogen are evolved and in traditional batteries these gases are exhausted to the atmosphere. This represents a loss of water from the electrolyte making it necessary to periodically top up' the cells with distilled water. Recently, battery manufacturers have been called upon to prevent this 'loss' and so produce what is sometimes termed a maintenance free battery. Incidentally this is not strictly true, batteries still need maintenance of connections, etc. The process whereby these gases are retained within the battery is called 'Recombination Electrolyte' technology.

Commonly RE technology is termed gas recombination and, although the process is not new, its application commercially has only just come about. It has needed the development of special microporous glass fibre separator materials and lead-calcium-tin composite plates. Coupled with this is the need to have a correct balance between the active materials in a cell; the volume of the electrolyte is critical. In order to effect gas recombination the cells must also work at a pressure slightly above ambient. A non-return valve is fitted as a safety precaution to prevent a build up of pressure to a dangerous level.

Other types of batteries, e.g. car batteries, employ pasted plates. These are developed from Faure's plates and the active material is held in lead-calcium-tin alloy grids. The grids are pasted with lead oxide-sulphuric acid pastes which are afterwards converted electrolytically to the required active materials. Plate thicknesses depend upon the work for which they are designed and vary from about 1.5 mm for aeroplane batteries to about 2 mm for car-starting and about 5 mm for starting diesel locomotive engines.

Because of the high currents required for starting engines the plates are arranged as close as possible to each other to keep the internal resistance to a minimum. This is achieved by using very thin diaphragm separators made from microporous PVC or rubber, occasionally with glass wool mats. Such batteries are assembled in ebonite or hard rubber containers and more recently polypropylene is being widely employed for car batteries.

Cells may be arranged in a monobloc unit having multiple cell compartments (for car batteries), or as individual cells housed in wooden crates. Because of the close assembly of the plates the space for the electrolyte is very limited which is consequently of a higher relative density, generally of the order of 1.280.

For motive power duties such as fork lift trucks, milk floats and coal mine locomotives, heavier plates combined with devices to assist in retaining the paste on the positive grid are employed. Cells containing heavy flat plate positives, up to approximately 6 mm thick, usually employ glass wool mats together with microporous PVC or rubber separators of substantial thickness and pasted negatives of up to approximately 5 mm thick in a compact assembly. However the vast majority of these batteries in the UK now use cells containing tubular positive plates up to 8.6 mm thick which are separated from flat pasted negative plates by microporous envelopes. The positive plate consists of a number of antimonial lead conducting spines held in a vertical grid by a top casting of antimonial lead and a bottom bar of either moulded polythene or cast antimonial lead. Over this grid is threaded a series of tubes, either a multi-tubular sleeve of terylene or individual tubes of glass fibre with an outer layer of perforated PVC. The space between the tube (retainer) and spine (conductor) is packed tight with positive material. This type of plate construction gives high mechanical strength and a high degree of porosity for acid penetration. Relative density is in the region of 1.280.

First charge.   Manufacturer's instructions must be adhered to when charging batteries, and it is only possible to consider the important points to be observed. Very pure acid must be used in this connection.

Dry and uncharged batteries have to be given a first charge in which the strength of acid for filling, time of soaking before commencing charge, charging rate and temperature on charge are always carefully specified. A faulty cell may show up in the first charge by its backwardness in reaching the gassing stage and by overheating. The fully charged readings and strength of electrolyte for final adjustment vary with different types, but first charge is never complete until the voltage (on charge) is in the region of 2.6–2.7 V per cell and all the plates are gassing freely.

Once the strength of the electrolyte has been adjusted at this stage, no further adjustment should be necessary throughout the life of the cell. Many automotive batteries are dry charged and only require filling with electrolyte of the appropriate relative density before being put into service. Vent plugs should be kept in place during first charging and also recharging and should only be removed for topping up and other maintenance operations.

Installation. When stands are supplied with a lead-acid battery they consist of two or more epoxy-coated steel end frames, together with longitudinal runners and tie bars. Before commencing erection, smear all the screw heads with petroleum jelly. Then follow the manufacturer's instructions regarding the method of erecting the stand. Usually rubber channels are provided upon which the individual cells stand and these should be placed in position and the stand height adjusted so that it is level.

Ensure all the boxes and lids of the cells are thoroughly clean and dry. Also clean the flat contact-making surfaces of the terminal pillars. If there is evidence that acid has been spilled the whole length of the pillars should be wiped down with a rag which has been dipped in a non-caustic alkali solution, preferably dilute ammonia. This will neutralise any acid on these parts. Do not allow any of this solution to get into the cells.

Wipe the pillars clean and dry and cover the whole length of each pillar down to the lid with petroleum jelly, e.g. Vaseline. Do not scrape the surface of any lead-plated copper pillar or connector because you may damage the plating.

For cells in self-contained cabinets installation is simply a matter of placing the cells on the plastic sheets on the shelves according to the makers instructions, starting with the lower shelves to ensure stability of the cabinet. When all the cells are in position the connecting straps may be installed.

Maintenance. Batteries and battery equipment should be kept scrupulously clean, as impurities inside and moisture outside the battery cause rapid self-discharge and deterioration. Vent plugs must be free from obstruction to avoid internal gas pressure and naked lights or sparking near batteries will ignite the evolved-gases, causing violent explosions. Never allow the electrolyte to get so low that either the separators or plate tops are uncovered. Mechanical damage follows the chemical changes that occur under these conditions. Topping up is best done when cells are on recharge or on float charge and only distilled or deionised water must be used. Lower than normal relative densities indicate that the cell or battery is not in a fully charged state. A refreshing charge should be given in these circumstances to ensure a fully charged condition. Low densities in odd cells are an indication of internal short-circuit, or the cells could have been left open-circuited for a long period of time. Many modern designs are of the sealed variety and no maintenance of the electrolyte is necessary.

Maintenance testing. Where the importance of the battery warrants, those responsible for its maintenance should periodically test its condition. In order to do this properly it should be tested on the equipment for which it is intended. A typical example of this would be where a battery is required to supply power for emergency lighting for a specific period or is required for starting up an engine. If a battery is supplying power for emergency lighting, then it should supply the necessary power for the appropriate standby time, and where it is required to start an engine, test starting should be carried out at regular intervals.

Cell voltages. 1. There is little value attached to cell voltage readings when no current is passing through the cells.

2. When a lead-acid cell voltage on charge does not rise above 2.3 or 2.4 after all the others are up at about 2.7 V per cell, that cell may have an internal short-circuit. These short-circuits are often hard to find, but the voltage shows it up at once. Rapid cell-voltage drop will occur if there is a short-circuit, particularly on discharge or in service, and this will invariably be accompanied by a fall in the relative density of the electrolyte. Such short-circuits may also be detected on batteries that are on float charge by a cell that is short-circuited having a lower floating voltage than the others. It will also be accompanied by a lower relative density.

3. If the voltage of any lead cell rises to 3 V when first put on charge, with recommended charge rate flowing, it could well be

sulphated and must be charged at a very low rate until its voltage has fallen and then commences to rise again.

4. Do not attempt to add acid to any lead cell that appears to contain only water. Put it on charge at about half normal rate for a time. If then the voltage is seen to rise slowly, the rate may be increased slowly, for the cell is only over-discharged. Only adjust the relative density of the acid after charging is ended; this will be indicated by the relative density and cell voltage remaining constant for three consecutive hourly readings.

Additional information.   For standby applications where many cells may be used for a battery, its best to have a battery room or cabinet in which the cells may be placed. Checks should be carried out at regular intervals to ensure that electrolyte levels are maintained, the battery is kept fully charged and the charging equipment is in good order. Each battery should have its own hydrometer and instruction booklet, together with a quantity of Vaseline or petroleum jelly to keep the cell connections well protected. Where the battery has to perform an important function, a thermometer, record book and cell-testing voltmeter are useful additions. Electrolyte examination and testing is not required or possible on sealed designs sometimes called 'maintenance free'. However, cell connections need checking as outlined above.

For motive power uses, the operator should be conversant with the maintenance operations required, i.e. the need for regular topping up and adequate recharging following the battery discharge. Again a hydrometer should be available and a cell testing voltmeter, thermometer and record book are valuable aids.

For automotive applications, it is necessary to keep the battery electrolyte level correct and occasionally check the relative density with a hydrometer.

For all applications, a supply of distilled water or deionised water should be available. Tap water should never be used for topping up cells as many domestic water supplies contain additives and impurities that may be harmful.

Where both lead-acid and alkaline cells are employed, it is best to have separate battery rooms and under no account should an alkali be added to a lead-acid cell or vice versa. No equipment such as an hydrometer or thermometer should be used on both lead-acid and alkaline batteries to avoid contamination.

the necessary conductivity. Sintered-plate cells are more costly than pocket-plate cells and are therefore used mainly for duties requiring a short-time high discharge rate. They occupy less volume than the equivalent capacity pocket-plate types.

Sealed nickel-cadmium cells are only available with relatively low ampere-hour capacities and find wide application for standby duties, particularly emergency lighting. Recharging of these cells without building up dangerous internal pressures requires the negative plate to have a larger capacity than the positive. Both pocket-plate and sintered plate and sintered-plate cells are available in sealed designs.

Characteristics. The basic characteristic of the nickel cadmium battery is extreme flexibility in regard to electrical and mechanical conditions. The battery will accept high or low rate charges and will deliver high or low currents, or work under any combination of these conditions. It can stand idle for long periods in any state of charge and will operate within very close voltage limits.

Types of battery. Two main types of battery are available.

1. The normal resistance type which is used where the full ampere hour capacity of the battery is regularly or occasionally required, e.g. traction and standby lighting.
2. The high performance type which is used where ampere hour capacity is seldom required but where the battery has to supply high currents for short periods, e.g. engine starting and switchgear operation.

Charging and discharging. The 'normal' charge rate for all types of nickel cadmium battery is C/5 for 7 hours where C is the rated capacity of the battery, which will restore a discharged battery in 7 hours. The charging rate and time can, however, be varied according to the duty.

Because of the all steel construction and the enclosed active materials batteries may be safely discharged at high rates up to short circuit without damage. The average discharge voltage is 1.2 V per cell at normal currents.

Maintenance. Very little maintenance is required for a nickel cadmium cell. It is only necessary to keep the cells clean and dry as with any battery and top up as required with distilled water.

Nickel cadmium alkaline cells

Chemical principle. The active material of the positive plate consists of nickel hydrate with a conducting admixture of pure graphite. The active material of the negative plate is cadmium oxide with an admixture of a special oxide of iron. The electrolyte is generally a 21% solution of potassium hydroxide of a high degree of purity.

For the purposes of practical battery operation the reaction may be written as follows:

Charge   *Discharge*

$$1Ni(OH)_3 + Cd \rightleftharpoons 2Ni(OH)_2 + Cl(OH)_2$$

In a fully-charged battery the nickel hydrate is at a high degree of oxidation and the negative material is reduced to pure cadmium; on discharge the nickel hydrate is reduced to a lower degree of oxidation and the cadmium in the negative plate is oxidised.

The reaction thus consists of the transfer of oxygen from one plate to the other and the electrolyte acts as an ionised conductor only and does not react with either plate in any way. The relative density does not change throughout charge or discharge. At normal temperature the cell is inert on open circuit.

Construction. There are a number of different forms of construction for nickel cadmium cells. In the nickel cadmium pocket plate cell the positive material consists of a mixture of nickel hydroxide and graphite. The mixture is compressed into flat pellets and supported in channels made from perforated nickel plated steel strips. The negative plate is similarly supported cadmium powder with an admixture of iron or nickel. The electrolyte is an aqueous solution of potassium hydroxide. With metal containers, the individual cells are insulated by spacers; many are available with plastics containers.

Nickel cadmium sintered plate cells use a highly porous nickel matrix produced by sintering special types of nickel powder. The porus plate, which is usually strengthened by a reinforcing grid of nickel or nickel-plated steel, contains about 80% voids into which the active materials, nickel and cadmium hydroxides respectively, are introduced. Intimate contact is thus maintained between the active material and a large surface of metallic nickel, to provide

BATTERY CHARGING

A battery on charge is not a fixed or static load. It has an extremely low internal resistance and unless something was done to control the charging current a very large current would flow if connected to a d.c. source of supply very much larger than the battery voltage. Because a battery has an output voltage the applied voltage from a charger has to be in excess of this voltage. The positive of the battery charger must be connected to the positive of the battery. The current which flows is the difference between the applied and battery voltages divided by the internal resistance of the battery.

As the battery becomes charged its voltage rises and the difference between this and the applied charger voltage falls and the charging current drops. Thus it will be seen that the requirements of a battery charger are that it must limit the charging current at the start of the charging cycle and yet maintain a sufficiently high current throughout the entire charging period to recharge the battery in an acceptable time.

Every charger has a definite characteristic called the 'slope characteristic'. This is the relationship between the output voltage and the current, throughout its complete range. At any given voltage it will always deliver a given current. This is inherent in the charger design and will not be affected by the capacity of the battery connected to it. The charger simply responds to the counter voltage of the battery connected to it.

This does not mean that any charger can be connected to any battery. If the charger voltage is less than the battery voltage no current will flow in the charge direction. Conversely, if the charger voltage is sufficiently high that the current drawn is beyond its capabilities, the charger will overheat and may even burn out unless protected by a fuse or other device.

The primary significance of the slope characteristic is its angle. This can vary with the application and is usually adjustable within a certain latitude. Slope A in Figure 21.1, referred to as a 'constant voltage' or 'constant potential' type, represents a charger which maintains a constant voltage throughout its load range. This would, at nominal voltage values, deliver an extremely high current to a discharged battery, as shown by line A in the diagram. A charger of this type therefore usually has some form of current limitation, shown as A_1 which causes the voltage to drop when the current exceeds rated load, thus preventing any

518

further current increase. This is the type of battery charger used in 'float' charging whereby the charger is connected permanently across the battery to maintain it in a fully charged condition.

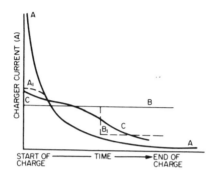

Figure 21.1   Currents produced by basic slopes

Slope B, referred to as a 'constant current' type, maintains the current at a constant value throughout the charge. This characteristic is frequently used for the recharging of nickel-iron batteries, as they can accept a comparatively high current throughout the entire charging period. It can also be used for lead acid batteries if sufficient time is available to charge throughout at the lower finish rate required by these types. A constant current charger can also be used to charge lead-acid batteries in a shorter time by conducting the early part of the charge at a constant high rate, and then shifting to a lower value, from B to B_1 on Figure 21.1, to complete the charge. This is referred to as 'two-rate' charging.

Slope C is a 'tapering'slope, with its angle designed to best suit the application to which it is put. This results in a moderately high charge to begin with, gradually decreasing during the charging period to a low rate at the end when the charge is complete.

Charging equipment – general.   Since all battery charging must be by direct current, the available power sources are: an existing d.c. supply; an a.c. to d.c. motor generator set; and a.c. to d.c. rectifier. As d.c. supplies direct from a supply company are no longer common, motor-generator sets or rectifiers now supply practically all power for battery charging purposes. The two basic groups which cover most chargers in industrial service are:

1. Single-circuit chargers adapted to charge only one battery at a time. These cover a wide range of applications.
2. Multi-current chargers designed to charge a number of batteries at the same time. The principal application of this type is the charging of industrial battery driven trucks or other batteries used for motive purposes.

Single-circuit rectifiers. The most elementary type of rectifier circuit consists of a transformer to step the mains voltage down to a value appropriate for the battery being charged with a rectifier on the secondary side to allow current to flow only one way in the battery circuit. This constitutes a half-wave rectifier. For greater efficiency and to provide a better wave form a number of rectifying elements may be arranged to provide full-wave rectification, either in single phase of three-phase systems, see Figure 21.2. Rectifiers are either of the selenium or silicon type. In most cases the silicon diode type is the more expensive of the two but has smaller space requirements, longer life and absence of ageing characteristics.

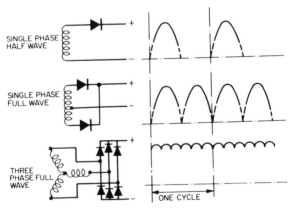

Figure 21.2   Basic rectifier circuits and waveforms produced

A basic rectifier (transformer and rectifying element only) has a drooping characteristic similar to a shunt generator, the droop being a function of the impedance of the component parts. This fundamental characteristic can be altered to obtain almost any desired slope by using special transformers, reactors, ballast resistors, transistorised control units or magnetic amplifiers.

For cycle charging a typical tapering slope is usually chosen for lead-acid batteries and a more nearly constant current slope for nickel alkaline types. Both types of characteristic can be achieved using a transistorised control unit. The other control methods can be used with reasonable success. With transistorised control units a compound slope is quite practicable with the angle of slope changing quite distinctly at the preset time in the charging period.

For float charging, precise control of the voltage is necessary and therefore transistorised control units are invariably employed. The basic characteristic is a constant voltage throughout the load range, with current limitation at or about full load.

When using these rectifiers on any system subject to surges or other sharp fluctuations in load, it is important to make a direct connection between the charger and battery terminals. Likewise connect the load directly to the battery terminals rather than to the charger. This enables the battery to absorb any surges or voltage peaks rather than the charger which could be damaged as a result. Control of load current is accomplished by reactors or magnetic amplifiers.

The rectifiers referred to so far have been selenium stacks or silicon diodes. More recently these diodes have been replaced by thyristors having gate control circuits. By applying a signal to the gate the thyristor can be 'fired' (made to conduct) and the current that will flow will be governed by the point in the cycle at which the thyristor is fired. Thus control of the current can be effected without the need for magnetic amplifiers.

Multi-circuit charging. Multi-circuit charging will be considered from its most common application – that of the recharging of battery-electric trucks. A typical charging system consists of one or more chargers supply d.c. to a common busbar and a number of panels, each one being designed to charge a single battery. This system is usually called a 'modified constant potential' system.

Source of power may either be a motor-generator set or a rectifier system. The characteristic must be the provision of a constant voltage throughout the load range within a tolerance of ±3%, and a current capacity equal to the sum of the currents drawn by all the individual charging panels. While there is no limit to the number of batteries which can be charged from one machine it is usual to use several chargers in parallel on one d.c. busbar, or separate the system into several groups, if more than 20 or 30 batteries are to be charged.

When using a single d.c. busbar it is desirable that all batteries consist of the same number of cells and be of the same general type, either lead-acid or alkaline. The ampere hour of the individual batteries however may vary.

The basic circuit of a panel is simply a ballast resistor in series with the battery across the d.c. busbar.

22

Cable management systems

Because of the dramatic increase in the electrical and electronic requirements of all types of buildings, domestic, commercial and industrial, has come the demand for the provision of safe yet adaptable wiring systems enabling sections of them to be easily altered without affecting the supply to the remainder. This is essential for those networks supplying so-called electronic offices which have to cater for data, voice and image communications facilities, building services, lighting, power, earthing and lightning protection.

Electronic aids are not only confined to offices but extend to include other buildings like hospitals, retail outlets, hotels and industrial establishments. Each building has different cabling requirements and the density of cabling and degree of versatility needed may vary considerably. This must be reflected in the design of the cable management system enabling it to be easily adapted to meet a wide variety of supply possibilities, with the minimum of effort and change of components.

One would expect frequent changes in modern commercial offices, particularly financial establishments containing many computer terminals, telecoms and data facilities but at the other end of the spectrum hotels may only require changes in the offices and reception area.

The growth of cabling systems has been brought about by the rapid increase of Information Technology (IT). To prevent the generation of masses of loose cables which become unmanageable, manufacturers of electrical distribution equipment have designed integrated systems using cable trunking, conduit, cable trays and ladders to house and support the cabling. These are known generically as 'cable management systems'. They take many forms and provide the user with a versatile facility, well able to cater for frequent changes of the supply network. At the same time they are safe even when being altered.

INTEGRATED SYSTEMS

There is no recognised definition of an integrated cable management system but the author suggests the following probably covers it adequately. An integrated cable management system is a combination of cable enclosure and/or support equipment (i.e. conduit, trunking, floorboxes or other outlets, ladders or trays), so designed as to form an integrated arrangement and being capable of minor changes without having to shut down the entire network or endangering personnel working on it.

One must stress at this point that a most important function is for the system to be 'integrated'. By this is meant that each piece of equipment joins onto the next without any gaps, loose joints, etc. so that the cabling is protected in every part of the network, from the point of supply to the point of use. Furthermore, should the user want to alter any part of the installation it should be a comparatively simple operation; for example if the position of computer terminals need to be changed around in an office, or part of an office, this should be possible without disconnecting all the other terminals being served by the same network.

The above definition does not cover some special systems like flat cable and cable harness methods which are being used to perform the same function equally well, and these are described later in the chapter.

Compliance with British Standards (or the IEC equivalent) for the products comprising an integrated system is extremely important, because there are many inferior items available on the British market and to introduce any, even as a small part of an installation, will jeopardise the safety of the whole. Furthermore, an integrated system should meet, wherever relevant, the requirements of the IEE Wiring Regulations recently designated BS 7671.

The different components forming an integrated system are described below.

Conduits

There are three types of conduit systems from which to choose; metal, insulating and composite. Metal conduit is generally steel; insulating conduit has no conductive components and is usually some form of uPVC; composite conduit comprises both metal and insulating material and is generally flexible. We do not

524

highlight all the advantages and disadvantages of these or other integrated systems components here. For fuller description of all the different types of cable management systems the reader is referred to *Cable Management Systems* by E. A. Reeves, available from Blackwell Science in Oxford, from which the information in this chapter has been extracted.

When considering the purchase of any conduit system, and the same goes for other cable management arrangements, it is wise to ensure that the manufacturer can supply all the necessary accessories required such as bends, junction boxes, outlet boxes, etc.

Metal conduit. This may be steel or stainless steel, the latter being used in sterile areas like food processing. Reputable suppliers generally belong to the British Electrical Systems Association (BESA) and also possibly the British Welded Steel Tube Manufacturers' Association. Where mechanical strength is required many users opt for steel, but there is always the risk of corrosion if the coating is damaged in any way.

Most manufacturers concentrate on supplying Class 2, black stoving enamel on the inside and outside, and Class 4 (Figure 22.1) which is hot-dipped galvanised on both inside and outside. Both can be provided in light gauge (about 1 mm thick walls) and heavy gauge (up to 2 mm thick walls).

Figure 22.1   Class 2 and Class 4 metal conduit has screwed ends, with one end protected by a coupler and the other by a cap (Burn Tubes)

Insulating conduit. This is often called plastics conduit and covers an enormous range of different materials, but here we confine the discussion to conduits made from unplasticised polyvinylchloride (uPVC). Round rigid conduit is available as light or heavy gauge in sizes from 16–50 mm diameter. No threading is required as with metallic varieties for joining adjacent sections together, a special adhesive being used. White or black finish is available. Being made from plastics there is no danger of corrosion. High impact types withstand the rough handling on site. Oval sections provide a shallower profile where necessary.

Channelling. Sometimes called capping, it can be employed for switch drops and cable buried in plaster to provide protection against damage by nails knocked in the wall during the lifetime of the installation.

Flexible conduit. Flexible conduit is defined in BS 6099 Part 1 as 'A pliable conduit which can be bent by hand with a reasonable small force, but without any assistance, and which is intended to flex frequently throughout its life.' Pliable conduit is defined as a conduit which can be bent by hand with a reasonable force but without other assistance. However, it is important to realise that pliable conduit must never be used where continual flexing is experienced.

Flexible conduit is available in steel or plastics forms, and particularly with steel, in many varieties. Liquidtight flexible conduit is usually of steel core with a PVC sheath, and employing specially designed connectors to provide an IP67 degree of protection.

Cable trunking systems

In this section we are dealing only with trunking that houses and protects cables linking points of an electrical network together. Perimeter, bench, mini, and poles and posts are dealt with under their own headings. Cable trunking under this definition is available in both metal and plastics materials. Three types of metal trunking are marketed, steel, stainless steel and aluminium and where a non-flammable material is specified the choice lies between these three types. Toxic gases are released by some plastics materials when subjected to flame but there are halogen-free designs of insulating trunking. Some plastics are flame retardant while others do not burn once the flame is removed. Heat resistance capability might be very important for trunking housing fire alarm circuit cables which need to operate for as long as possible in the event of a fire. Aluminium trunking is claimed to protect data cables from interference.

In any long vertical trunking runs, particularly where they pass through floors or exceed about 5 m, then some kind of fire barrier must be incorporated inside. This is dealt with in more detail in the book referred to earlier.

Steel trunking is available in 23 sizes from 38×38 mm up to 300×300 mm as laid down in BS 4678 Part 1. Plastics trunking generally follows the same size range. One must bear in mind when comparing plastics and metal cable management systems and their component parts that one big advantage of plastics designs is their lightness compared to metal. Trunking and conduit are also easier to cut to length and join together.

Multi-compartment trunking. Both metal and plastics trunking are available in multi-compartment designs (Figure 22.2) to provide complete segregation of cables in the individual

Figure 22.2 Ega range of uPVC cable trunking (MK Cable Management)

compartments. Metal trunking has the added advantage of shielding cables in adjacent compartments.

Some designs of multi-compartment trunking can accept outlets on the lids, transforming them to perimeter trunking systems suitable for mounting at either floor or dado level.

Components and fittings. Lids for plastics trunking are usually of the clip-on type while for metal designs some kind of mechanical fixing like a turnbuckle is employed. It is important when ordering trunking to ensure that a full range of fittings is available including angles, tees, corners, crossover units, cable retainers, connectors, flange couplers, reducers, adaptors etc.

Earth continuity. One of the claimed advantages of metallic cable management systems is that the housing can be used as the earth continuity conductor (ecc) thus saving on cabling space. BS 7671 permits this but there is always the danger of a break in the circuit due to corrosion or a faulty joint at a fitting and many installers prefer to run separate ecc inside the trunking.

Perimeter trunking systems

Perimeter trunking systems include skirting and dado (sometimes called undersill) and are made from plastics (Figure 22.3), steel or aluminium. Steel and plastics systems are covered by BS 4678: Parts 1 and 4, respectively, although neither specifically deals with perimeter systems *per se*. Perimeter systems are suitable for domestic, commercial and industrial installations and are ideal for refurbishment applications.

Figure 22.3   Sterling multi-compartment plastics perimeter trunking is available in six profile variations and three colours and is suitable for skirting and dado mounting (Marshall-Tufflex)

528

Basically the systems are designed to house power, lighting, telecommunications and data cables but there are some hybrid arrangements that incorporate other services as well such as compressed air or heating pipes. One design combines ducting with a protection rail. Multi-compartment arrangements are the norm with the different services occupying separate, segregated compartments. This enables work to be carried out on one of the services without interfering with the others. Some multiple-socket outlets can be wall-mounted and are therefore, in a sense, perimeter systems, although they are of limited length and are not in the same category as true perimeter systems.

Generally trunking systems are wired with cable but there are some designs which incorporate busbars used to power the socket-outlets.

Cornice trunking is available and provides an ideal solution to wiring in buildings with concrete floors and ceilings. Manufacturers tend to design it to complement a particular range of their own standard trunking and this needs to be recognised.

Screened compartments. For some computer and data circuits screened compartment systems are available for plastics systems. They may be built into the trunking body or supplied as an extra metallic compartment which fixes onto a standard design. Metallic systems do not need special treatment except where the compartment dividers do not form a complete enclosure.

Socket-outlet positions. One of the advantages of perimeter systems is the facility to fit as many outlet points as required, anywhere along the length, allowing for a minimum distance between adjacent outlets. The position of the outlet is either on the trunking lid itself or on an extension piece mounted above the trunking, which raises it to a level from the floor recommended by the IEE Wiring Regulations.

Shapes. Two design shapes are popular, either a rectangular section or one incorporating a chamfered cover plate. The chamfer may be symmetrical, i.e. both the top and bottom of the cover is chamfered or the angle may be confined to the top part of the trunking only. All are of narrow depth with increased cable capacity provided by increasing the height. Three-compartment design is popular with the centre compartment being the largest and the top and bottom compartments generally having the same dimensions.

Miniature-trunking

Miniature-trunking systems (shortened to mini-trunking) were originally introduced for the refurbishment of domestic dwellings and light commercial establishments. Being surface-mounted and made from plastics material, enabled quick installation, it not being unknown to rewire a flat or small house within one day. But since then its application has spread to new buildings, not only for housing power and lighting cable but also telecommunications, alarms and data systems. All mini-trunking is characterised by a low profile, some products being as shallow as 7.5 mm but the usual being of the order of 12.5 to 16 mm. A typical product is shown as Figure 22.4.

Two methods of fixing are available either by screws (or tacks) or by adhesive, the latter having a peel-off backing paper. Once the adhesive design is fixed to the wall or ceiling it cannot be adjusted as is possible with mechanical fixing methods. One design has clip-on components which further speed-up installation.

Figure 22.4   A mini-trunking system showing joints, tee-pieces and other fittings, together with a switchbox accessory. Close-fitting joints provide a high level of safety (Legrand Electric)

530

To give some idea of the sizes of this trunking one manufacturer provides eight different sections ranging from 16 mm square up to 50 mm wide by 32 mm deep. Although most are of single-compartment construction there are a limited number of two-compartment designs on the market.

All lids are of the clip-on variety and are either flush with the trunking sides or overlap them. At least one supplier has a double-locking clip-on lid to provide security if it is needed.

Bench trunking systems

Bench trunking is not defined by BS 4778 although the trunking standard is used as a guide by manufacturers. The term describes a triangular-shaped metal or plastics trunking with the sloping face carrying socket-outlets (Figure 22.5), although a rectangular trunking section with multiple socket-outlets, when fitted to a bench, performs the same function. Where necessary the trunking should also comply with BS 5490 dealing with ingress of foreign bodies, i.e. an installation may require an IP number.

Bench trunking can be hard wired or may incorporate busbars, and usually comes in specific lengths although these may not be the same from individual manufacturers. Some designs can be compartmented to segregate different services and others may come prewired.

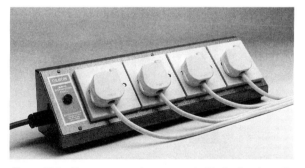

Figure 22.5   From an extensive range suitable for bench, wall or floor mounting, this unit is a steel trunking finished in two-tone colour, with four socket-outlets, 2 m of cable and a 13 A fused plug (Olsen Electronics)

531

Residual current devices. To protect against earth faults many manufacturers fit residual current devices as standard in their products thus meeting the requirements of the IEE Wiring Regulations.

Clean supplies. A special design of bench trunking protects sensitive equipment from transients, spikes, switching surges, radio frequency interference and lightning surges. It can be used to supply computers and other sensitive data equipment. The trunking system may be called a power conditioner and as such is covered by BS 6204.

Service poles and posts

Service poles and posts have been developed to meet the demand for concentrated power, telecommunications and data outlet facilities in open-plan offices, isolated workstations, reception areas and financial institutions (Figure 22.6). Power requirements are usually quite small and so they may be supplied from a floor

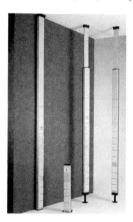

Figure 22.6   Single-sided telescopic aluminium multi-compartment service poles (one post also shown), are suitable for mounting against walls. The Prefadis C9000 range is also available in a double-sided version. Most poles are fitted with a 16 A miniature circuit-breaker as standard. (Telemecanique)

outlet box of a platform floor installation. Basically poles and posts are floor-standing multi-compartment trunking sections carrying outlets on one or more faces. Service poles may be fed from the floor (as indicated) or from the ceiling, while posts, being truncated versions of poles, are generally supplied from a floor box.

Virtually all designs consist of an extruded aluminium spine to provide the necessary stiffness, with clip-on covers over formed channels, either of metal or plastics, carrying the various outlets. Some poles may also include one or more of the following, lighting switches, circuit-breakers, luminaires, fire alarm break glass units or other items, increasing their versatility.

Although square or rectangular sections are the norm there are part-circular and elliptical designs with special provision for carrying the socket-outlets.

Cavity floor systems

Several terms are used to describe cabling systems that are installed in the space underneath a false floor, or sometimes flush with the floor. Access, raised access, platform, raised and cavity floor are some examples. They may be used to describe the complete system, i.e. cabling and housing, or either the cabling or the trunking. There is no British Standard covering these systems but the Property Services Agency document MOB PF2 PS *Platform floors (raised access floors). Performance specification*, is the standard that reputable suppliers work to. In this section we use the term cavity floor system to describe the complete electrical distribution network serving floor outlet boxes or other facilities providing power, data, telecom or computer services to terminal equipment.

Adjustable jacks are used to support the false floor and provide the space between the screed and the underside of the flooring for the services cabling. This system is often called a pedestal design, but there are flooring systems that do not rely on adjustable jacks to provide the space for the cabling and these are termed non-pedestal designs.

Non-pedestal designs. Floor panels are located close to the structural floor and individual panels are often meant to be permanent fixtures so that service outlet positions are therefore usually decided at the design stage thereby restricting the versatility of the installation. The depth of the floor box is also critical and this will restrict the choice available to the user. Fluted deck and batten type are two arrangements of non-pedestal design, the latter providing some degree of versatility after installation.

One of the biggest advantages of these systems is where there is a limited floor to ceiling space which cannot be altered, i.e. for refurbishment projects.

Pedestal designs. Any depth of floor void can be provided within reason, from 40 mm up to 1 000 mm or even greater in special cases, where perhaps other services such as central heating or air conditioning are also to be contained in the space. Rectangular or square panels are generally employed, a popular size being 1 200 mm by 600 mm for the fixed type and 600 mm square for the removable design.

Electrical systems. Basically there are three distinct cable management systems used with cavity floors, with quite wide variations in each. The first is where the cables are contained in some form of enclosure fixed to the subfloor (Figure 22.7). Alternatively the cable enclosure may be fitted flush with the floor panel. Secondly the cables may be laid on the subfloor to

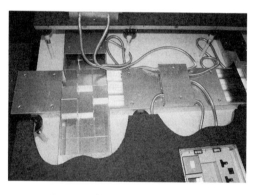

Figure 22.7   This cavity floor metal trunking system shown fitted to the sub-floor is the same as the company's surface trunking but usually has conduit knockouts in the sides and dividers for flexible conduit connections between the trunking and the floor outlet-boxes (Cableduct)

join components together, sometimes called plug and socket systems. The third method is to suspend the enclosure close up to the underside of the floor panels.

Subfloor systems.   In some plug and socket systems cables are linked together by multiple outlet feeder boxes to the subfloor and these boxes supply the service outlets fitted in the floor panels. If these feeder boxes have spare capacity it is a simple matter to reposition or install another floor outlet box, by using one of the spare sockets. Segregation of services is maintained within the feeder boxes. A variation is where the cables are replaced by busbar systems and services outlet boxes fed by plugging direct into them.

The most common cavity floor cable management system is where the compartmented trunking (either plastics or metallic) is fixed to the subfloor which carry the services cables. Feeds to outlet boxes are taken out through flexible conduit so preserving the segregation of services. There are many variations of the subfloor systems which are described more fully in *Cable Management Systems*.

Flushfloor systems.   These are usually in the form of metallic trunking, the cover of which is fixed flush with the surface of the floor panels. Outlet boxes fit into the lid at any desired position along the length of the run.

Under-panel systems.   The multi-compartment metallic trunking is lifted into position by a jacking system so that the trunking top just touches the underside of the panel, so that no cover for the trunking is needed. Like flushfloor systems they leave the structural slab free for other services and present the cable at a convenient working height.

Outlet boxes.   Basically an outlet box consists of three components, the base, an inner section carrying the accessory plates, and a top part usually carrying the lid. Trims are supplied as a separate item. The base is provided with partitions to maintain the segregation between compartments in multi-compartment designs. Up to four compartments are provided as normal although some manufacturers offer more. Knockouts on the sides of the base unit provide facilities for fitting 20 mm or 25 mm flexible conduit. Rectangular or circular boxes are available.

Floor distribution systems

This section covers all types of floor distribution systems except cavity described above, including both metal and plastics flushfloor and undersurface screeded systems, undercarpet wiring and pedestal boxes associated with undersurface systems.

Screeded systems. There are two types of screeded systems. The first is a cable enclosure buried in the screed with access at finished floor level only through junction and services outlet boxes, with interconnection between the junction boxes being by single or multiple ducts through which the cables are drawn. The other has the trunking fixed to the floor with the screed up to the lid which is taken as the datum (Figure 22.8). Full access is provided by taking off the lid which is used to house the outlet boxes, or for a modular design the service outlet replaces a section of the lid.

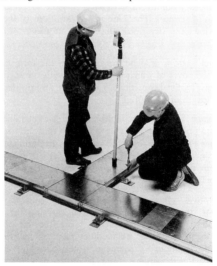

Figure 22.8   Metal flushfloor system being installed on the sub-floor prior to screeding. The top of the trunking is taken as the datum line for screeding purposes. Outlet boxes fit direct into the trunking (Walsall Conduits)

536

Undercarpet systems. There are basically two ways in which power, telecoms and data services can be taken to equipment, either in a very shallow profile multi-compartment trunking or by very thin specially designed cables. These cables are not to be understood in the conventional sense.

Trunking systems, usually plastics, employ conventional cables. Typically a three-compartment arrangement has dimensions of 60 mm wide by 14 mm deep. Another supplier quotes dimensions of 70 mm wide by 9.6 mm deep. These trunking systems feed shallow profile pedestal boxes which can be fitted with the required outlets, and generally rest on the floor with the carpet brought up to them, i.e. they are visible.

Flat cable designs lie under the carpet and usually have single service capability which can be integrated to give all three services. Special outlet boxes are required.

Pedestal boxes. These are available as standard and non-standard designs, the former being employed for both carpet trunking systems and screeded underfloor arrangements. The non-standard design is for use with flat cables.

Cable trays

Cable trays have traditionally been made from metal but increasingly plastic is being used because of its non-corrosive qualities (Figure 22.9). Trays of both materials are characterised

Figure 22.9 Powercomponents PVC cable tray. The special ventilated structure ensures high rigidity combined with light weight. Because the tray cannot corrode it is suitable for use in many hostile environments (Mita)

by perforations in their bases and sometimes on the sides, these being used to hold the cables in position. Some of the plastics products do not have perforations and then they are like trunking without lids. Heights of the longitudinal side members depend on the width and strength of the tray. Different methods are adopted for fastening sections together.

Perforations in the base are usually in the form of elongated slots although a few circular holes are sometimes included in the pattern. Popular widths start at 50 mm and 75 mm; side depths also range from 24 mm to 100 mm, while standard lengths are 2.44 m, 3 m and 4 m.

Cable ladders

Because both cable ladders and cable trays are used to carry heavy loads of cable over long distances and with large spans they should be treated more like structural support systems than the many forms of cable trunking already described. They find their applications in power stations, chemical and shopping complexes, factories, leisure centres and similar establishments where a great many cables have to be accommodated and routed over long distances.

Cable ladders are generally made of pre-galvanised sheet steel for dry environments but of post-galvanised metal where there is a possibility of corrosion. Post-galvanising is generally known as hot-dip galvanising although there are other methods of protection against corrosion (Figure 22.10). Stainless steel and extruded aluminium alloys are alternative materials for ladders. There is also a growing interest in plastics designs.

There is a great variety in the design of both sides and rungs of ladders, manufacturers claiming advantages with their particular products. Ladder sides carry perforations while some rungs also carry holes. Flat, oval, and T-section rungs indicate the wide variety on the market. Spacings between rungs also varies but 300 mm is a popular choice.

Wire tray system

The wire tray system is neither a cable ladder or cable tray although it performs the function of both to some degree. It is produced from high mechanical strength steel wire to form a welded mesh construction, with sides, to keep the cable in position. It is therefore much lighter than its counterparts (Figure 22.11).

538

Figure 22.10 KHZ hot-dip galvanised cable ladder having oval rungs and closed rail profile meets Cadbury's stringent requirements at its Bournville factory (Wibe)

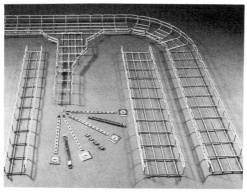

Figure 22.11 The Hi-Way wire tray system can be fabricated on site to the exact size by simply cutting the steel wires, preferably using a bolt cutter with offset cutting jaws. Bends, tees, crosses and level changes can also be formed from the wire tray itself (Vantrunk Engineering)

539

Busbar trunking sytems

This section relates to 440 V busbar trunking systems, usually mounted overhead, and fitted with tap-off units to feed equipment on the floor below (Figure 22.12). They should comply with BS 5486 Parts 1 and 2, particularly Part 2.

Basic components of a busbar trunking system are a set of copper or aluminium busbars complete with supports and insulators, factory assembled within insulated or sheet steel trunking. Ratings range from 25 A up to 5000 A or even greater. Trunking units are provided in straight sections and are usually for three-phase networks although there are some single-phase arrangements at the lower end of the range. Five-busbar designs allow for three-phase and neutral together with a protective conductor. Centre and end feed facilities are available particularly with the higher ratings. Good joints between adjacent sections are essential if hot spots are to be avoided.

Tap-off units are easily plugged into the busbars and can be altered quite simply should the need arise. They may be of the all-insulated or metal-enclosed design which can be safely connected and disconnected onto live busbars. For busbar systems over 800 A tap-off units are manufactured with ratings from 40 A to 1000 A, and they may be fixed or unpluggable.

There is at least one cast resin design of busbar system with ratings up to 6300 A at 17.5 kV for a.c. networks and 16 000 A at 1.5 kV for d.c. circuits.

Figure 22.12 Power-Bar 800 busbar trunking has an IP42 enclosure, five tap-off openings per 3 m length and ratings of 250 A, 400 A and 630 A. Fire barriers are positioned to suit (Square D)

Special systems

One should draw attention to cable management systems that do not fit into the normal category but it is not proposed to describe them fully. Mineral insulated cable can be employed using a radial network and junction boxes.

An American modular system, while accommodating cables in trunking systems, can be adapted to provide voice, data and/or video systems by utilising surface-mounted boxes, wall plates and connector modules which snap onto the trunking as required.

Nurse call systems employ trunking at the hospital bedhead to house electrical facilities such as radio, television, sound distribution, power socket-outlets, computer outlets and telecommunications.

Zip-up cable jacket and sleeving are ideal for protecting cable supplies to motors, control gear and switchgear.

Wired office furniture enables one to bring the services as close to the point of use as possible, dispensing with any trailing cables on floors.

Harnesses and pre-wired assemblies are made up for specific applications and can be economical where many identical arrangements are required.

23

Hazardous area electrical work

In areas where explosive gas–air mixtures can be present, electrical apparatus that gives off sparks during operation or in the event of a fault, and equipment with hot surfaces constitute a potential source of a hazard. The consequences of an explosion in such an area can be quite horrific, as disasters such as occurred on the Piper 'Alpha' platform, or before that at Flixborough have proved.

In the chemical, mining, many other branches of industry, and even in everyday life, combustible materials which can form explosive atmospheres and which could under certain circumstances cause an explosion are manufactured, stored and processed. The concept of 'explosion protection' of electrical apparatus has been developed and formalised to prevent accidents occurring in hazardous areas during normal operation of the electrical apparatus.

Historically, the major impetus to the development of special equipment, procedures, standards and codes for particularly hazardous working environments was probably the large-scale development of the coal mining industry in Britain and the rest of the world at the time of the steam power revolution. Davy's miners' safety lamp is an example of a piece of equipment designed specifically for use in hazardous areas, and in the early years of this century, commercial competitions were held for the design of safe electrical equipment for mines. The concept of 'intrinsic safety' is thought to date from a mining accident in South Wales in 1913. An explosion killing 439 men was probably caused by the ignition of methane from the spark created as a miner touched his shovel between the two bare wires which formed the signalling system to call for the mine's winch.

The legacy of mining's importance remains today—the British coal mining industry is governed by slightly different regulations than those applying to other hazardous areas (see sections on gas groupings and certification later in this chapter), but hazardous areas are also found, for example in offshore oil and gas installations and onshore petrochemical installations and it is quite likely that at some point in his career an electrical engineer will need to know something of the subject of explosion protection.

The numerous sets of standards and codes of practice covering the manufacture, suitability, installation and maintenance of electrical equipment in potentially hazardous areas, combined with differences in these standards and codes of practice around the world, make the subject complex and sometimes intimidating, and in a single chapter of a book such as this it is impossible to give more than a brief background to the subject and to indicate where full and detailed information can be found. It should also be pointed out that the information contained in this chapter is not intended to replace published standards, codes of practice or other relevant publications.

A worldwide perspective is important because of the increasingly international nature of the electrical system design and building industry. For instance, Britain's technological expertise in the offshore oil and gas industry has meant work for British designers and builders of control systems destined for similar use overseas, but those systems must be suitable for use in the country in which they will be used.

In common with many areas of trade, steps towards harmonisation of standards across the European Union have been and continue to be taken, a subject covered further in the section below on standards.

This chapter gives information on the definitions of hazardous areas, equipment design concepts for explosion protection and the standards governing the manufacture and certification of electrical equipment designed for use in hazardous areas which will help the electrical engineer in selecting, installing and using such equipment. In addition, information is given on working practices, in general and in specific areas.

HAZARDOUS AREAS

The first requirement is to know what a hazardous area is. The principal factors relevant to the classifications of a hazardous area are the nature of the gases present in the potentially explosive atmosphere and the likelihood of that atmosphere being present.

The concept of 'zone classification' has been developed to summarise these factors. The nature of the atmosphere is characterised by the chemical composition of the gas and its ignition temperature. The notions of 'gas grouping' and 'temperature classification' have been developed to formalise this.

Before we look in more detail at these definitions, it is instructive to consider how explosions occur. A useful concept is that of the 'hazard triangle', Figure 23.1. The sides of the triangle represent fuel, oxygen and a source of ignition, all of which are required to create an explosion. The *fuel* considered in this chapter is a flammable gas, vapour or liquid although dust may also be a potential fuel. *Oxygen* is present in air at a concentration of approximately 21 per cent. The *ignition source* could be a spark or a high temperature. Given that a hazardous area may contain fuel and oxygen, the basis for preventing explosion is ensuring that any ignition source is either eliminated or else does not come into contact with the fuel-oxygen mixture.

Figure 23.1   The 'hazard triangle'

Zone classification. Table 23.1 shows the IEC 79 zone classification used in Europe and most other parts of the world. The relevant British Standard is BS5345 Pt 2. The table also indicates which types of explosion protection is suitable for use within each zone. These explosion protection concepts are described later in the chapter.

The American system of hazardous area classification is structured in a different way, according to the National Electrical

544

Table 23.1 IEC 79 classification of hazardous area zones

	Suitable protection
Zone 0	
Areas in which hazardous explosive gas atmospheres are present constantly or for long periods, for example in pipes or containers	Ex 'ia' Ex 's' (where specially certified Zone 0)
Zone 1	
Areas in which hazardous explosive gas atmospheres are occasionally present, for example in areas close to pipes or draining stations	Ex 'd'; Ex 'ib'; Ex 'p'; Ex 'e'; Ex 's'; Ex 'o'; Ex 'q'; Ex 'm'; Equipment suitable for Zone 0
Zone 2	
Areas in which hazardous explosive gas atmospheres are rare or only exist for a short time, for example areas close to Zones 0 and 1	Ex 'N'/Ex 'n'; Equipment suitable for Zones 1 & 0

Code. In brief, hazardous areas are classified as either 'Division 1', where ignitable concentrations of flammable gases or vapours may be present during normal operation, or 'Division 2', where flammable gases or vapours occur in ignitable concentrations only in the event of an accident or a failure of a ventilation system.

In the UK, the Factories Act states that where there is a risk of a flammable dust cloud, explosion protection and measures to reduce the risk of ignition will be required, although this is not dealt with in this chapter.

Gas grouping and temperature classification. Different gases require different amounts of energy to ignite them and the two concepts of gas grouping and temperature classification are used in Europe to classify electrical apparatus according to its suitability for use with explosive atmospheres of particular gases.

Table 23.2 lists common industrial gases in their appropriate gas groups. Gas group I is reserved for classifying equipment suitable for use in coal mines, and the differences between electrical equipment certified for use in mines is covered in more detail later. Gas group II—which contains gases found in other industrial applications—is subdivided according to the relative flammability of the most explosive mixture of the gas with air.

Table 23.2   CENELEC/IEC gas grouping

Group	Representative gases
I	Methane
IIA	Acetone, ethane, ethyl acetate, ammonia, benzol, acetic acid, carbon monoxide, methanol, propane, toluene, ethyl alcohol, l-amyl acetate, N-hexane , N-butane, N-butyl alcohol, petrol, diesel, aviation fuel, heating oils, acetaldehyde, ethyle ether
IIB	Town gas, ethylene (ethene)
IIC	Hydrogen, acetylene (ethyne), hydrogen disulphide

Table 23.3 defines each temperature class according to the maximum allowed apparatus surface temperature, and indicates common gases for which these classifications are appropriate.

North American practice defines hazardous materials in classes. Flammable gases and vapours are Class 1 materials, combustible dusts are Class 2 materials and 'flyings' (such as sawdust) are Class 3 materials. Class 1 is subdivided into four groups depending on flammability: A (e.g. acetylene), B (e.g. hydrogen), C (e.g. ethylene) and D (e.g. propane, methane). Note that when compared with the IEC gas groupings, the subgroup letters are in opposite order of flammability.

North American temperature classification is similar to IEC standards, but further subdivides the classes to give more specific temperature data.

Standards and Codes of Practice for UK and Europe

Standards. As in many areas of industry, European Norms (EN) exist alongside equivalent British Standards (BS). The origin of these norms are the European Directives concerning electrical equipment for use in potentially explosive atmospheres
546

Table 23.3 CENELEC/IEC temperature classification

Class	Highest permissable surface temperature (°C)	Representative gases
T1	450	Acetone, ethane, ethyle acetate, ammonia, benzol, acetic acid, carbon monoxide, methanol, propane, toulene, town gas, hydrogen
T2	300	Ethyl alcohol, (-amyl acetate, N-hexane, N-butane, N-butyl alcohol, ethylene
T3	200	Petrol, diesel, aviation fuel, heating oils
T4	135	Acetaldehyde, ethyl ether
T5	100	
T6	85	Hydrogen disulphide

published in 1975 and 1979. A separate directive relating to mines was published in 1982. As a result of moves towards the removal of hindrances to free trade within Europe, CENELEC published European Norms (EN) which constitute harmonised standards. Work is underway on a new European Directive, due to come into effect in 1996, which will cover the conformity assessment procedures for all such electrical and mechanical equipment, protective systems and components. Until the new Directive is in place, and until 2003, the structure of standards and codes described in this chapter is likely to be relevant.

As with all standards, review and revision is a continual process, and the European Norms which relate to hazardous areas are at present being updated and reissued in their second editions.

It is important for the electrical engineer to be aware of the European Norm standards for the manufacture of explosion protected equipment (EN 50014 and others) and equivalents in their latest editions because they determine the types of equipment available for use, and also impinge on installation procedure. Other relevant standards include those covering cables (see later).

Equipment certified as meeting EN standards carries the distinctive hexagonal conformity mark (Figure 23.2).

Figure 23.2   The EN conformity mark for certified explosion-protected equipment

Codes of Practice.   Regulations and guidance for the installation and maintenance of electrical equipment in hazardous areas vary from country to country because of differences in installation practice. It is essential, of course, to take into account the national Wiring Regulations or general codes of practice for electrical installations. In addition in the UK, British Standard BS 5345 covers the selection, installation and maintenance of electrical apparatus for use in potentially explosive atmospheres, although it is not applicable to mining. BS5345 has nine parts, some of which have already been mentioned (e.g. Part 2, which covers zone classification). More details of the code of practice is given towards the end of this chapter.

Area assessment procedure.   Companies using flammable materials should carry out an area assessment exercise in accordance with BS 5345 Part 2 (zone classification). Other relevant industry codes also exist, such as those relating to petroleum or chemical industries. In general, the assessment procedure would result in a written report identifying and listing the flammable materials which are used, recording all potential hazards and their source and type, identifying the extent of the zone by taking factors such as type of potential release of flammable material, ventilation and so on, and including other relevant data. Because the area assessment process is complex, various specialist companies and organisations offer a commercial hazardous area assessment service.

ELECTRICAL EQUIPMENT

The manufacturing and other processes within designated hazardous areas are similar to those in non-hazardous areas, and so there is a need for electrical power for motors, lighting, control and instrumentation. It is not surprising that as well as complete electrical systems built and certified to Ex standards, all manner of electrical equipment, from motors (Figure 23.3) and control stations (Figure 23.4), through luminaires (Figure 23.5) and switchgear to plugs and sockets (Figure 23.6), and even handlamps and torches, can be found in Ex-certified versions from specialist manufacturers.

Figure 23.3   Flameproof motor type EEx 'd', frame size 355 S4, 250 kW 4-pole (ABB Motors)

Figure 23.4   Explosion-protected control station (ABB Control Ltd)

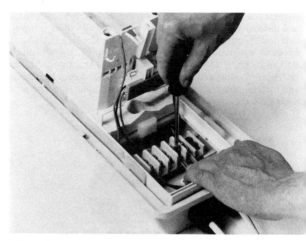

Figure 23.5   Wiring an EEx edm luminaire (ABB Control Ltd)

Figure 23.6   Ex 'de' plug and socket designed for hazardous area use (ABB Control Ltd)

Equipment designed and manufactured in Europe for use in hazardous areas will typically carry a string of codes designating information about the product's suitability for use and the certifying authority's mark. As an example, we will consider the markings on a luminaire unit certified to harmonised European standards: 'ABB CEAG ellk 92 EEx ed IIC T4 18/91. PTB No. Ex-92.C.1028', together with the hexagonal EN conformity symbol. The marks refer to:

1. the manufacturer's name (ABB CEAG);
2. model designation (ellk 92);
3. European Norm code (EEx);
4. protection code (ed);
5. gas group (IIC);
6. temperature classification (T4);
7. serial number (18/91); and
8. certifying authority code (PTB No. Ex-92.C.1028).

Some of these markings are self-explanatory (for example, manufacturer's name), and others have been covered earlier in this chapter (gas group and temperature classification). This section will explain the other markings, and give further information on how electrical equipment is designed and certified for use in hazardous areas.

Apparatus certified as suitable for use in British coal mines will carry slightly different marks. The gas group mark will be I, and there will be no temperature class, because gas group I contains only one hazardous gas and this has a relatively high ignition temperature; thus the T class rating is considered unnecessary. The conformity mark consists of the letters MEx in a distinctively shaped box. The differences in the certification process for equipment for the mining industry are dealt with later.

Types of protection. According to EN 50014, electrical apparatus for use in explosive atmospheres can be designed with various protection concepts, Table 23.4. The apparatus is explosion protected (that is, it will not create or transmit the explosion of a surrounding hazardous atmosphere during normal operation) if it has been certified as meeting a suitable standard by a recognised test authority. Equipment is designated EEx if it has been certified to harmonised European standards; the more general Ex designation is used in the rest of this chapter.

Table 23.4 Types of protection and relevant standards

Protection concept	Designation (European where appropriate)	Relevant standards		
		Harmonised European	Equivalent British Standard	International IEC
General requirements		EN 50014	BS 5501 Pt 1	IEC 79-0
Oil immersion	EEx 'o'	EN 50015	BS 5501 Pt 2	IEC 79-6
Pressurised	EEx 'p'	EN 50016	BS 5501 Pt 3	IEC 79-2
Quartz-sand filled	EEx 'q'	EN 50017	BS 5501 Pt 4	IEC 79-5
Flameproof enclosure	EEx 'd'	EN 50018	BS 5501 Pt 5	IEC 79-1
Increased safety	EEx 'e'	EN 50019	BS 5501 Pt 6	IEC 79-7
Intrinsic safety	EEx 'i'	EN 50020 EN 50039	BS 5501 Pt 7 BS 5501 Pt 9	IEC 79-11
Encapsulation	EEx 'm'	EN 50028	BS 5501 Pt 8	
Non-sparking/ incendive	EX 'N'/'n'		BS 4683 Pt 3	
Special protection	Ex 's'		SFA 3009	

The electrical system design engineer will have to make a choice as to the method of protection to use for the system being designed, and to select apparatus and components accordingly. In practice, the majority of electrical equipment for use in hazardous areas will be designated according to Ex 'd', Ex 'e', 'N/n' or Ex 'i' concepts. Detailed descriptions of these types follow, along with brief details on the other explosion-protection design concepts.

The concept of ingress protection ('IP rating') is relevant to the design of explosion-protected electrical equipment as standards for some Ex apparatus also require a suitable IP rating. Most type tests for certification include more stringent ingress protection tests as well. Table 23.5 defines ingress protection ratings.

Flameproof enclosure – Ex 'd'. Those parts of electrical apparatus that can ignite an explosive gas–air mixture are contained in an enclosure which can withstand the pressure created in the event of ignition of explosive gases *inside* the

Table 23.5 Ingress protection ratings

First digit	Description of protection against solid objects	Second digit	Description of protection against liquids
0	No protection	0	No protection
1	Up to 50 mm, e.g. accidental touch by hands	1	Drops of water, e.g. condensation
2	Up to 12 mm, e.g. fingers	2	Direct sprays of water up to 15° from vertical
3	Up to 2.5 mm (tools and wires)	3	Direct sprays of water up to 60° from vertical
4	Up to 1 mm (tools, wires and small wires)	4	Splashing from all directions
5	Dust (no harmful deposit)	5	Low-pressure jets of water from all directions
6	Totally protected against dust	6	Strong-pressure jets of water, e.g. on ship decks
		7	Immersion in water
		8	Long periods of immersion under pressure

A third digit indicates degree of protection against impact, from no protection (0) to 20 joule (9).

enclosure, and which can prevent the communication of the explosion to the atmosphere surrounding the enclosure.

The rationale of the Ex 'd' concept is the *containment* of any explosion which may be created by the equipment, and thus the concept is applicable to virtually all types of electrical apparatus given that the potential sparking or hot elements can be contained in a suitably sized and strong enclosure. The factors taken into account by equipment manufacturers and system designers include arc and flame path lengths and types, surface temperature, internal temperature with regard to temperature classification, and distance from components in the enclosure to the enclosure wall (12 mm minimum is specified by the standard).

Some component types are unsuitable for use with flameproof enclosures, for example rewireable fuses, and components containing flammable liquids.

A major consideration in the use of Ex 'd' enclosures is the making of flameproof joints, which must be flanged, spigotted or screwed. Maximum gaps and minimum widths of any possible flame path through a joint are defined by the standard, which also lays down requirements for screw thread pitch, quality and length.

The certification process for Ex 'd' equipment involves examination of the mechanical strength of the enclosure, and explosion and ignition tests under controlled conditions.

Increased safety – Ex 'e'. Measures have been taken to prevent sparks or hazardous temperatures in internal or external parts of the electrical apparatus during normal operation. The *prevention* of explosion from normally non-sparking/arcing or hot equipment is the guiding philosophy behind the Ex 'e' concept. No sparking devices can be used and various electrical, mechanical and thermal methods are used to increase the level of safety to meet the certification test requirements.

Examples of equipment designed and constructed to the Ex 'e' protection philosophy include luminaires, terminal boxes and motors. One advantage of Ex 'e' as a means of protection is that boxes and enclosures can be made of plastics and other materials that are easier to work with than using Ex 'd' flameproof enclosures.

Key design concepts for Ex 'e' equipment are the electrical, physical and thermal stability of the materials, and the compatibility of different materials, used for items such as electrical terminations. Manufacturers of specialist explosion-protected electrical apparatus have been ingenious in the design of electrical terminations for Ex 'e' equipment to ensure firm, positive and maintenance-free connection of conductors. Ex 'e' certified enclosures require IP54 ingress protection although in practice, enclosures are available to meet a very wide range of operating conditions.

Ex 'n'. 'Type N' protection (non-sparking) is a British designation for electrical equipment suitable for use in Zone 2 but not Zone 1; the concept is covered by BS 6941 and as yet is not incorporated into harmonised European standards. When it is, the designation will become EEx 'n'.

Equipment manufactured with Type N protection includes terminal boxes, luminaires and motors, which, because of the less hazardous nature of locations assessed as Zone 2 and the consequently less demanding requirements of the standard, can be manufactured more simply and at less cost than other types of explosion-protected equipment for Zone 1 locations. The Type N protection concept combines certain aspects of Ex 'e' and Ex 'i' concepts in its use of 'non-incendive' circuit elements and design for greater safety.

Intrinsic safety – Ex 'i'. An electrical device is intrinsically safe (IS) if all the circuits it contains are intrinsically safe. The EN 50 020 standard stipulates that under normal operation and certain specified fault conditions, no sparks or thermal effects are produced which can cause the ignition of a specified gas atmosphere. Sparks or thermal effects are not produced because the energy in the IS circuits is very low; intrinsically safe circuits are control and instrumentation circuits rather than power circuits. According to Code of Practice British Standard BS 5345 Part 4, to be defined as 'simple apparatus' the maximum stipulated voltage which a field device can generate is 1.2 V, with a current not exceeding 0.1 A, an energy of 20 microjoules and power of 25 mW, although IS circuitry exceeding these ratings can still meet the requirements of EX 'i', but requires certification.

There are two Ex 'i' protection types. Ex 'ia' equipment will not cause ignition (i) in normal operation, (ii) with a single fault, and (iii) with any two faults; a safety factor of 1.5 applies in normal operation and with one fault, and a safety factor of 1.0 applies with two faults. Ex 'ia' equipment is suitable for use in all zones, including Zone 0.

Ex 'ib' equipment is incapable of causing ignition (i) in normal operation, and (ii) with a single fault; a safety factor of 1.5 applies in normal operation and with one fault, and a safety factor of 1.0 applies with one fault if the apparatus contains no unprotected switch contacts in parts likely to be exposed to the potentially explosive atmosphere, and the fault is self-revealing. Ex 'ib' equipment is suitable for use in all zones except Zone 0.

Components for intrinsically safe circuits contain barriers to prevent excessive electrical energy from entering the circuit. The two principal barrier types are Zener barriers (used when an intrinsically safe earth is available) and galvanically isolated barriers (used where an IS earth is not available, Figure 23.7). An IS earth must be provided by a clearly marked conductor of not less than 4 mm^2 with impedance not greater than 1 ohm from the barrier earth to the main power supply earth.

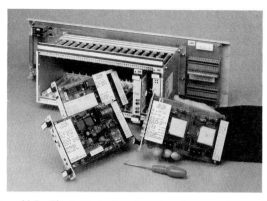

Figure 23.7   The 'intrinsically safe engineering system' – a rack to accommodate various galvanically isolated IS interface cards for control and instrumentation circuits (ABB Control Ltd)

The specified ingress protection rating of an enclosure for IS apparatus is IP20 but for hostile exterior environments, IP66 is recommended by the Code of Practice.

Other types of protection.   Ex 'p' protection uses air or inert gas to maintain positive pressure to prevent the entry of flammable gas or vapour into the enclosure or room. Another method is to reduce the volume of gas or vapour within the enclosure or room below the explosive gas–air mixture level by dilution from a clean external source.

Ex 'o' protection refers to apparatus in which ignition of a gas–air mixture is prevented by immersing the live or sparking apparatus in a specified minimum depth of oil, determined by type testing. Ex 'o' equipment is rarely encountered.

Ex 'q' protected apparatus has the live or sparking elements immersed in granular quartz or other similar material. The required protection rating is IP54.

Ex 'm' equipment has potential ignition sources encapsulated to prevent them coming into contact with explosive atmospheres.

Ex 's' equipment has been shown by test to be suitable for use in the appropriate zone although the apparatus does not comply with the standard of the established concepts described above.

Cables. As in any electrical system, cables are required in hazardous areas for instrumentation, communication, lighting, power and control. Such cables are the subject of various regulations.

In intrinsically safe (IS) electrical systems, the inductive and capacitive nature of cables means that they can store energy and this energy must be taken into account when designing IS circuits. This is achieved by strict control of capacitance and L/R ratios in conjunction with Zener barriers or galvanic isolators. British Standards relevant to IS circuit cabling include BS 5308 Parts 1 and 2 which cover polyethylene-insulated cables for use in petroleum refineries and related applications, and PVC-insulated cables widely used in chemical and industrial applications. Although BS5308 cables are frequently used in IS systems for hazardous-area use, the words 'intrinsically safe' do not appear in the standard's title, and the foreword states that the cables 'may be suitable for Group II intrinsically safe systems'. The foreword refers to the code of practice BS 5345 on selection, installation and maintenance of electrical apparatus for use in potentially explosive atmospheres.

Other cable standards which are relevant to hazardous–area work and which may be specified include BS 6883/IEC 92-3 (covering wiring for ships and topside on offshore oil and gas installations), BS 6425/IEC 754-1 (smoke and halogen emission), BS 6387/IEC 331 (fire resistance), BS 4066/IEC 332 (flame retardance) and BS 6207 (mineral-insulated cables), Figure 23.8. Cable glands also have to be flameproof, Figure 23.9.

Certification. Within Europe, there are national authorities which issue certification documents for electrical equipment to prove that equipment meets a specific standard for explosion protection. Table 23.6 gives details of these authorities in each country. In the UK there are two accredited test authorities: BASEEFA (a unit of EECS) and SIRA. As well as being accepted throughout the European Union, certification by BASEEFA/SIRA is also recognised in other parts of the world, particularly in the Middle and Far East.

In the UK, certification to the standards relevant for use in mines is carried out by MECS (a unit of EECS). The procedure is similar to that for certification to standards for other hazardous areas, but in addition to explosion protection requirements, electrical equipment must also provide the high degree of electrical,

Figure 23.8   Flex-Flame HCF cable for use on offshore oil and gas platforms can withstand temperatures of 1100°C for one hour – exceeding IEC 331/BS6387 standards for fire resistance (ABB Control Ltd)

Figure 23.9   An EEx 'd' flameproof cable gland, used primarily for off-shore applications (CMP Products)

mechanical and operational safety (pitworthiness requirements) demanded by the Mines Inspectorate.

Once a product has been tested by an authorised test house and certified as meeting a specific standard, the equipment manufacturer can then certify that product under licence from the test authority. Individual components usually have conditions for safe use attached to their certificate. 'Certificates of conformity' for a complete piece of electrical apparatus allow installation in hazardous areas without further verification.

Table 23.6 Certifying authorities

Country	Authority
UK	BASEEFA/EECS and SIRA
	(MECS for mining applications)
Germany	PTB
France	CERCHAR and LCIE
Spain	LOM
Italy	CESI
Belgium	INIEX
Denmark	DEMKO
Sweden	SEMKO
Norway	NEMKO
USA	FM
Canada	CSA

In brief, the certification process involves an assessment of the conformity of the equipment to the specific standard sought, examination of a prototype to ensure that it complies with the design documents, and testing. The certified design is defined in a set of approved drawings listed on the certificate.

INSTALLATION, INSPECTION AND MAINTENANCE PRACTICE

As well as information on selection of electrical apparatus for use in hazardous areas, the British Standard code of practice BS 5345 contains guidance on installation, inspection and maintenance of equipment. The code is divided into nine parts; Part 1 gives general guidance, Part 2 covers hazardous-area classification, and each remaining part is specific to one of the types of protection concept. The code contains much useful information of a general nature concerning electrical work in hazardous areas. It does not cover work in mines or areas where explosive dusts may be present, and it makes clear that the code is not intended to replace recommendations produced for specific industries or particular applications.

In this section we mention some of the areas of recommendation of the code, in particular from those parts relating to Ex 'd', Ex 'e' and Ex 'i' equipment. This is in no way intended to replace the code itself.

559

In general, operation and maintenance should be taken into account when designing process equipment and systems in order to minimise the release of flammable gases. For, example, the requirement for routine opening and closing of parts of a system should be borne in mind at the design stage. No modifications should be made to plant without reference to those responsible for hazardous-area classification, who should be knowledgeable in such matters. Whenever equipment is reassembled, it should be carefully examined. The code gives recommended inspection schedules for equipment of each type of protection concept, which sets out what should be inspected on commissioning and at periodic intervals. For all equipment, the protection type, surface temperature class and gas group should all be checked to ensure that the equipment is suitable for its zone of use, and also that no unauthorised modifications have been carried out. Circuit identification should also be checked.

For installations making use of flameproof enclosures, the code recommends preventing solid obstacles such as steelwork, walls or other electrical equipment from approaching near to flanged joints or openings. Minimum clearance distances of up to 40 mm are given. Gaps should be protected against the ingress of moisture with approved non-setting agents. Extreme care should be taken in the selection of these non-setting agents to avoid potential separation of joint surfaces. The code specifies the type of threads for entry tappings into flameproof enclosures, and draws attention to the requirement to follow any directions contained in the certification documents for cable systems and terminations. Types of cables suitable for use are also specified, along with their appropriate methods of entry. Ex 'd' apparatus with integral cables where the cable terminations are encapsulated should not be tampered with but returned to the manufacturer if maintenance is required.

The section of the code on inspection, maintenance and testing of increased safety Ex 'e' equipment also includes a recommendation to check that ratings of lamps used—which may have been replaced—are correct. An appendix contains guidance on the chemical influence of combustible gases on certain mechanical and electrical properties of any insulating materials, for example panels, gaskets or encapsulation material used in Ex 'e' equipment. An appendix covering Ex 'e' cage motors and associated protection equipment gives recommendations intended to ensure that all parts of the motor do not rise above a safe temperature.

560

The part of the code dealing with Ex 'i' equipment pays particular attention to the interconnecting cables used in IS systems, recommending, for instance, minimum conductor sizes to ensure temperature compliance in fault conditions, specific separations between individual IS circuits and earth, insulation thicknesses, screening and mechanical protection of cables. The use of multicore cable is considered, as is the siting of cables to avoid potential induction problems. In general, cable entries should be designed to minimise mechanical damage to cables. With all IS equipment, the need to follow the specific requirements of certification documents is emphasised, and during inspection, it is recommended that attention be paid to lamps, fuses, earthing and screens, barriers and cabling. Certain specified on-site testing and maintenance of energised IS circuits is permitted inside the hazardous area provided any test equipment used is certified as intrinsically safe in itself, and that conditions on certification documents are followed.

It is impossible in this section to give detailed, specific guidance on the installation, inspection and maintenance of electrical equipment in hazardous areas. The electrical engineer is advised to consult the BS 5345 code of practice, or other codes relevant to specific industries or applications.

SOURCES OF FURTHER INFORMATION

BSI Enquiries, 389 Chiswick High Road, London, telephone 0181-996 7000.
SIRA Test and Certification, Saighton Lane, Saighton, Chester, telephone (01244) 332200.
EECS (BASEEFA and MECS), Harpur Hill, Buxton, Derbyshire, telephone (01298) 28000.

This chapter was supplied by ABB Control of Coventry.

Index

564